T0353895

Mathematical Modelling

This book investigates human–machine systems through the use of case studies such as crankshaft maintenance, liner piston maintenance, and biodiesel blend performance. Through mathematical modelling and using various case studies, the book provides an understanding of how a mathematical modelling approach can assist in working out problems in any industrial-oriented activity.

Mathematical Modelling: Simulation Analysis and Industrial Applications details a data analysis approach using mathematical modelling sensitivity. This approach helps in the processing of any type of data and can predict the result so that based on the result, the activity can be controlled by knowing the most influencing variables or parameters involved in the phenomenon. This book helps to solve field and experimental problems of any research activity using a data-based modelling concept to assist in solving any type of problem.

Students in manufacturing, mechanical, and industrial engineering programs will find this book very useful. This topic has continued to advance and incorporate new concepts so that the manufacturing field continues to be a dynamic and exciting field of study.

Mathematical Modelling

Simulation Analysis and Industrial Applications

Pramod Belkhode
Prashant Maheshwary
Kanchan Borkar
J.P. Modak

CRC Press is an imprint of the
Taylor & Francis Group, an informa business

First edition published 2023
by CRC Press
6000 Broken Sound Parkway NW, Suite 300, Boca Raton, FL 33487-2742

and by CRC Press
4 Park Square, Milton Park, Abingdon, Oxon, OX14 4RN

CRC Press is an imprint of Taylor & Francis Group, LLC

ISBN: 978-1-032-33198-0 (hbk)
ISBN: 978-1-032-33201-7 (pbk)
ISBN: 978-1-003-31869-9 (ebk)

DOI: 10.1201/9781003318699

Typeset in Times
by SPi Technologies India Pvt Ltd (Straive)

Contents

Author Biographies ... xi

Chapter 1 Evaluation of the System ... 1

1.1 Introduction ... 1
1.2 Causes and Effects Relationships ... 2
1.3 Ergonomics ... 2
1.4 Anthropometry ... 2
1.5 Approach to Formulate the Mathematical Model ... 3

Chapter 2 Concept of Field Data-Based Modelling ... 5

2.1 Introduction ... 5
2.2 Formulation of Mathematical Model ... 7
2.3 Limitations of Adopting Field Database Model ... 8
2.4 Identification of Causes and Effects of an Activity ... 8
2.5 Dimensional Analysis ... 9
2.6 Dimensional Equation ... 9
2.6.1 Rayleigh's Method ... 10
2.6.2 Buckingham Π Theorem Method ... 11

Chapter 3 Design of Experimentation ... 13

3.1 Introduction ... 13
3.2 Limitations of Adopting Field Data-Based Model Formulation for Man Machine System ... 13
3.3 The Approach for Formulating a Field or Experimental Data-Based Model ... 14
3.4 Identification of Variables ... 14
3.5 Problems Associated with Crankshaft/Liner Piston Maintenance Activities of LocoShed ... 15
3.5.1 Independent and Dependent Variables of Crankshaft Maintenance Activity ... 16
3.5.2 Dimensional Analysis of Crankshaft Maintenance Operation ... 17
3.5.3 Establishment of Dimensionless Pi Terms for Crankshaft Maintenance Activity ... 19
3.5.4 Formulation of a Field Data-Based Model for Response Variables of Crankshaft Maintenance Activity ... 22
3.5.5 Model Formulation by Identifying the Curve Fitting Constant and Various Indices of Pi Terms of Crankshaft Maintenance Activity ... 22
3.5.6 Independent and Dependent Variables of Liner Piston Maintenance ... 24
3.5.7 Dimensional Analysis of Liner Piston Maintenance ... 24
3.5.8 Establishment of Dimensionless Pi Terms for Liner Piston Maintenance ... 28

3.5.9 Formulation of a Field Data-Based Model for Response Variables of Liner Piston Maintenance 29
3.5.10 Model Formulation by Identifying the Curve Fitting Constant and Various Indices of Pi Terms of Liner Piston Maintenance 30
3.6 Problem Associated with Fossil Fuels 32
3.6.1 Diesel Blending 32
3.6.2 Independent and Dependent π Term 33
3.6.3 Establishment of Dimensionless Group of π Terms 36
3.6.4 Creation of Field Data-Based Model 36
3.6.5 Model Formulation by Identifying the Curve Fitting Constant and Various Indices of π Terms 37
3.7 Problem Associated with Conventional Power Generation 38
3.7.1 Identification of Variables Affecting the Phenomenon 39
3.7.2 Formation of Pi (π) Terms for All Dependent and Independent Variables Affecting the Phenomenon 41
3.7.3 Formulation of Experimental Data Base Model for Solar Updraft Tower 44

Chapter 4 Experimentation 47

4.1 Introduction 47
4.2 Instrumentation and Data Collection 47
4.2.1 Instrumentation for Crankshaft Maintenance Activity 47
4.2.2 Data Collection from Field for Crankshaft Maintenance Activity 49
4.2.3 Instrumentation Used for Liner Piston Maintenance Activity 49
4.2.4 Data Collection from the Field for Liner Piston Maintenance Activity 54
4.2.5 Basis for Arriving at Number of Observations 54
4.2.6 Calculation of Field Human Energy Consumed in Maintenance Activity 57
4.2.7 Calculation of Human Energy Consumed in Crankshaft Maintenance Activity 70
4.2.8 Instrumentation and Collection of Data for Solar Updraft Tower 71
4.2.9 Instrumentation and Collection of Data for the Engine Performance by Using the Alternative Fuels 74
4.2.10 Establishment of Dimensionless Group of π Terms 76
4.2.11 Creation of Field Data-Based Model 76
4.2.12 Model Formulation by Identifying the Curve Fitting Constant and Various Indices of π Terms 77
4.2.13 Basis for Arriving at Number of Observations 78

Chapter 5 Formulation of Mathematical Model 83

5.1 Formulation of Field Data-Based Model for Crankshaft Maintenance Operation 83
5.1.1 Model Formulation for π_{D1}, Overhauling Time of Crankshaft Maintenance Activity by Identifying the Curve Fitting Constant and Various Indices of Pi Terms 86

5.1.2 Model Formulation for π_{D2}, Human Energy Consumed in Crankshaft Maintenance Activity by Identifying the Curve Fitting Constant and Various Indices of Pi Terms ... 88
5.1.3 Model Formulation for π_{D3}, Productivity of Crankshaft Maintenance Activity by Identifying the Curve Fitting Constant and Various Indices of Pi Terms ... 91
5.1.4 Models Developed for the Dependent Variables – Crankshaft Maintenance Activity ... 94
5.2 Formulation of Field Data-Based Model for Liner Piston Maintenance Activity ... 95
5.2.1 Model formulation for π_{D1}, Overhauling Time of Liner Piston Maintenance Activity by Identifying the Curve Fitting Constant and Various Indices of Pi Terms ... 97
5.2.2 Model Formulation for π_{D2}, Human Energy Consumed in Liner Piston Maintenance Activity by Identifying the Curve Fitting Constant and Various Indices of Pi Terms ... 100
5.2.3 Model Formulation for π_{D3}, Productivity of Liner Piston Maintenance Activity by Identifying the Curve Fitting Constant and Various Indices of Pi Terms ... 103
5.2.4 Models Developed for the Dependent Variables – Liner Piston Maintenance Activity ... 106
5.3 Formulation of Field Data-Based Model for Diesel Blending ... 106
5.3.1 Model Formulation by Identifying the Curve Fitting Constant and Various Indices of π Terms ... 107
5.3.2 Model Formulation for Brake Thermal Efficiency (Z_1) ... 108
5.3.3 Model Formulation for Brake-Specific Fuel Consumption (Z_2) ... 109
5.4 Formulation of Field Data-Based Model for Solar Updraft Tower ... 109
5.4.1 Model Formulation for Turbine Speed Developed by Identifying the Constant and Various Indices of π Terms ... 109
5.4.2 Model Formulation for Turbine Power Developed by Identifying the Constant and Various Indices of π Terms ... 111

Chapter 6 Artificial Neural Network Simulation ... 113
6.1 Introduction ... 113
6.2 Procedure for Formulation of ANN Simulation ... 113
6.3 Ann Program for Crankshaft Maintenance Activity ... 114
6.3.1 ANN Program for Overhauling Time of Crankshaft Maintenance Activity (z_{1C}) ... 114
6.3.2 ANN Program For Human Energy Consumed in Crankshaft Maintenance Activity (z_{2C}) ... 122
6.3.3 ANN Program for Productivity of Crankshaft Maintenance Activity (z_{3C}) ... 127
6.4 ANN Program for Liner Piston Maintenance Activity ... 131
6.4.1 ANN Program for Overhauling Time of Liner Piston Maintenance Activity (z_{1P}) ... 131
6.4.2 ANN Program for Human Energy Consumed in Liner Piston Maintenance Activity (z_{2P}) ... 139
6.4.3 ANN Program for Productivity of Liner Piston Maintenance Activity (z_{3P}) ... 140

6.5 ANN Program for Brake Thermal Efficiency and Brake-Specific Fuel Consumption ... 151
6.5.1 ANN Program for Brake Thermal Efficiency ... 151
6.5.2 ANN Program for Brake-Specific Fuel Consumption ... 154
6.6 ANN Program for Solar Updraft Tower ... 157
6.6.1 ANN Program for Turbine Speed ... 157
6.6.2 ANN Program for Turbine Power ... 162

Chapter 7 Sensitivity Analysis ... 165

7.1 Sensitivity Analysis of Crankshaft Maintenance Activity ... 165
7.1.1 Effect of Introduced Change on the Dependent π Term – Overhauling Time of a Maintenance Activity of Crankshaft Maintenance Activity (z_{1C}) ... 165
7.1.2 Effect of Introduced Change on the Dependent π Term – Human Energy Consumed in Crankshaft Maintenance Activity (z_{2C}) ... 169
7.1.3 Effect of Introduced Change on the Dependent π Term – Productivity of Crankshaft Maintenance Activity (z_{3C}) ... 170
7.2 Sensitivity Analysis of Liner Piston Maintenance Activity ... 171
7.2.1 Effect of Introduced Change on the Dependent π Term – Overhauling Time of a of Liner Piston Maintenance Activity (z_{1P}) ... 171
7.2.2 Effect of Introduced Change on the Dependent Pi Terms – Human Energy Consumed in Liner Piston Maintenance Activity (z_{2P}) ... 176
7.2.3 Effect of Introduced Change on the Dependent π Term – Productivity of Liner Piston Maintenance Activity (z_{3P}) ... 176
7.3 Optimization of Models for Crankshaft Maintenance Activity ... 177
7.4 Optimization of the Models for Liner Piston Maintenance Activity ... 181
7.5 Reliability of Models ... 185
7.5.1 Reliability of Crankshaft Maintenance Activity ... 185
7.5.2 Reliability of Liner Piston Maintenance Activity ... 186
7.6 Sensitivity Analysis of Brake Thermal Efficiency and Brake-Specific Fuel Consumption ... 187
7.6.1 Effect of Introduced Change on the Dependent π Term: Brake Thermal Efficiency ... 189
7.6.2 Effect of Introduced Change on the Dependent π Term: Brake Thermal Efficiency ... 191
7.7 Sensitivity Analysis of Turbine Speed and Turbine Power ... 191
7.7.1 Effect of Introduced Change on the Dependent π Term: Turbine Speed ... 191
7.7.2 Effect of Introduced Change on the Dependent π Term: Power Developed ... 191

Chapter 8 Interpretation of Mathematical Models ... 195

8.1 Models Developed for Dependent Variables of Crankshaft Maintenance Activity ... 195
8.1.1 Interpretation of Model of Crankshaft Maintenance Activity ... 195

8.1.1.1 Analysis of the Model for Dependent Pi Term – Overhauling Time of Crankshaft Maintenance Activity (z_{1C})....195
8.1.1.2 Analysis of the Model for Dependent Pi Term – Human Energy Consumed in Crankshaft Maintenance Activity (z_{2C})....197
8.1.1.3 Analysis of the Model for Dependent Pi Term – Productivity in Crankshaft Maintenance Activity (z_{3C})....198
8.1.2 Analysis of Performance of Models by ANN Simulation of Crankshaft Maintenance Activity....199
8.2 Models Developed for Dependent Variables of Liner Piston Maintenance Activity....200
8.2.1 Interpretation of Models of Liner Piston Maintenance Activity....201
8.2.2 Analysis of Performance of Models by ANN Simulation of Liner Piston Maintenance Activity....204
8.3 Analysis of the Mathematical Model for the Dependent Pi Term Brake Thermal Efficiency....206
8.4 Analysis of the Mathematical Model for the Dependent Pi Term Brake-Specific Fuel Consumption....206
8.5 Models Developed for Dependent Variables Turbine Speed....208
8.5.1 Analysis of the Model for Dependent π Term Z_1....208
8.6 Analysis of Performance of the Models of Power Developed....209
8.6.1 Interpretation of the Model....209
8.6.2 Analysis of the Model for Dependent π Term Z_2....209
8.6.2.1 Analysis of Performance of the Models....210

Bibliography....211
Index....217

Author Biographies

Dr. Pramod Belkhode did his doctoral research degree in Mechanical Engineering from Rashtrasant Tukadoji Maharaj Nagpur University, Nagpur. He is currently working as an Assistant Professor at Laxminarayan Institute of Technology since 2009. He has published more than thirty research papers in international journals and has delivered more than 20 guest lectures on various topics. Four students have been awarded doctoral degrees under his supervision.

Dr. Prashant Maheshwary did his double doctoral research degree in two different specializations (Machine Design and Thermal Engineering) of Mechanical Engineering. An accomplished teaching professional with more than 3 decades of teaching experience and 10 years of research experience, he has published more than twenty research papers in SCI and Scopus (A+) indexed international journals and has in-depth administrative experience which he gained as the Director of an Educational Institute since 2008. He is a self-driven, result-oriented person with flexibility and the ability to connect to all levels in an organization. Continuous self-development and continuous learning have added to his knowledge and are an asset to the Institution he is associated with.

Dr. Kanchan Borkar is currently an Assistant Engineer in the Department of Military Engineer Services under the Ministry of Defence for the last 12 years in Nagpur. She did her doctorate in Mechanical Engineering, master in Design Engineering, and graduated from Rashtrasant Tukadoji Maharaj Nagpur University, Nagpur (MS) India. She has published more than fifteen research papers in international journals and has delivered 5 guest lectures on various topics.

Dr. J.P. Modak is currently a professor of Mechanical Engineering at J. D. College of Engineering, Nagpur. With a keen interest in the areas of automation, the theory of experimentation, ergonomics, man-machine system, applied robotics, rotor dynamics and kinematics, and dynamics of mechanism, Dr. Modak has contributed more than 800 research paper publications, including more than 100 journals of international repute. He has won the national award for the "Best Research Paper" twice during his eminent career. He has been instrumental in developing various products and applications in the field of Mechanical Engineering and has turned out around 90 Ph.D. students and 12 postgraduate degrees by research candidates.

1 Evaluation of the System

1.1 INTRODUCTION

Most industrial activities are executed manually due to limitations of mechanization, such as technological constraints and costs. Industrial activities such as maintenance operations, loading and unloading operations on process machines, and similar such other operations are performed manually. Management attitudes are often conservative and traditional, and with predominantly spontaneous judgement-based decision-making. Hence, most operations are carried out manually. Operators work with different types of machine tools and process machines under different environmental conditions. Ergonomic design of the workstation varies according to suitability for operators with different constraints. Ergonomics deals in an integrated way, considering the working environment, tools, materials, and processes. Work can be completed efficiently with human comfort if designed based on principles of ergonomics. If ergonomic principles are not applied in the design of man–machine systems, it results in low efficiency, poor health, and increased accident rates.

Solar power plants in use in the world are equipped to transform solar radiation into electrical energy via natural phenomena. The solar chimney power plant produces electricity from solar energy. Sunshine heats the roofed collector sheets surrounding the central base of a tall chimney tower. The chimney effect is caused by a hot air updraft resulting from convection. This airflow drives wind turbines placed in the chimney updraft or around the chimney base to produce electricity. The three essential elements of solar updraft tower, viz. solar air collector, chimney/tower, and wind turbines, have been familiar for centuries.

The mathematical model is formed with all the variables involved in the design of the solar updraft tower. These independent variables are grouped thus: variables related to collector, variables related to the chimney, variables related to atmospheric condition, and variables related to heating condition. The indices of each grouped pie terms predict the performance of the dependent variables, such as speed and power produced by the turbine. Based on the mathematical model, performance of the solar updraft is optimized.

Due to the inevitable exhaustion of petroleum, as well as its rising price, plus environmental problems generated by the burning of fossil fuels, means that the hunt for alternative fuels has gained prominence. The use of alternative fuels not only evades the petroleum crises, but also reduces pollutant gases emitted by engines. Hence the blends of various proportions, such as 10%, that being 10% treated transformer oil and 90% diesel fuel; likewise, blends of 20%, 25%, 30%, and 40% are also made, and then compared with the properties of pure diesel fuel.

A diesel engine run was conducted with a single-cylinder diesel engine. The result obtained was fuelled with blends of treated transformer oil and diesel fuel varying in proportions, such as 10:90, 20:80, 25:75, 30:70, and 40:60. The runs were covered under varying loads of 10 kg, 15 kg, and 20 kg. The performance of the engine was evaluated on the basis of brake thermal efficiency (BTE) and brake specific fuel consumption (BSFC).

The variables affecting the effectiveness of the phenomenon under consideration are blends of treated transformer oil with diesel, and performance characteristics are optimized using the formulation of mathematical models, and then the performance evaluated using indices of the formulated model based on the independent and dependent variables involved in the experiment.

Various factors of the operations are identified so as to optimize productivity and conserve human energy needed for operations. There are many approaches to develop or upgrade industrial activities,

DOI: 10.1201/9781003318699-1

such as method study (motion study), work measurement (time study), and productivity. A field or experimental data-based modelling approach is proposed to study man–machine systems.

1.2 CAUSES AND EFFECTS RELATIONSHIPS

Formulation of logic-based model correlating causes and effects is not possible for these types of complex phenomenon. Only approach appropriate for this type of study of phenomenon i.e. man–machine system is by field or experimental data-based modelling. Field or experimental data-based model correlates the inputs or causes (in other words, outputs of such activity) by formulating the quantitative mathematical modelling. The indices of the causes of the mathematical indicate the most influencing inputs. Such correlation indicates the deficiency and the strength of the man–machine system which helps to improve the performance of the system. Hence, for improving system/activity performance it is essential to form such analytical cause–effect relationships conceptualized as field data-based models or experimental data-based models.

1.3 ERGONOMICS

Ergonomics deals with the interaction of the human operator and the physical system of works. Performance of man–machine systems are optimized by designing the system with the help of principles of ergonomics, data, and methods. Ergonomists contribute to the design and evaluation of the system based on the type of job, environment, product, and task to be performed in order to make them compatible. These principles are used for enhancing safety, reducing fatigue, increasing comfort with improved job satisfaction while enhancing effectiveness, i.e. productivity in carrying out the tasks.

1.4 ANTHROPOMETRY

Anthropometry deals with the measurement of the size and proportions of the human body as well as parameters, such as reach and visual range capabilities. The application of anthropometry in design of tools and equipment is to incorporate the relevant human dimensions, aiming to accommodate at least 90% of potential users, taking into account both static and dynamic (functional) factors. Static factors include height, weight, and shoulder breadth. Dynamic factors consist of body movement, distance reach, and movement pattern. Anthropometry is the branch of ergonomics that deals with different human body dimensions of operators, who are accommodated by providing adjustability in machines used. Hence, the anthropometric data of operators is collected.

Postural discomfort is happens for the operator when muscular discomfort is experienced in attempting to maintain body posture during work. The interaction of operator with machine during operation leads to ugly postures; therefore, it is necessary to study the specifications of machines from the viewpoint of postural comfort. In physiological response studies, human energy consumed while performing the task using tools of different designs is recorded. Normally, the selection of tools is made on the basis of the constructional features of the machine, tools available, operating conditions, and the environment. Theses form the independent activity variables. The physiological cost (dependent/response variable) incurred in operation is recorded for different conditions of these independent variables. Productivity of operation is considered as other dependent/response variables. These variables are also recorded in order to study the effect of independent variables on quality of operation.

Hilbert suggested the experimentation theory to know the output of any activity in terms of various inputs of any phenomenon. In fact, it is felt that such an approach is not yet seen towards correctly understanding the operation performed by human being. This approach finally establishes a field data-based model for the phenomenon. The various inputs in the industrial activity are i) body specifications of the operator (viz. the anthropometric measurements); ii) specifications of machine;

iii) specifications of tools; iv) other process related parameters; and v) specifications of environmental factors such as ambient temperature, humidity, and air circulation at the place of work. The response variables of the phenomenon are i) time of operation; ii) productivity; iii) human energy consumed during the operation; and iv) quality of operation. A quantitative relationship is established amongst the responses and inputs. The inputs as well as the corresponding responses are measured. Such quantitative relationships are known as mathematical models. Two model types are established, viz. the model using the concept of least-square multiple regression curve (here in after referred as mathematical model) and the "Artificial Neural Network" based model. The interest of the operator lies in arranging inputs so as to obtain targeted responses. Once the models are formed, they are optimized. The optimum conditions are deduced for which the independent variables should be set to achieve maximum productivity, minimum human energy expenditure, and accepted quality. From the analysis of models, the intensity of influence of various independent variables on the dependent variables and the nature of relationship between independent and dependent variables is determined. Finally, some important conclusions are drawn based on the analysis of models.

1.5 APPROACH TO FORMULATE THE MATHEMATICAL MODEL

The main objective of this book is to explain the approach to formulate a mathematical model for the man–machine system. In order to form this model, the most critical industrial activities – maintenance activities of the Locoshed Industry, performance of solar updraft towers, and performance of diesel blends, are identified and studied. Locoshed industries involve operations such as liner piston maintenance and crank piston maintenance. The formulation of mathematical models for maintenance activities of the Locoshed Industry, performance of solar updraft towers, and performance of diesel blends, are identified and studied, and selected as the case studies. In the present method, productivity is less, and human energy requirement is substantial. The variables related to the maintenance operation in the locoshed are identified to enhance the productivity with minimum human energy. In this book, the approximate generalized mathematical models have been established applying concepts of Theories of Experimentation for the operations in locoshed. The general procedure adopted is as follows.

1. Review the existing literature on industrial operations. A general overview in relation with ergonomic aspects in which the interaction between man–machine system.
2. Study of existing workstation including man–machine system. This study includes maintenance types, schedules, problems, functions, purpose, and benefits.
3. Possibility of Formulation of model for improving industrial activity. The industrial activity is exceedingly difficult to plan and incurs high costs. There is therefore a need to develop a model that helps reduce human energy and repair times. Data is collected based on sequence of industrial activity by direct measurement. From this data, input and output variables are decided, and a model is generated by forming dimensionless equations using regression analysis.
4. Possibility to validate output and model of the system – the aim is to find out the utility and effectiveness of model. The effectiveness of the model is decided by the ANN simulation, sensitivity analysis, and optimization technique.

2 Concept of Field Data-Based Modelling

2.1 INTRODUCTION

This chapter discusses the concept of field data-based modelling with the help of examples seen in everyday life. Detailed concepts of man–machine systems are discussed, with effects and causes. Activities dominated by human beings and their influences are explained. The accuracy of models depends on the magnitude of curve fitting constants, and the importance of indices is explained.

In life we come across many activities. These activities have some environmental systems in which the activities take place. The environment or system can be defined in terms of its parameters some of which are always constant in their magnitudes whereas some are variable. The activities are set in action by some parameters which are considered as causes. These causes interact with parameters of the system as a result of this interaction some effects are produced. The above said matter in a diagrammatic form can be presented as shown in Figure 2.1.

Figure 2.1 shows one rectangular block within which is written activity along with its nature, i.e. activity may be totally physical, or it may be a combination of human directed/operated physical activity stated as a man–machine system, or it may be an activity mainly dominated by human beings stated in the block as "Totally Human System". The functioning of an activity is influenced by two sets of parameters one set is characterizing the features of environment of an activity and other set is planned parameters or causes which influence the functioning of the system.

Accordingly, Figure 2.1 shows these parameters. The parameters characterizing features of an environment are E1, E2, E3, E4 etc. Some of which are permanently fixed, say, E1 and E2, whereas the remaining are time variant, say, E3 and E4, over which one has no control. The planned parameters are known as causes shown as A, B, C, D, E, etc. and effects of an activity are shown as Y1, Y2, Y3, Y4, etc.

If one analyses any activity of society, one would be able to identify the causes A to E etc, system parameters E1 to E4 etc and the effects Y1 to Y4 etc. This may be treated as qualitative analysis of the societal activity. This can be demonstrated by one example from everyday life of a man–machine system.

Let us consider a gardener preparing flowerbeds in a kitchen garden of a conventional house of an upper-middle-class family.

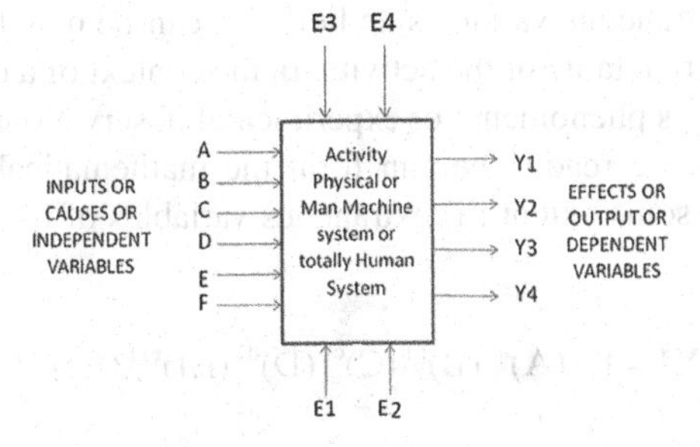

FIGURE 2.1 Block diagrammatic representation of an activity.

The relationship between inputs or causes or independent variables and effects or output or dependent variables.

DOI: 10.1201/9781003318699-2

Supposing the house owner has to prepare around six to eight flower beds. Each is of the size 3 m in length, 1 m wide, 0.5 m in depth. The house owner instructs his gardener accordingly. Let us say that gardener decides to start the work from specific day along with his team of 2–3 helpers. The tools necessary for this operation are i) a Kudali (axe); ii) a phawada (a spade); and iii) a soil collector (a ghamela). The work would take place in a shift of 8 hours: from 8.00 am in the morning till 5.00 pm in the afternoon, with a lunch break from 12.30 pm to 1.30 pm. Let us say that it is a team of three members, with one supervisor performing this task. One member, say, A, digs the soil, B collects the dry soil with a spade and puts it in a ghamela, and C then carries the soil to the where a heap of soil is accumulated.

The planned sequence of working is A digs for 10 mins with a rest for 3 mins at the end of every 7 mins. During this rest of 3 mins, B collects this soil in the ghamela, and C carries this soil to the heap. Like this, workers A, B, and C will work for 30 mins in a sequence. During these 30 mins, the leader of the team with observe activity performed in 7 mins of digging and 3 mins of carrying soil to heap and measuring the quantity in kgf. The measurements taken are i) Initial pulse rate of A, B, C; ii) Pulse rate at the end of 7 mins of digging for A; iii) Pulse rate of B and C at the end of minute 10; iv); the measurement of rise in body temperature would also be noted; (v) the measurement of soil dug in kg at the end of every 10 mins will also be recorded; and (vi) the surface finish of sides of rectangular space created by A will also be noted.

The complete set of observations of the above listed parameters at the end of every 30 mins would be recorded. Along with this, a record would also be maintained of the enthusiasm of workers at the beginning and end of every 30 mins. The specifications of the tools include terms of their geometry, weight, sharpness of digging point, and edge of the spade. At the end of the shift every day, the total of all causes and effects will be generated. Referring to Figure 2.1 for this operation cause A may be anthropometric dimensions of operators including their number, B experience, qualifications of operators, E enthusiasm and attitude of operators, C may be geometric dimensions of the tools, their weight, their condition i.e. (sharpens of edges, tips), D soil condition at the spot before digging, i.e. at the beginning of every 10 mins, F time of operation, whereas E1, E2 could be general features of kitchen garden, E3 ambient temperature, pressure, humidity, E4 noise level. E1, E2 be could constant parameters, and E3, E4 extraneous variables of the system, i.e. activity under consideration in this case. Obviously, the responses of the activity could be total soil dug out in every 10 mins, say Y1, human energy input in terms of pulse rate or blood pressure rise Y2, the quality of operation performed Y3.

In this case, one might say that E1 and E2 are constant parameters of the system, whereas E3 and E4 are system extraneous variables; A, B, C, D, E, and F are planned, and/or actual unplanned, but measured causes or inputs, and Y1, Y2, and Y3 would be responses/outputs of the activity.

In the list of variables above there are some variables which are difficult to measure, e.g. enthusiasm and attitude of the operator. These are categorized as abstract inputs or causes, and sometimes effects; i.e. how the worker feels psychologically at the end of every 10 min. However, these abstract quantities can be measured using the concept of weightages. In short, all causes, constant parameters of the system, and some extraneous variables and effects, can be quantified. In that case, one can arrive at a complete observation table of the activity. In the context of a completely physical system it could have been specified as phenomenal or experimental observations.

Once these observations are ready one can form the mathematical co relationships amongst i) causes; ii) effects; and to some extent iii) extraneous variables of the form as far as this activity is concerned as

$$Y1 = K1(A)^{a1}(B)^{b1}(C)^{c1}(D)^{d1}(E1)^{e11}(E2)^{e21} \tag{2.1}$$

$$Y2 = K2(A)^{a2}(B)^{b2}(C)^{c2}(D)^{d2}(E1)^{e12}(E2)^{e22} \tag{2.2}$$

$$Y3 = K3(A)^{a3}(B)^{b3}(C)^{c3}(D)^{d3}(E1)^{e13}(E2)^{e23} \tag{2.3}$$

TABLE 2.1
Types of the Derived Quantities

Ser	Variables/Constants	Distinguishing Feature of the Physical Quantity	Example
1	Dimensional variables	Have dimension but no physical value	Force, velocity, power
2	Dimensional variables	Neither have dimension nor fixed value	Specific gravity
3	Dimensional constants	Have fixed dimensions and fixed value	Gravitational constant
4	Dimensional constants	Have no dimension but have fixed values	1, 2, 3

Equations 2.1–2.3 can be formed based on information presented in observation Table 2.1 and using the already available mathematical treatments specially matrix algebra. All exponents of Equations 2.1–2.3 can be obtained. The quantities K1, K2, and K3 are known as curve fitting constants. They represent collectively what is not identified as causes logically and/or as a result of action of extraneous variables quantitatively but collectively.

Once the Equations 2.1–2.3 are formed they can be looked upon as design tools for planning similar activities in the future, and not necessarily with the same conditions, but potentially applicable for variations of all causes and constant parameters of the system within, say, 80%–120% range of their variation. Equations 2.1–2.3 can be considered as field data-based models; the concept first launched by one of the authors of this book, Dr. J.P. Modak, in the body of knowledge. It is the field data-based model because it is formed from actual field studies.

2.2 FORMULATION OF MATHEMATICAL MODEL

The experimental studies to be carried for such investigations needs proper planning. Normally, large numbers of variables are involved in such experimental studies. It is expected that the influence of all the variables and parameters be studied economically, without sacrificing the accuracy by reducing the number of variables to a few dimensionless terms through the technique of dimensional analysis (Bansal, 2005; De Felice and Petrillo 2011). Theory of engineering experimentation is a study of scientific phenomenon, which includes the analysis and synthesis of the scientific phenomenon. The study may include a theoretical approach and an experimental approach.

In a theoretical approach, we apply the laws of mechanics and physics, which include i) force balance; ii) momentum balance; iii) energy balance; and iv) quantity balance. Effects of several independent variables on complex process are studied to formulate the process. In an experimental approach, various steps are involved in formulating the model for such a complex phenomenon:

a. **Identify the Causes and Effects:** Performing qualitative analysis of process to identify various physical quantities. These are the causes (inputs). The experiment takes more time and becomes complex if a large number of independent variables is involved. By deducing dimensional equation for the phenomenon, numbers of variables are reduced.
b. **Perform Test Planning:** This involves deciding test envelope, test points, test sequence and plan of experimentation (Fulder et al. 2005; Kotus and Szwarc 2011; Belkhode et al. 2021; Atmanli et al. 2015).
    i. **Test Envelope:** To decide the range of variation of individual independent π terms.
    ii. **Test Points:** To decide and specify values of independent π terms at which experimental set up be set during experimentation.
    iii. **Test Sequence:** To decide the sequence in which the test points be set during experimentation. Sequence may be of ascending order, descending order, or random order. Usually ascending or descending order depending on nature of the phenomenon adopted for irreversible experiment, while random order is adopted for reversible experiments.

iv. **Plan of Experimentation:** In planning, a decision is reached regarding how to vary the independent variable. Planning may be the classical plan or factorial plan. In a classical plan, only one variable is varied at a time, maintaining all other variables constant. In a factorial plan, more than one (or even all) independent variables are varied at a time.

c. **Physical Design of an Experimental set up:** Here, it is necessary to work out physical design of an experimental set up including deciding specifications and procurement of Instrumentation and Experimentation. The next step would be to execute experimentation as per test planning. This will generate experimental data regarding causes (inputs) and effects (responses).

d. **Checking and Rejection of Test Data:** Based on the experimental results, it is necessary to check it for its reliability. The erroneous data is identified and removed from the gathered data for this purpose some statistics-based rules are to be adopted.

e. **Formulation of the Model:** Quantitative relationship is formulated in terms of the dimensional equation between the dependent and independent π terms. This establishes the relationship between Outputs (Effects) and Inputs (Causes).

2.3 LIMITATIONS OF ADOPTING FIELD DATABASE MODEL

For experimental systems in complex activities. Unfortunately, in many such systems, it is not possible to implement the test planning component of the experimental technique (Simsek 2020; Belkhode 2018; Belkhode et al. 2020). For a variety of reasons, it is necessary to enable the activity to proceed as planned. This happens when one wishes to formulate a model for any activity in engine performance, civil construction activities, human assembly operations, industry manufacturing activities etc.

However, the links between the dependent and independent factors are defined qualitatively, based on existing published studies, and the generalized quantitative correlations are not always understood. As a result, it is impossible to establish the quantitative relationship. Since there is no potential to develop a scientific model (logic-based), the only option is to develop a facts-based, or, to be more exact, experimental-data-based, model. As a result, it is recommended that such a model be developed in the current study.

The technique given by Schenck H. Jr. for building generalized experimental data-based models has been presented in the current investigation, which entails the following stages:

- Considering a phenomenon.
- Identification of variables or parameters affecting the phenomenon.
- Reduction of variables through Dimensional analysis.
- Selecting sufficient number of cities with variation of causes and extraneous variables.
- Executing experimental work for data collection.
- Rejection of absurd data.
- Formulation of the model.

Depending upon that filtered data, a quantitative link in between dependent and independent components of the dimensional equation must be developed.

2.4 IDENTIFICATION OF CAUSES AND EFFECTS OF AN ACTIVITY

Identification of causes and effects is the first step. As these causes and effects vary as time elapses. Selection of dependent and independent variables must be done as the phenomena takes place based on observed qualitative analysis of the process. There are four types:

1. Independent variables
   Independent variables can be defined as that influences the activity. This can be changed independently of other variables of the activity.

2. Dependent variables
   The phenomenal quantity or parameters which changes due to variation in the values of the independent variables are called the response or dependent variables.
3. Extraneous variables
   Any parameter which influences the process, but its magnitude cannot be changed or altered at our wish, such as ambient pressure, humidity, temperature, or operator related parameters, such as enthusiasm and attitude.
4. Controlled variables
   Controlled variables are the phenomenal quantities which remain constant all through the duration of activity which are independent variables but due to practical reasons they are not alterable, such as acceleration due to gravity.

2.5 DIMENSIONAL ANALYSIS

The dimensional analysis was rightly used primarily as experimental tool to combine any experimental variable into one (or to reduce the number of experimental variables) this technique was then applied mainly in fluid mechanics and heat transfer for almost all experiments.

a. Dimensions/Quantities
   There are two types of quantities or dimensions.
   i. Fundamental Quantities or Fundamental Units
      Mass (M), Length (L) and Time (T) are three constant dimensions. If heat is involved, then temperature (θ) is also taken as fundamental quantity.
   ii. Derived Dimensions (Secondary Quantities)
      If the physical quantity is designed based on some fundamental, then it is quantities known as derived quantities. The derived quantities classified as shown in the Tables 2.1 and 2.2.

2.6 DIMENSIONAL EQUATION

If in an equation containing physical quantity, each quantity is represented by its dimensional formula, and the result is known as the dimensional equation (Yilmaz et al. 2018).

$$\text{Kinetic energy} = \tfrac{1}{2}\text{mv}^2$$

Here, m is the mass of the body, v is velocity. Writing the formula for kinetic energy in the dimensional equation form, we have

$$[\text{M}]\times[\text{MLT}^{-1}]^2 = [\text{ML}^2\text{T}^{-2}]$$

This is known as the dimensional equation.

The connection between the dependent and independent variables is evaluated in terms of its fundamental dimensions in dimensional analysis of a physical phenomenon to get insights into the underlying link between the dimensionless factors that influence the event. There are numerous ways to reduce the complexity of dimensionless parameters. The most common methods are Rayleigh's and the Buckingham Π theorem. Sometimes, even by observations and preliminary qualitative analysis, a dimensional equation can be formed.

TABLE 2.2
Dimensional Formulae of the Derived Quantities

Ser	Physical Quantity	Relation with Other π Quantities	Dimensional Formula	SI Unit[a]
1	Area	Length × breadth	$[L] \times [L] = [L^2]$	m^2
2	Volume	Length × breadth × height	$[L] \times [L] \times [L] = [L^3]$	m^3
3	Density	Mass/volume	$[M]/[L^3] = [M^1L^{-3}T^0]$	kgm^{-3}
4	Velocity	Distance/time	$[L]/[T] = [M^0L^1T^{-1}]$	ms^{-1}
5	Acceleration	Velocity/time	$[M^0L^1T^{-1}]/[T] = [M^0L^1T^{-2}]$	ms^{-2}
6	Force	Mass × acceleration	$[M] \times [M^1L^1T^{-2}] = [M^1L^1L^{-2}]$	N
7	Momentum	Mass × velocity	$[M] \times [M^0L^1T^{-1}] = [M^1L^1T^{-1}]$	$kgms^{-1}$
8	Work	Force × distance	$[MLT^{-2}] \times [L] = [ML^2T^{-2}]$	Nm
9	Power	Work/time	$[ML^2T^{-2}]/[T] = [ML^2T^{-3}]$	W
10	Pressure	Force/area	$[MLT^{-2}]/[L^2] = [ML^{-1}T^{-2}]$	Nm^{-2}
11	Kinetic energy	½ × mass × (velocity)²	$[M] \times [MLT^{-1}]^2 = [ML^2T^{-2}]$	Nm
12	Potential energy	Mass × g × distance	$[M] \times [M^0LT^{-2}] \times [L] = [ML^2T^{-2}]$	Nm
13	Impulse	Force × time	$[MLT^{-2}] \times [T] = [MLT^{-1}]$	Ns
14	Torque	Force × distance	$[MLT^{-2}] \times [L] = [ML^2T^{-2}]$	Nm
15	Stress	Force/area	$[MLT^{-2}]/[L^2] = [ML^{-1}T^{-2}]$	Nm^{-2}
16	Strain	Extension in length/original length	$[L]/[L] = [M^0L^0T^0]$	Number
17	Elasticity	Stress/strain	$[ML^{-1}T^{-2}]/[M^0L^0T^0] = [ML^{-1}T^{-2}]$	Nm^{-2}
18	Surface tension	Force/length	$[MLT^{-2}]/[L^1] = [ML^0T^{-2}]$	Nm^{-1}
19	Force constant of spring	Applied force/extension in length	$[MLT^{-2}]/[L^1] = [ML^0T^{-2}]$	Nm^{-1}
20	Gravitational constant	Force × (distance)²/(mass)²	$[MLT^{-2}] \times [L^2]/[M^2] = [M^{-1}L^3T^{-2}]$	Nm^2kg^{-2}
21	Frequency	1/time period	$1/[T] = [M^0L^0T^{-1}]$	s^{-1}
22	Angle	Arc/radius	$[L]/[L] = [M^0L^0T^0]$	Rad
23	Angular velocity	Angle/time	$[M^0L^0T^0]/[T] = [M^0L^0T^{-1}]$	Rad s^{-1}
24	Angular acceleration	Angular velocity/time	$[M^0L^0T^{-1}]/[T] = [M^0L^0T^{-2}]$	Rad s^{-2}
25	Moment of inertia	Mass × (distance)²	$[M]/[L^2] = [M^1L^{-2}T^0]$	Kgm^2
26	Angular momentum	Moment of inertia × Angular velocity	$[M^1L^{-2}T^0] \times [M^0L^0T^{-1}] = [ML^2T^{-1}]$	Kgm^2s^{-1}
27	Heat	Energy	$[ML^2T^{-2}]$	J
28	Planck's constant	Energy/frequency	$[ML^2T^{-2}]/[M^0L^0T^{-1}] = [ML^2T^{-1}]$	Js
29	Velocity gradient	Change in velocity/distance	$[M^0LT^{-1}] \times [L] = [M^0L^0T^{-1}]$	s^{-1}
30	Radius of gyration	Distance	$[M^0LT^0]$	m

2.6.1 Rayleigh's Method

Let Y be an independent variable which depends on x_1, x_2, x_3, x_4 etc. According to Rayleigh's method, Y is a function of x_1, x_2, x_3, x_4... etc. and mathematically it can be written as

$$Y = f(x_1, x_2, x_3, x_4)$$

The equation can be written as follows:

$$Y = k\, x_1^{\,a}, x_2^{\,b}, x_3^{\,c}, x_4^{\,d}$$

Where k is constant and a, b, c, and d are arbitrarily indices. The values of a, b, c, and d are obtained by comparing the powers of the fundamental dimension on both the side.

2.6.2 Buckingham π Theorem Method

The Buckingham Π theorem asserts that if the factors determining a physical event have m main dimensions, then the phenomenon may be represented by (n–m) independent dimensionless groups. This theorem can be used for reducing the number of variables affecting the process. According to the theorem, if an equation is dimensionally homogeneous, it may be reduced to a connection between a full set of dimensionless products. In this procedure, m numbers of repeating factors are chosen, and dimensional lower groups are generated one at a time by each of the remaining variables. Rayleigh's approach is also known as the repeating variable method. The recurring variables must be chosen with care. They must have all of the essential dimensions involved in the situation.

1. The dependent variable must not be chosen as a repeating variable.
2. The repeating variables should be chosen in such a way that one variable contain geometric property, other variable contain flow properly and third variable contain fluid property.
3. Usually a length parameter (D or H); a typical velocity V and the fluid density are convenient set of repeating variables.

When the number of variables exceeds the number of fundamental dimensions, Rayleigh's approach of dimensional analysis becomes more arduous (MLT). Buckingham's Π theorem is used to solve this problem.

Using this idea, current tests may significantly enhance their working procedures and be shortened, needing less time, while maintaining control. Using Buckingham's Π theorem, deducing the dimensional equation for a problem minimizes the number of variables in the experiments. It is evident that, if we take the product of the π terms, it will also be a dimensionless number, and hence a π term. This idea is used to achieve further reduction in the number of independent π term variables, which further forms fewer π terms. An attempt is made to apply the above discussed matter to form the mathematical model for brake thermal efficiency and brake specific fuel consumption, maintenance activities of liner piston, crank shaft of locomotive, and solar updraft tower to improve the performance in terms of production, time consumption, and human energy consumption.

3 Design of Experimentation

3.1 INTRODUCTION

It is not possible to plan such activities, on the lines of design of experimentation (H. Schenck, Jr.), when one is studying any complete physical phenomenon, but the phenomenon is very complex to the extent that it is not possible to formulate a logic-based model correlating causes and effects of such a phenomenon, that one is required to apply field data-based models. Formulation of relationships amongst causes and effects (in other words, inputs and outputs), however, is essential. This is because it is only after formulation of such relationships that strengths and weaknesses of present methods are revealed. Once, the weaknesses are known, improvements become possible. Hence, from the point of view of improving system/activity performance, it is essential to form such analytical cause–effect relationships conceptualized in this research work as field or experimental data-based models. In such a situation, the various steps involved in formulating model for such a complex phenomenon are as follows:

1. **Identify the Causes and Effects:** Causes and effects of the phenomenon are identified by performing qualitative analysis of physics of such a phenomenon, and establishing dimensional equations for them. Once a dimensional equation is formed, it is a confirmation that all involved physical quantities are considered. Finding out the dimensionless input quantities due to physical quantities is the first step in this direction. These are the causes (or inputs).
2. **Perform Test Planning:** This involves deciding Test Envelope, Test Points, Test Sequence, and plan of experimentation.
   i. **Test Envelope:** To decide range of variation of an individual independent pi term.
   ii. **Test Points:** To decide & specify values of independent pi terms at which experimental setup be set during experimentation.
   iii. **Test Sequence:** To decide the sequence in which the test points be set during experimentation.
   iv. **Plan of Experimentation:** Whether to adopt Classical Plan or Factorial Plan.
3. **Physical Design of an Experimental setup:** Here, it is necessary to work out physical design of an experimental setup, including deciding specifications and procuring instrumentation, and then to consider the setup.
4. **Next step** would be to execute experimentation as per test planning, and gather data regarding causes (Inputs) and effects (Responses) and to purify the gathered data using statistical methods.
5. **Final step** is to establish the relationship between outputs (Effects) and inputs (Causes).

3.2 LIMITATIONS OF ADOPTING FIELD DATA-BASED MODEL FORMULATION FOR MAN MACHINE SYSTEM

For a man–machine system in complex industrial activities, it is only partially possible to plan experimentation. However, in many such systems, test planning part of experimentation approach is not feasible. One must allow the activity (i.e. phenomenon) to take place as it is, or as planned by others. This happens when one wishes to formulate a model for a maintenance activity of locoshed such as crankshaft maintenance and liner piston maintenance.

DOI: 10.1201/9781003318699-3

A theoretical approach can be adopted in a case where known logic is applicable by correlating the various dependent and independent parameters of the system. Qualitatively, the relationships between the dependent and independent variables are known based on the available literature references, but sometimes the generalized quantitative relationships are not known. Whatever quantitative relationships are known, are pertaining to a specific anthropometric data and a specific task. The relevant quantitative data of locoshed workers is not available. Hence formulating the quantitative relationship based on the logic is not possible, as it is not possible to formulate a theoretical model (logic-based), so one is left with the alternative of formulating an experimental data-based method, or to be very specific in this case, a field data-based model. To that end, it is proposed that such a model is formulated in the present investigation.

3.3 THE APPROACH FOR FORMULATING A FIELD OR EXPERIMENTAL DATA-BASED MODEL

The approach for formulating a field data-based model suggested by Schenck H. Jr. has been proposed in the present investigation, and it involves following steps:

- Identification of variables or parameter affecting the phenomenon.
- Reduction of variables through dimensionless analysis.
- Test planning.
- Rejection of absurd data.
- Formulation of model.

Identification of dependent and independent variables of the phenomenon is to conducted based on known qualitative physics of the phenomenon. If the system involves many independent variables, the experimentation can become tedious and time consuming, and furthermore, costly. So, by deducing the dimensional equation for the phenomenon, we can reduce the number of independent variables and it gets confirmed that all variables are considered. The exact mathematical form of this equation can be obtained based on the field data. This mathematical form is the targeted model. Upon obtaining field data results, adopting the appropriate method for test data checking and rejection of erroneous data is conducted from gathered data. Based on this purified data, one must formulate quantitative relationships between the dependent and independent pi terms of the dimensional equation.

3.4 IDENTIFICATION OF VARIABLES

The first step in this process is identification of variables. The parameters of the phenomenon are known as variables. The term variable is used in a general sense to imply any physical quantity that undergoes change. Identification of variables is based on known qualitative physics of the phenomenon. These variables are of three types:

1. Independent.
2. Dependent.
3. Extraneous.

The term variable is used in a general sense to imply any physical quantity that undergoes change. Any aspect or a physical quantity, due to variation in it, which influences the final outcome of an activity (largely man–machine system) or a phenomenon are known as input or independent variable of the activity. If a physical quantity can be changed independent of other quantities, then, it is also an independent variable. If a physical quantity can be changed in response to the variation of one or more number of variables, then it is treated as a dependent or response variable. If a physical quantity that affects our test is changing in random and in uncontrolled manner in the phenomenon, then it is called as an extraneous variable.

3.5 PROBLEMS ASSOCIATED WITH CRANKSHAFT/LINER PISTON MAINTENANCE ACTIVITIES OF LOCOSHED

The diesel locomotive workshop is a place where repair and maintenance of locomotives is carried out. During the maintenance work in a diesel locomotive workshop, there is a high probability of an accident on any working day. Various hazards are faced during maintenance and overhauling of locomotives, involving components such as the engine block, cylinder head, under frame, fuel injection pump, machines, and equipment used during the maintenance work, potentially resulting in minor or major fatality. Loss of man hours and trained manpower may result from this, which affects the maintenance schedule. The physiological and biomechanical demands of manual work in such environments are greater with the above constraints. Further, locoshed workers have to work in humid, poorly illuminated, and noisy environment. So, due to the present maintenance method, productivity is adversely affected, and there is a substantial requirement of human energy and time.

The engine block of locomotive is one of its key components, and is used to generate and transmit power through the transmission unit. The engine crankshaft and piston liner assembly play vital roles in converting one motion into another. In order to form a mathematical model, the most critical crankshaft and liner piston activities are identified and studied. For this, existing maintenance schedule methods, past failure data, and their experiences are considered. The maintenance of crankshaft and liner piston assembly are carried out in heavy schedule (preventive maintenance strategy). Maintenance of crankshafts and liner pistons involves disassembling assembling of all other connected parts of the engine, which can affect engine alignment, and improper maintenance can result in various engine problems such as lubrication issues, heat generation, vibration, and many other potential problems, reducing engine efficiency and productivity. Again, the power generation of a locomotive depends on the proper working of its crankshaft.

It is one of the primary tasks, and makes up 50–60% of the total time and effort, compared to other activities. It is a repetitive task, involving awkward postures of workers, and can be physically demanding on the neck, shoulders, back, and forearms, compared to other industrial workers.

Great concern is therefore needed to minimize the occurrences of such hazards, and for this purpose it is necessary to form a generalized mathematical model using theories of experimentation for crankshaft and liner piston assembly maintenance activities in locoshed. From this model, present method can be improved with optimized techniques based on field data-based modelling in which dependent and independent variables of an activity can be compared, and the most effective method for improving present practice can be understood.

The effectiveness of crankshaft and liner piston maintenance depends on human factors, workstation data, tools used, specifications of crankshafts and liner pistons, solvents used, and extraneous variables such as temperature, humidity, light, and noise. Therefore, we shall study the effect of these inputs (independent variables) on the outputs, i.e. overhauling time of crankshaft and liner piston maintenance activity, human energy used in crankshaft and liner piston maintenance, and productivity of crankshaft and liner piston activity. Numerous studies have been carried out for the assessment of maintenance operations, but none have applied the Theory of Experimentation suggested by Hilbert (1961) in maintenance systems. No research has been carried out so far to quantitatively specify the influence of an individual independent variable on a response variable. This is only possible by establishing a mathematical model. Hence, we shall conduct that here. Systematically applying this approach will offer new insight to optimization of parameters for crankshaft and liner piston maintenance, in order to optimize productivity for these operations. Once the model is formulated, we will obtain clear idea of the variation of dependent variables in terms of interaction of the various independent variables. Then, applying optimization techniques, determination of the optimum conditions for the execution of the phenomenon becomes possible. Hence it is always felt that any experimental research should ultimately be precipitated to formulation of a model.

A theoretical approach can be adopted when known logic is applicable correlating the various dependent and independent parameters of the system. Though qualitatively, the relationships between the dependent and independent variables are known based on the available literature references, but sometimes the generalized quantitative relationships are not known. Hence formulating the quantitative relationship based on the known logic is not possible. When it is not possible to formulate a theoretical model, one is left only with the alternative of formulating a field data-based model. Hence, it is proposed to formulate such a model here.

Field observations, as regards the response of the phenomenon, demonstrate variation of the inputs of the phenomenon, or that independent variables are varied. Sometimes, quantitative relationships correlating with response variables with a single input are established. In a some investigations, the influence of simultaneous actions of various inputs in order to create the response in the nature of phenomenon have been studied. However, a quantitative representation of the interaction between inputs and response is rare. The main objective of this work is to establish the quantitative relationship of the interaction of inputs on the response variables, and optimize maintenance of the following:

1. Crankshaft
2. Liner piston

It is therefore intended to study the interactive influence of inputs on response variables for crankshaft maintenance activity.

3.5.1 Independent and Dependent Variables of Crankshaft Maintenance Activity

The crankshaft maintenance activity is influenced by following variables (Table 3.1).

TABLE 3.1
Independent and Dependent Variable-Crankshaft Maintenance Activity

Ser	Description of Variables	Variable Type	Symbol	Dimension
1	Overhauling time of crankshaft maintenance activity	Dependent	T_{cs}	$[M^0L^0T^1]$
2	Human energy consumed in crankshaft maintenance activity	Dependent	HE_{cs}	$[M^1L^2T^{-2}]$
3	Productivity of crankshaft maintenance activity	Dependent	Pd_{cs}	$[M^0L^0T^{-1}]$
4	Age of worker	Independent	Ag_s	$[M^0L^0T^1]$
5	Experience of worker	Independent	Ex_s	$[M^0L^0T^1]$
6	Skill of worker	Independent	Sk_s	$[M^0L^0T^0]$
7	Enthusiasm of worker	Independent	En_s	$[M^0L^0T^0]$
8	Habits of worker	Independent	Hb_s	$[M^0L^0T^0]$
9	Health of worker	Independent	hl_s	$[M^0L^0T^0]$
10	Anthropometric data of worker	Independent	Ant_s	$[M^0L^1T^0]$
11	Temperature of workstation	Independent	Tmp_s	$[ML^2T^{-2}]$
12	Humidity of workstation	Independent	Hmd_s	$[M^0L^0T^0]$
13	Noise of workstation	Independent	Noi_s	$[M^0L^0T^0]$
14	Illumination of workstation	Independent	ilm_s	$[M^1L^0T^{-3}]$
15	Diameter of split pin	Independent	Dsp	$[M^0L^1T^0]$

(Continued)

TABLE 3.1 (Continued)

Ser	Description of Variables	Variable Type	Symbol	Dimension
16	Length of split pin	Independent	Lsp	$[M^0L^1T^0]$
17	Diameter of saddle nut	Independent	Dsn	$[M^0L^1T^0]$
18	Length of saddle nut	Independent	Lsn	$[M^0L^1T^0]$
19	Diameter of saddle bolt	Independent	Dsb	$[M^0L^1T^0]$
20	Length of saddle bolt	Independent	Lsb	$[M^0L^1T^0]$
21	Diameter of small Allen bolt	Independent	Dsmab	$[M^0L^1T^0]$
22	Length of small Allen bolt	Independent	Lsmab	$[M^0L^1T^0]$
23	Diameter of side Allen bolt	Independent	Dsiab	$[M^0L^1T^0]$
24	Length of side Allen bolt	Independent	Lsiab	$[M^0L^1T^0]$
25	Diameter of saddle cap	Independent	Dsc	$[M^0L^1T^0]$
26	Length of saddle cap	Independent	Lsc	$[M^0L^1T^0]$
27	Diameter of main journal of crankshaft	Independent	Dmj	$[M^0L^1T^0]$
28	Length of main journal of crankshaft	Independent	Lmj	$[M^0L^1T^0]$
29	Diameter of CP journal of crankshaft	Independent	Dcpj	$[M^0L^1T^0]$
30	Length of CP journal of crankshaft	Independent	Lcpj	$[M^0L^1T^0]$
31	Diameter of bend ring	Independent	Dbr	$[M^0L^1T^0]$
32	Length of bend ring	Independent	Lbr	$[M^0L^1T^0]$
33	Diameter of small Allen spanner	Independent	Dsmsp	$[M^0L^1T^0]$
34	Length of small Allen spanner	Independent	Lsmsp	$[M^0L^1T^0]$
35	Diameter of side Allen spanner	Independent	Dsisp	$[M^0L^1T^0]$
36	Length of side Allen spanner	Independent	Lsisp	$[M^0L^1T^0]$
37	Diameter of barring rod	Independent	Dbrr	$[M^0L^1T^0]$
38	Length of barring rod	Independent	Lbrr	$[M^0L^1T^0]$
39	Diameter of main journal bearing	Independent	Dmjb	$[M^0L^1T^0]$
40	Thickness of main journal bearing	Independent	Tmjb	$[M^0L^1T^0]$
41	Diameter of crank pin journal bearing	Independent	Dcpjb	$[M^0L^1T^0]$
42	Thickness of crank pin journal bearing	Independent	Tcpjb	$[M^0L^1T^0]$
43	Length of frame	Independent	Lfrm	$[M^0L^1T^0]$
44	Width of frame	Independent	Wfrm	$[M^0L^1T^0]$
45	Height of frame	Independent	Hfrm	$[M^0L^1T^0]$
46	Kerosene (solvent)	Independent	Ker	$[M^1L^0T^0]$
47	Lube oil	Independent	Loil	$[M^1L^0T^0]$
48	Emery belt	Independent	Eb	$[M^1L^0T^0]$
49	Compressed air	Independent	Cmar	$[M^1L^0T^0]$
50	Axial clearance of main journal	Independent	Ax mj	$[M^0L^1T^0]$
51	Axial clearance of crank pin	Independent	Axcp	$[M^0L^1T^0]$
52	Saddle bolt elongation	Independent	Elsb	$[M^0L^1T^0]$

3.5.2 Dimensional Analysis of Crankshaft Maintenance Operation

1. π_1 – Anthropometric data of workers performing crankshaft maintenance activity

$$\pi_1 = \left[\frac{(a \times c \times e \times g)}{(b \times d \times f \times h)}\right]$$

Where a, b, c, d, e, f, g, and h are anthropometric dimensions of operators.

2. π_2 – Workers data performing crankshaft maintenance activity
   The ratio of various variables measures of an operator is taken and dimensionless pi term π_2 is calculated as:

$$\pi_2 = [(Ags/Exs), (sks/Ens), (hls/Hbs)]$$

   Where Ags, Exs, sks, Ens, hls, Hbs is the workers data performing crankshaft maintenance activity, i.e. age, experience of worker, skill of worker, enthusiasm, health, and habits.
3. π_3 – Specification of crankshaft
   The ratio of various linear dimensions of the crankshaft is taken as shown below and dimensionless pi term π_3 is calculated as:

$$\begin{aligned}\pi_3 = [&(Dsps/Dsns)\times(Lsps/Dsns)\times(Dsmab/Dsns)\times(Lsns/Dsns)\times(Dsbs/Dsns)\\&\times(Lsbs/Dsns)\times(Lsmab/Dsns)\times(Dsiab/Dsns)\times(Lsiab/Dsns)\times(Dsc/Dsns)\\&\times(Lsc/Dsns)\times(Dmj/Dsns)\times(Lmj/Dsns)\times(Dcpj/Dsns)\times(Lcpj/Dsns)\\&\times(tmjb/Dsns)\times(Dmjb/Dsns)\times(Dcpjb/Dsns)\times(tcpjb/Dsns)]\end{aligned}$$

4. π_4 – Tools used in crankshaft maintenance activity
   The ratio of various geometric dimensions of tools used in crankshaft maintenance is shown below, and dimensionless pi term π_4 is calculated as:

$$\begin{aligned}\pi_4 = [&(Dsmsp/Dbr),(Lbr/Dbr),(Lsmsp/Dbr),(Dsisp/Dbr),\\&(Lsisp/Dbr),(Dbrr/Dbr),(Lbrr/Dbr)]\end{aligned}$$

5. π_5 – Solvent used in crankshaft maintenance activity
   The ratio of various linear dimensions of solvents used in the crankshaft maintenance is shown below, and dimensionless pi term π_5 is calculated as:

$$\pi_5 = [(ker/l\,oil),(Ca/E\,belt)]$$

6. π_6 – Axial clearance of crankpin and main journal of crankshaft
   The ratio of various dimensions of axial clearance of crankshaft pin of the crankshaft maintenance activity is shown below, and dimensionless pi term π_6 is calculated as:

$$\pi_6 = [(Axcp/Elsb),(Axmj/Elsb)]$$

7. π_7 – Workstation data of crankshaft maintenance activity
   The ratio of various linear dimensions of workstation of the crankshaft maintenance activity is shown below, and dimensionless pi term π_7 is calculated as:

$$\pi_7 = [(Lfrm/hfrm),(Wfrm/hfrm)]$$

8. π_8 – Temperature at workplace

$$\pi_8 = temp/100$$

9. π_9 – Humidity at workplace
   Humidity measured in percentage. It is a dimensionless term, and hence it is one of the pi terms.

$$\pi_9 = [(Humd)]$$

10. π_{10} – Light at workplace

$$\pi_{10} = \left[\frac{(\text{illmus} \times (\text{Ags})^3)}{\text{Wt}}\right]$$

11. π_{11} – Noise at workplace
Decibel is a logarithmic unit that indicates ratio or gain. It is a dimensionless unit, and hence it is one of the pi terms.

$$\pi_{11} = [(\text{Noise})]$$

12. π_{D1} – Pi term relating to overhauling time of crankshaft maintenance activity

$$\pi_{D1} = (\text{field time})^2 \times (\text{g/length of saddle bolt})$$

$$\left[M^0L^0T^0\right] = \left[M^0L^0T^2\right]^a \left[M^0L^0T^{-2}\right]^b \left[M^0L^{-1}T^0\right]^c$$

$$M = 0 = 0 + 0 + 0,\ a = 0$$

$$L = 0 = 0 + b - c,\ b - c = 0$$

$$T = 0 = 2a - 2b + 0, a - b = 0, 0 - b = 0, b = 0 \text{ and } c = 0$$

13. π_{D2} – Pi term relating to human energy consumed in crankshaft maintenance activity

$$\pi_{D2} = (\text{field energy})/(\text{field time} \times 1.341 \times 10^{-3}\,\text{HP})$$

$$\left[M^0L^0T^0\right] = \left[M^1L^2T^{-2}\right]^a \left[M^0L^0T^{-1}\right]^b \left[M^{-1}L^{-2}T^3\right]^c$$

$$M = 0 = a + 0 - c,\ a - c = 0$$

$$L = 0 = 2a + 0 - 2c, a - c = 0, a = c = 0$$

$$T = 0 = -2a - b + 3c, -2a - b + 3c = 0, 0 - b + 0 = 0$$

14. π_{D3} – Pi term relating productivity of crankshaft maintenance activity

$$\pi_{D3} = (\text{crankshaft overhauling time/total engine overhauling time})$$

$$\left[M^0L^0T^0\right] = \left[M^0L^0T^1\right]^a \left[M^0L^0T^{-1}\right]^b$$

$$M = 0 = 0 + 0, a = 0$$

$$L = 0 = 0 + 0, b = 0$$

$$T = 0 = a - b, 0 - b = 0$$

3.5.3 Establishment of Dimensionless Pi Terms for Crankshaft Maintenance Activity

The independent variables have been reduced into a group of pi terms. Following is the list of dimensionless independent and dependent pi terms of crankshaft maintenance activity (Tables 3.2, 3.3 and 3.4).

It is also decided to establish the quantitative relationships of the interaction of these inputs (a, b, c, d, e, f, g, h, i, j, and k) mentioned above on the response variable time of overhauling of crankshaft, and human energy consumed as well as productivity therein. Similarly interactive influences of inputs on response variables are studied for liner piston activity (Table 3.5).

TABLE 3.2
Independent Dimensionless Pi Terms of Crankshaft Maintenance Activity

Ser	Nature of Basic Physical Quantities	Independent Dimensionless Pi Term
1	Pi term relating anthropometric data of worker	π_1 = [(a × c × e × g)/(b × d × f × h)]
2	Pi term relating aptitude data of worker	π_2 = [(Ags/Exs), (sks/Ens), (hls/Hbs)]
3	Pi term relating specification of crankshaft	π_3 = [(Dsp/Dsn) × (Lsp/Dsn) × (Dsmab/Dsn) × (Lsn/Dsn) × (Dsb/Dsn) × (Lsb/Dsn) × (Lsmab/Dsn) × (Dsiab/Dsn) × (Lsiab/Dsn) × (Dsc/Dsn) × (Lsc/Dsn) × (Dmj/Dsn) × (Lmj/Dsn) × (Dcpj/Dsn) × (Lcpj/Dsn) × (Tmjb/Dsn) × (Dmjb/Dsn) × (Dcpjb/Dsn) × (Tcpjb/Dsn)]
4	Pi term relating specification of tools	π_4 = [(Lbr/Dbr), (Dsmsp/Dbr), (Lsmsp/Dbr), (Dsisp/Dbr), (Lsisp/Dbr), (Dbrr/Dbr), (Lbrr/Dbr)]
5	Pi term relating specification of solvents	π_5 = [(ker/l oil), (Cmar/Eb)]
6	Pi term relating specification of Axial clearance of crank pin and Saddle bolt elongation	π_6 = [(Axcp/Elsb), (Axmj/Elsb)]
7	Pi term relating specification of workstation	π_7 = [(Lfrm/Hfrm), (wfrm/Hfrm)]
8	Pi term relating specification of temperature at workplace	π_8 = [temp/100]
9	Pi term relating specification of humidity at workplace	π_9 = [humidity in %]
10	Pi term relating specification of illumination at workplace	π_{10} = [(ilms × (Ags) 3/wt)]
11	Pi term relating specification of noise at workplace	π_{11} = [noise]

TABLE 3.3
Dependent Dimensionless Pi Terms of Crankshaft Maintenance Activity

Ser	Nature of Basic Physical Quantities	Dependent Pi Terms
1	Pi term relating to overhauling time of crankshaft maintenance activity	z_{1C} = [(field time of crankshaft)^2 × g/length of saddle]
2	Pi term relating Human Energy consumed in crankshaft maintenance activity	z_{2C} = [field energy consumed in crankshaft/(field time of crankshaft × 1.341 × 10^{-3})]
3	Pi term relating productivity of crankshaft maintenance activity	z_{3C} = [overhauling time of crankshaft/total engine overhauling time] = field overhauling time of crankshaft/21,600 min

It is also decided to establish the quantitative relationships of the interaction of these inputs (a, b, c, d, e, f, g, h, i, j, and k) mentioned above on the response variable overhauling time of liner piston, human energy consumed in maintenance activity of liner piston and productivity of liner piston maintenance activity.

The advantage of such quantitative relationships is to ascertain the relative influence of inputs on the responses. This is possible if such a quantitative relationship, such as a model, is established. Once the model is developed, the exercise of optimization could be taken up subsequently. The interest of locoshed workers lies in arranging optimized inputs so as to acheive targeted responses.

TABLE 3.4
The Interactive Influence of Inputs on Response Variables for Crankshaft Maintenance Activity

Types of Variables		Description
Independent variables	a	Anthropometric data of worker
	b	Worker data, including attitude and aptitude to work
	c	Specification of crankshaft
	d	Specification of tools
	e	Solvents used for crankshaft maintenance
	f	Axial clearance of crankpin
Independent variables	g	Workstation data
	h	Temperature at workplace
	i	Humidity at workplace
	j	Light at workplace
	k	Noise at workplace
Dependent variables	A	Overhauling time of crankshaft maintenance activity (z_{1C})
	B	Human Energy consumed in crankshaft maintenance activity (z_{2C})
	C	Productivity of crankshaft maintenance activity (z_{3C})

TABLE 3.5
The Interactive Influence of Inputs on Response Variables for Liner Piston Activity

Types of Variables		Description
Independent variables	a	Anthropometric data of worker
	b	Workers data, including attitude to do and aptitude for work
	c	Specification of liner piston
	d	Specification of tools used in liner piston maintenance activity
	e	Solvents used in liner piston maintenance activity
	f	Axial clearance of big end bearing
	g	Workstation data of liner piston maintenance activity
	h	Temperature at workplace
	i	Humidity at workplace
	j	Light at workplace
	k	Noise at workplace
Dependent variables	A	Overhauling time of liner piston maintenance activity (z_{1P})
	B	Human energy consumed in liner piston maintenance activity(z_{2P})
	C	Productivity of liner piston maintenance activity (z_{3P})

When it comes to the formulation of the model, there is no other substitute but to adopt the methodology of experimentation, more suitably the one suggested by Hilbert (Schenck 1961). This method is applied to a complex phenomenon for which logic-based modelling is highly improbable. The inputs to the phenomenon, the extraneous variables, and response variables are identified. The inputs can be varied over a broad yet in practically possible range, and response data is captured. On the basis of the gathered data, models are formed, and the effect of extraneous variables is minimized by applying the suggested method properly.

3.5.4 Formulation of a Field Data-Based Model for Response Variables of Crankshaft Maintenance Activity

For formulation of model, 11 independent pi terms and 3 dependent pi terms have been identified for crankshaft maintenance activity and each dependent pi term is a function of available independent pi terms:

$$z_{1C} = f(\pi_{D1}) = f(\pi_1, \pi_2, \pi_3, \pi_4, \pi_5, \pi_6, \pi_7, \pi_8, \pi_9, \pi_{10}, \pi_{11})$$
$$z_{2C} = f(\pi_{D2}) = f(\pi_1, \pi_2, \pi_3, \pi_4, \pi_5, \pi_6, \pi_7, \pi_8, \pi_9, \pi_{10}, \pi_{11})$$
$$z_{3C} = f(\pi_{D3}) = f(\pi_1, \pi_2, \pi_3, \pi_4, \pi_5, \pi_6, \pi_7, \pi_8, \pi_9, \pi_{10}, \pi_{11})$$

f stands for "function of".
Where

$z_{1C} = \pi_{D1}$, first dependent pi term, overhauling time of crankshaft maintenance
$z_{2C} = \pi_{D2}$, second dependent pi term, human energy consumed in crankshaft maintenance
$z_{3C} = \pi_{D3}$, third dependent pi term, productivity of crankshaft maintenance

The probable exact mathematical form for a dimensional equation of the phenomenon could be relationships assumed to be of exponential form
Dependent pi term:

$$\begin{aligned} z_C = K \times \{ &[(a \times c \times e \times g)/(b \times d \times f \times h)]^a, \\ &[(Ags/Exs), (sks/Ens), (hls/Hbs)]^b, \\ &[(Dsps/Dsns) \times (Lsps/Dsns) \times (Dsmab/Dsns) \times (Lsns/Dsns) \times (Dsbs/Dsns) \\ &\times (Lsbs/Dsns) \times (Lsmab/Dsns) \times (Dsiab/Dsns) \times (Lsiab/Dsns) \\ &\times (Dsc/Dsns) \times (Lsc/Dsns) \times (Dmj/Dsns) \times (Lmj/Dsns) \times (Dcpj/Dsns) \\ &\times (Lcpj/Dsns) \times (tmjb/Dsns) \times (Dmjb/Dsns) \times (Dcpjb/Dsns) \times (tcpjb/Dsns)]^c, \\ &\quad [(Dsmsp/Dbr), (Lbr/Dbr), (Lsmsp/Dbr), (Dsisp/Dbr), (Lsisp/Dbr), (Dbrr/Dbr), \\ &\quad (Lbrr/Dbr)]^d, \\ &\quad [(ke\, r/loil), (Ca/E\, belt)]^e \\ &\quad [(Axcp/Elsb), (Axmj/Elsb)]^f \\ &\quad [(Lfrm/hfrm), (Wfrm/hfrm)]^g \\ &\quad [temp/100]^h \\ &\quad [(Humd)]^i \\ &\quad [(illmus \times Ags)^3/Wt]^j \\ &\quad [(Noise)]^k \} \end{aligned}$$

3.5.5 Model Formulation by Identifying the Curve Fitting Constant and Various Indices of Pi Terms of Crankshaft Maintenance Activity

The multiple regression analysis helps to identify the indices of the different pi terms in the model aimed at, by considering 11 independent pi terms and 1 dependent pi term. Let the model be of the form:

$$z_{1C} = K_1 \times [(\pi_1)^a{}_1 \times (\pi_2)^b{}_1 \times (\pi_3)^c{}_1 \times (\pi_4)^d{}_1 \times (\pi_5)^e{}_1 \times (\pi_6)^f{}_1 \times (\pi_7)^g{}_1 \times (\pi_8)^h{}_1 \times (\pi_9)^i{}_1 \times (\pi_{10})^j{}_1 \times (\pi_{11})^k{}_1]$$
$$z_{2C} = K_2 \times [(\pi_1)^a{}_2 \times (\pi_2)^b{}_2 \times (\pi_3)^c{}_2 \times (\pi_4)^d{}_2 \times (\pi_5)^e{}_2 \times (\pi_6)^f{}_2 \times (\pi_7)^g{}_2 \times (\pi_8)^h{}_2 \times (\pi_9)^i{}_2 \times (\pi_{10})^j{}_2 \times (\pi_{11})^k{}_2]$$
$$z_{3C} = K_3 \times [(\pi_1)^a{}_3 \times (\pi_2)^b{}_3 \times (\pi_3)^c{}_3 \times (\pi_4)^d{}_3 \times (\pi_5)^e{}_3 \times (\pi_6)^f{}_3 \times (\pi_7)^g{}_3 \times (\pi_8)^h{}_3 \times (\pi_9)^i{}_3 \times (\pi_{10})^j{}_3 \times (\pi_{11})^k{}_3]$$

This is known as an exponential form of a model.

To determine the values of K_1, a_1, b_1, c_1, d_1, e_1, f_1, g_1, h_1, i_1, j_1, and k_1 and to arrive at the regression hyper-plane, the above equations are presented as follows.

Taking log on both sides of Equation 4.2 for overhauling time of crankshaft maintenance (z_{1C}) we get 12 unknown terms in the equations:

$$\log z_{1C} = \log k_1 + a_1 \log \pi_1 + b_1 \log \pi_2 + c_1 \log \pi_3 + d_1 \log \pi_4 + e_1 \log \pi_5 + f_1 \log \pi_6 + g_1 \log \pi_7 + h_1 \log \pi_8 + i_1 \log \pi_9 + j_1 \log \pi_{10} + k_1 \log \pi_{11}$$

Let:

$$z_{1C} = \log z_{1C}, \quad B = \log \pi_2 \quad E = \log \pi_5, \quad H = \log \pi_8,$$
$$K_1 = \log K_1, \quad C = \log \pi_3, \quad F = \log \pi_6, \quad I = \log \pi_9,$$
$$A = \log \pi_1, \quad D = \log \pi_4, \quad G = \log \pi_7, \quad J = \log \pi_{10},$$
$$k_1 = \log \pi_{11}$$

Putting these same values in log equations, they can be written as:

$$Z_{1C} = K_1 + a_1A + b_1B + c_1C + d_1D + e_1E + f_1F + g_1G + h_1H + i_1I + j_1J + k_1K$$

This is a regression equation of Z on A, B, C, D, E, F, G, H, I, J, K in an n dimensional coordinate system this represents a regression hyper-plane as:

$$\Sigma z_{1C} = nK_1 + a_1\Sigma A + b_1\Sigma B + c_1\Sigma C + d_1\Sigma D + e_1\Sigma E + f_1\Sigma F + g_1\Sigma G + h_1\Sigma H + i_1\Sigma I + j_1\Sigma J + k_1\Sigma K$$

$$\Sigma z_{1C}A = K_1\Sigma A + a_1\Sigma A^2 + b_1\Sigma AB + c_1\Sigma AC + d_1\Sigma AD + e_1\Sigma AE + f_1\Sigma AF + g_1\Sigma AG + h_1\Sigma AH + i_1\Sigma AI + j_1\Sigma AJ + k_1\Sigma AK$$

$$\Sigma z_{1C}B = K_1\Sigma B + a_1\Sigma AB + b_1\Sigma B^2 + c_1\Sigma BC + d_1\Sigma BD + e_1\Sigma BE + f_1\Sigma BF + g_1\Sigma BG + h_1\Sigma BH + i_1\Sigma BI + j_1\Sigma BJ + k_1\Sigma BK$$

$$\Sigma z_{1C}C = K_1\Sigma C + a_1\Sigma AC + b_1\Sigma CB + c_1\Sigma C^2 + d_1\Sigma C + e_1\Sigma CE + f_1\Sigma CF + g_1\Sigma CG + h_1\Sigma CH + i_1\Sigma CI + j_1\Sigma CJ + k_1\Sigma CK$$

$$\Sigma z_{1C}D = K_1\Sigma D + a_1\Sigma AD + b_1\Sigma BD + c_1\Sigma CC + d_1\Sigma D^2 + e_1\Sigma DE + f_1\Sigma DF + g_1\Sigma DG + h_1\Sigma DH + i_1\Sigma DI + j_1\Sigma DJ + k_1\Sigma DK$$

$$\Sigma z_{1C}E = K_1\Sigma E + a_1\Sigma AE + b_1\Sigma BE + c_1\Sigma CE + d_1\Sigma DE + e_1\Sigma E^2 + f_1\Sigma FE + g_1\Sigma GE + h_1\Sigma HE + i_1\Sigma IE + j_1\Sigma JE + k_1\Sigma EK$$

$$\Sigma z_{1C}F = K_1\Sigma F + a_1\Sigma AF + b_1\Sigma FB + c_1\Sigma FC + d_1\Sigma FD + e_1\Sigma FE + f_1\Sigma F^2 + g_1\Sigma FG + h_1\Sigma FH + i_1\Sigma FI + j_1\Sigma FJ + k_1\Sigma FK$$

$$\Sigma z_{1C}G = K_1\Sigma G + a_1\Sigma AG + b_1\Sigma GB + c_1\Sigma GC + d_1\Sigma GD + e_1\Sigma GE + f_1\Sigma GF + g_1\Sigma G^2 + h_1\Sigma GH + i_1\Sigma GI + j_1\Sigma GJ + k_1\Sigma GK$$

$$\Sigma z_{1C}H = K_1\Sigma H + a_1\Sigma AH + b_1\Sigma HB + c_1\Sigma HC + d_1\Sigma HD + e_1\Sigma HE + f_1\Sigma HF + g_1\Sigma HG + h_1\Sigma H^2 + i_1\Sigma HI + j_1\Sigma HJ + k_1\Sigma HK$$

$$\Sigma z_{1C}I = K_1\Sigma I + a_1\Sigma AI + b_1\Sigma IB + c_1\Sigma IC + d_1\Sigma ID + e_1\Sigma IE + f_1\Sigma IF + g_1\Sigma IG + h_1\Sigma IH + i_1\Sigma HI^2 + j_1\Sigma IJ + k_1\Sigma IK$$

$$\Sigma z_{1C}J = K_1\Sigma J + a_1\Sigma AJ + b_1\Sigma JB + c_1\Sigma JC + d_1\Sigma JD + e_1\Sigma JE + f_1\Sigma JF + g_1\Sigma JG + h_1\Sigma JH + i_1\Sigma JI + j_1\Sigma J^2 + k_1\Sigma JK$$

$$\Sigma z_{1C}K = K_1\Sigma K + a_1\Sigma AK + b_1\Sigma KB + c_1\Sigma KC + d_1\Sigma KD + e_1\Sigma KE + f_1\Sigma KF + g_1\Sigma KG + h_1\Sigma KH + i_1\Sigma KI + j_1\Sigma KJ + k_1\Sigma K^2$$

In the above set of equations, the values of the term Z on LHS and the multipliers of K_1, a_1, b_1, c_1, d_1, e_1, f_1, g_1, h_1, i_1, j_1, and k_1 on RHS side are substituted to calculate the values of the unknowns (K_1, a_1, b_1, c_1, d_1, e_1, f_1, g_1, h_1, i_1, j_1, and k_1). After substituting these values in the equations, we will get a set of 12 equations, which are to be solved simultaneously to get values of unknowns (viz. K_1, a_1, b_1, c_1, d_1, e_1, f_1, g_1, h_1, i_1, j_1, and k_1). The above equation can be verified in the matrix form and values of unknown K_1, a_1, b_1, c_1, d_1, e_1, f_1, g_1, h_1, i_1, j_1, and k_1 can be obtained by using matrix analysis.

$$X = \text{inv}(w) \times Z$$

The matrix obtained by matrix method for solving these equations is:

$$z_{1C} \times \begin{bmatrix} 1 \\ A \\ B \\ C \\ D \\ E \\ F \\ G \\ H \\ I \\ J \\ K \end{bmatrix} = \begin{bmatrix} n & A & B & C & D & E & F & G & H & I & J & K \\ A & A^2 & BA & CA & DA & EA & FA & GA & HA & IA & JA & KA \\ B & AB & B^2 & CB & DB & EB & FB & GB & HB & IB & JB & KB \\ C & AC & BC & C^2 & DC & EC & FC & GC & HC & IC & JC & KC \\ D & AD & BD & CD & D^2 & ED & FD & GD & HD & ID & JD & KD \\ E & AE & BE & CE & DE & E^2 & FE & GE & HE & IE & JE & KE \\ F & AF & BF & CF & DF & EF & F^2 & GF & HF & IF & JF & KF \\ G & AG & BG & CG & DG & EG & FG & G^2 & HG & IG & JG & KG \\ H & AH & BH & CH & DH & EH & FH & GH & H^2 & IH & JH & KH \\ I & AI & BI & CI & DI & EI & FI & GI & HI & I^2 & JI & KI \\ J & AJ & BJ & CJ & DJ & EJ & FJ & GJ & HJ & IJ & J^2 & KJ \\ K & AK & BK & CK & DK & EK & FK & GK & HK & IK & JK & K^2 \end{bmatrix} \times \begin{bmatrix} K_1 \\ a_1 \\ b_1 \\ c_1 \\ d_1 \\ e_1 \\ f_1 \\ g_1 \\ h_1 \\ i_1 \\ j_1 \\ k_1 \end{bmatrix}$$

A = 12 × 12 matrix of multipliers K_1, a_1, b_1, c_1, d_1, e_1, f_1, g_1, h_1, i_1, j_1, and k_1
B = 12 × 1 matrix of terms on LHS
C = 12 × 1 matrix of solutions of K_1, a_1, b_1, c_1, d_1, e_1, f_1, g_1, h_1, i_1, j_1, and k_1

Then, C = inverse (A) × B gives the unique values of K_1, a_1, b_1, c_1, d_1, e_1, f_1, g_1, h_1, i_1, j_1, and k_1. In the above equations, n is the number of sets of readings. A, B, C, D, E, F, G, H, I, J, and K represent the independent pi terms π_1, π_2, π_3, π_4, π_5, π_6, π_7, π_8, π_9, π_{10}, and π_{11}. While z represents a dependent pi term. Next, calculate the values of independent pi terms for corresponding dependent pi term, which helps to form the equations in matrix form. Solving, the above matrix by using MATLAB software for this purpose for making this process of model formulation precise, quick, and less cumbersome.

3.5.6 Independent and Dependent Variables of Liner Piston Maintenance

The liner piston maintenance activity is influenced by following variables (Table 3.6).

3.5.7 Dimensional Analysis of Liner Piston Maintenance

1. π_1 – Anthropometric data of workers performing liner piston maintenance

$$\pi_1 = [(a \times c \times e \times g)/(b \times d \times f \times h)]$$

Where a, b, c, d, e, f, g, and h are anthropometric dimensions of operators

TABLE 3.6
Independent and Dependent Variable-Liner Piston Maintenance Activity

Ser	Description of Variables	Variable Type	Symbol	Dimension
1	Overhauling time for liner piston maintenance activity	Dependent	To	$[M^0L^0T^1]$
2	Human energy consumed in liner piston activity	Dependent	H E	$[M^1L^2T^{-2}]$
3	Productivity of liner piston overhauling	Dependent	Pd	$[M^0L^0T^{-1}]$
4	Age of worker	Independent	Aw	$[M^0L^0T^1]$
5	Experience of worker	Independent	Exw	$[M^0L^0T^1]$
6	Skill of worker	Independent	Skw	$[M^0L^0T^0]$
7	Enthusiasm of worker	Independent	Ew	$[M^0L^0T^0]$
8	Habits of worker	Independent	Hw	$[M^0L^0T^0]$
9	Health of worker	Independent	HlW	$[M^0L^0T^0]$
10	Anthropometric data of worker	Independent	Ad	$[M^0L^1T^0]$
11	Temperature of workstation	Independent	Tws	$[ML^2T^{-2}]$
12	Humidity of workstation	Independent	Hws	$[M^0L^0T^0]$
13	Noise of workstation	Independent	Nws	$[M^0L^0T^0]$
14	Illumination of workstation	Independent	IlWs	$[M^1L^0T^{-3}]$
15	Diameter of liner	Independent	Dlin	$[M^0L^1T^0]$
16	Length of liner	Independent	Llin	$[M^0L^1T^0]$
17	Diameter of piston	Independent	Dpis	$[M^0L^1T^0]$
18	Length of piston	Independent	Lprng	$[M^0L^1T^0]$
19	Diameter of piston rings	Independent	Dprng	$[M^0L^1T^0]$
20	Thickness of piston rings	Independent	Tprng	$[M^0L^1T^0]$
21	Length of piston stroke	Independent	Lpstrk	$[M^0L^1T^0]$
22	Length of piston pin	Independent	Lppin	$[M^0L^1T^0]$
23	Diameter of piston pin	Independent	Dppin	$[M^0L^1T^0]$
24	Length of connecting rod	Independent	Lcr	$[M^0L^1T^0]$
25	Diameter of big end bearing	Independent	Dbgbr	$[M^0L^1T^0]$
26	thickness of big end bearing	Independent	Tbgbr	$[M^0L^1T^0]$
27	Diameter of little end bearing	Independent	Dltbr	$[M^0L^1T^0]$
28	Length of little end bearing	Independent	Lltbr	$[M^0L^1T^0]$
29	Diameter of big end nut	Independent	Dbnt	$[M^0L^1T^0]$
30	Length of big end nut	Independent	Lbnt	$[M^0L^1T^0]$
31	Diameter of big end bolt	Independent	Dbbl	$[M^0L^1T^0]$
32	Length of big end bolt	Independent	Lbbl	$[M^0L^1T^0]$
33	Small end bearing axial clearance	Independent	Scl	$[M^0L^1T^0]$
34	Big end bearing axial clearance	Independent	Bcl	$[M^0L^1T^0]$
35	Bolt elongation	Independent	Bel	$[M^0L^1T^0]$
36	Diameter of bracket compressor to set piston rings	Independent	Dbrcm	$[M^0L^1T^0]$
37	Length of bracket compressor to set piston rings	Independent	Lbrcm	$[M^0L^1T^0]$
38	Diameter of bracket expander to remove piston and liner ring	Independent	Dbrex	$[M^0L^1T^0]$
39	Length of bracket expander to remove piston and liner ring	Independent	Lbrex	$[M^0L^1T^0]$
40	Diameter of socket for big end nut bolt	Independent	Dst	$[M^0L^1T^0]$
41	Length of socket for big end nut bolt	Independent	Lst	$[M^0L^1T^0]$
42	Length of hammer	Independent	Lhmr	$[M^0L^1T^0]$
43	Diameter of hammer	Independent	Dhmr	$[M^0L^1T^0]$
44	Length of nose plier	Independent	Lnpl	$[M^0L^1T^0]$

(Continued)

TABLE 3.6 (Continued)

Ser	Description of Variables	Variable Type	Symbol	Dimension
45	Length of liner puller	Independent	Llipl	$[M^0L^1T^0]$
46	Diameter of liner puller	Independent	Dlipl	$[M^0L^1T^0]$
47	Length of bush puller	Independent	Lbpl	$[M^0L^1T^0]$
48	Diameter of bush puller	Independent	Dbpl	$[M^0L^1T^0]$
49	Kerosene (solvent)	Independent	Ke	$[M^0L^1T^0]$
50	Emery belt	Independent	Eb	$[M^1L^0T^0]$
51	Lube oil	Independent	Loil	$[M^1L^0T^0]$
52	Orient 11 solvent	Independent	O11	$[M^1L^0T^0]$
53	Height of connecting rod workstation	Independent	Hcrt	$[M^0L^1T^0]$
54	Length connecting rod workstation	Independent	Lcrt	$[M^0L^1T^0]$
55	Breath of connecting rod workstation	Independent	Bcrt	$[M^0L^1T^0]$

2. π_2 – Workers data performing liner piston maintenance
   The ratio of various variables measured of an operator is shown as below, and dimensionless pi term π_2 is calculated as:

$$\pi_2 = [(\text{Ags/Exs}), (\text{sks/Ens}), (\text{hls/Hbs})]$$

   Where Ags, Exs, sks, Ens, hls, Hbs is the workers data, i.e. age, experience of worker, skill of worker, enthusiasm, health, and habits
3. π_3 – Specification of liner piston
   The ratio of various specifications of dimensions of liner piston is shown below, and dimensionless pi term π_3 is calculated as:

$$\begin{aligned}\pi_3 = [&(\text{Dlins/Lcr}) \times (\text{l.lin/L.cr}) \times (\text{D.pis/L.cr}) \times (\text{L.pis/L.cr}) \times (\text{D.prng/Lc.r}) \times (\text{t.prng/L.crs})\\ &\times (\text{l.pstrk./L.cr}) \times (\text{D.ppin/L.cr}) \times (\text{L.ppin./L.cr}) \times (\text{Dbg.br./L.cr}) \times (\text{t.bg.br/L.crs})\\ &\times (\text{Dlt.brj./L.cr}) \times (\text{t.lt.brj/L.cr}) \times (\text{D.b.nj./L.cr}) \times (\text{L.b.nt/L.cr}) \times (\text{d.b.bl/L.cr})\\ &\times (\text{l.b.bl/L.cr})]\end{aligned}$$

4. π_4 – Specification of tools used in liner piston maintenance activity
   The ratio of various geometric dimensions of tools used in linear piston maintenance is shown below, and dimensionless pi term π_4 is calculated as:

$$\begin{aligned}\pi_4 = [&(\text{Dbr/Lbr}), (\text{Dsmsp/Lbr}), (\text{Lsmsp/Lbr}), (\text{Dsisp/Lbr}), (\text{Lsisp/Lbr}), (\text{Dbrr/Lbr}),\\ &(\text{Lbrr/Lbr})]\end{aligned}$$

5. π_5 – Specification of solvent used in liner piston maintenance activity
   The ratio of various linear dimensions of solvents used in the liner piston maintenance activity is shown below, and dimensionless pi term π_5 is calculated as:

$$\pi_5 = [(\text{ker /Ebelt}), (\text{orient11/L oil})]$$

6. π_6 – Axial clearance of big end bearing of liner piston maintenance activity
The ratio of linear dimensions of axial clearance of big end bearing and saddle bolt elongation is taken to calculate dimensionless pi term π_6 in liner piston maintenance.

$$\pi_6 = [(A \times \text{bigend/Elsb}),\ (A \times \text{smallend/Elsb})]$$

7. π_7 – Workstation data liner piston maintenance
The ratio of various linear dimensions of workstation of the liner piston maintenance is shown below, and dimensionless pi term π_7 is calculated as:

$$\pi_7 = \{[(h(\text{crodtable}))/(L(\text{crodtable}))],[(b(\text{crodtable}))/(L(\text{crodtable}))]\}$$

8. π_8 – Temperature at workplace

$$\pi_8 = [\text{temp}/100]$$

9. π_9 – Humidity at workplace
Humidity measured as a percentage. It is a dimensionless term, hence it is one of the pi terms:

$$\pi_9 = [(\text{Humd}\%)]$$

10. π_{10} – Illumination at workplace

$$\pi_{10} = \left[\frac{(\text{illmus} \times (\text{Ags})^3)}{\text{Wt}}\right]$$

11. π_{11} – Noise at workplace
Decibel is a logarithmic unit that indicates ratio or gain. It is a dimensionless unit and hence it is one of the pi terms.

$$\pi_{11} = [(\text{Noise})]$$

12. π_{D1} – Pi term relating to overhauling time of liner piston maintenance activity

$$\pi_{D1} = [(\text{field time of liner piston overhauling})^2 \times \text{g/length of liner}]$$
$$[M^0L^0T^0] = [M^0L^0T^2]^a \quad [M^0L^0T^{-2}]^b \quad [M^0L^{-1}T^0]^c$$
$$M = 0 = 0 + 0 + 0,\ a = 0$$
$$L = 0 = 0 + b - c,\ b - c = 0$$
$$T = 0 = 2a - 2b + 0, a - b = 0, 0 - b = 0, b = 0 \text{ and } c = 0$$

13. π_{D2} – Pi term relating to human energy consumed in liner piston maintenance

$$\pi_{D1} = [(\text{field time of liner piston over hauling})^2 \times (\text{g/length of liner})]$$
$$[M^0L^0T^0] = [M^1L^2T^{-2}]^a \quad [M^0L^0T^{-1}]^b \quad [M^{-1}L^{-2}T^3]^c$$
$$M = 0 = a + 0 - c, a - c = 0$$
$$L = 0 = 2a + 0 - 2c, a - c = 0, a = c = 0$$
$$T = 0 = -2a - b + 3c, -2a - b + 3c = 0, 0 - b + 0 = 0$$

14. π_{D3} – Pi term relating productivity of liner piston maintenance activity
    π_{D3} = (liner overhauling time/total engine overhauling time)

$$[M^0L^0T^0] = [M^0L^0T^1]^{a\cdots}[M^0L^0T^{-1}]^{b\cdots}$$
$$M = 0 = 0 + 0, a = 0$$
$$L = 0 = 0 + 0, b = 0$$
$$T = 0 = a - b, 0 - 0 = 0$$

3.5.8 Establishment of Dimensionless Pi Terms for Liner Piston Maintenance

The independent variables have been reduced to a group of pi terms. Following is the list of dimensionless independent and dependent pi terms of linear piston maintenance activity (Table 3.7 and 3.8).

TABLE 3.7
Independent Dimensionless Pi Terms of Liner Piston Maintenance Activity

Ser	Nature of Basic Physical Quantities	Independent Dimensionless Pi Term
1	Pi term relating anthropometric data of worker	$\pi_1 = [(a \times c \times e \times g)/(b \times d \times f \times h)] = Ad$
2	Pi term relating data of worker	$\pi_2 = [(Ags/Exs), (sks/Ens), (hls/Hbs)] = wd$
3	Pi term relating specification of liner piston	π_3 = [(Dlins/Dbnt) × (l.lin/Dbnt) × (D.pis/Dbnt) × (L.pis/Dbnt) × (D.prng/Lc.Dbnt) × (T.prng/Dbnt) × (Lpstrk./Dbnt) × (D.ppin/Dbnt) × (L.ppin/Dbnt) × (Lcr/Dbnt) × (Dbg.br/Dbnt) × (Tbg.br/Dbnt) × (Dlt.br/Dbnt) × (T.lt.br/Dbnt) × (L.b.nt/Dbnt) × (Dbbl/Dbnt) × (Lbbl/Dbnt)]
4	Pi term relating specification of tools used in liner piston maintenance activity	π_4 = [(Dbrcm/Lbrcm) × (Dbrex/Lbrcm) × (Tbrex/Lbrcm) × (Dst/Lbrcm) × (Lst/Lbrcm) × (Dhmr/Lbrcm) × (Lhmr/Lbrcm) × (Lnpl/Lbrcm) × (Dlipl/Lbrcm) × (Llipl/Lbrcm) × (Dbpl/Lbrcm) × (Lbpl/Lbrcm)] .
5	Pi term relating specification of solvent, lube oil, and compressed air	π_5 = [(Ker/Eb) × (o1 1/l oil)]
6	Pi term relating specification of Axial clearance of big end bearing and Saddle bolt elongation	π_6 = [(bcl/Bel) × (Scl/Bel)]
7	Pi term relating specification of workstation of liner piston	π_7 = [(h.crt/l.crt), (b.crt/l.crt)]
8	Pi term relating specification of temperature at workplace	π_8 = [temp]
9	Pi term relating specification of humidity	π_9 = [humidity %]
10	Pi term relating specification of illumination	π_{10} = [(ilms × (Ags)3)/wt)]
11	Pi term relating specification of noise	π_{11} = [noise]

TABLE 3.8
Dependent Dimensionless Pi Terms of Liner Piston Maintenance Activity

Ser	Nature of Basic Physical Quantities	Dependent Pi Terms
1	Pi term relating to overhauling time of liner piston maintenance activity	z_{1P} = [(field time of liner piston overhauling)^2 × g/length of liner]
2	Pi term relating Human Energy consumed in liner piston maintenance activity	z_{2P} = [field energy consumed in liner piston/ (field time of liner piston × 1.341 × 10^{-3})]
3	Pi term relating productivity of liner piston maintenance activity	z_{3P} = [overhauling time of liner piston/total engine overhauling time] = field overhauling time of liner piston/14,400 min

3.5.9 Formulation of a Field Data-Based Model for Response Variables of Liner Piston Maintenance

For formulation of model, 11 independent pi terms and 3 dependent pi terms have been identified for liner piston maintenance activity, and each dependent pi term is a function of available independent pi terms:

$$Z_{1P} = f(\pi_{D1}) = f(\pi_1, \pi_2, \pi_3, \pi_4, \pi_5, \pi_6, \pi_7, \pi_8, \pi_9, \pi_{11})$$
$$Z_{2P} = f(\pi_{D2}) = f(\pi_1, \pi_2, \pi_3, \pi_4, \pi_5, \pi_6, \pi_7, \pi_8, \pi_9, \pi_{11})$$
$$Z_{3P} = f(\pi_{D3}) = f(\pi_1, \pi_2, \pi_3, \pi_4, \pi_5, \pi_6, \pi_7, \pi_8, \pi_9, \pi_{11})$$

f stands for "function of".
Where

$z_{1P} = \pi_{D1}$ first dependent pi term, overhauling time of liner maintenance activity
$z_{2P} = \pi_{D2}$ second dependent pi term, Human Energy consumed in liner piston maintenance activity
$z_{3P} = \pi_{D3}$ third dependent pi term, productivity of liner piston maintenance activity

The probable exact mathematical form for dimensional equation of the phenomenon could be relationships assumed to be of exponential form

Dependent pi term:

$$\begin{aligned}
Z_p = K \times & [(a \times c \times e \times g)/(b \times d \times f \times h)]^a, [(Ags/Exs), (sks/Ens), (hls/Hbs)]^b, \\
& [(Dlins/Lcr) \times (1.lin / L.cr) \times (D.pis/L.cr) \times (L.pis/L.cr) \times (D.prng/Le.r) \\
& \times (t.prng/L.crs) \times (l.pstrk/L.cr) \times (D.ppin./L.cr) \times (L.ppin/L.cr) \times (Dbg.br/L.cr) \\
& \times (tbg.br/L.cr) \times (Dlt.brj/L.cr) \times (t.lt.brj/L.cr) \times (D.b.nj/L.cr) \times (L.b.nt/L.cr) \\
& \times (d.b.bl/L.cr) \times (l.b.bl/L.cr)]^c, \\
& [(Dbr/Lbr), (Dsmsp/Lbr), (Lsmsp/Lbr), (Dsisp/Lbr), (Lsisp/Lbr), (Dbrr/Lbr), \\
& (Lbrr/Lbr)]^d, \\
& [(ker /E.belt), (orient11/L.oil)]^e, \\
& [(bcl/Bel)^* (Scl/Bel)]^f, \\
& [(h.crt/l.crt), (b.crt/l.crt)]^g, \\
& [temp/100]^h, \\
& [(Humd)]^i, \\
& [((illmus \times Ags)^3)/Wt]^j, \\
& [(Noise)]^k\}
\end{aligned}$$

3.5.10 Model Formulation by Identifying the Curve Fitting Constant and Various Indices of Pi Terms of Liner Piston Maintenance

The multiple regression analysis helps to identify the indices of the different pi terms in the model aimed at, by considering 11 independent pi terms and 1 dependent pi term. Let the model be of the form:

$$Z_{1P} = K_1 \times [(\pi_1)^{a}{}_1 \times (\pi_2)^{b}{}_1 \times (\pi_3)^{c}{}_1 \times (\pi_4)^{d}{}_1 \times (\pi_5)^{e}{}_1 \times (\pi_6)^{f}{}_1 \times (\pi_7)^{g}{}_1 \times (\pi_8)^{h}{}_1 \times (\pi_9)^{i}{}_1 \times (\pi_{10})^{j}{}_1 \times (\pi_{11})^{k}{}_1]$$
$$Z_{2P} = K_2 \times [(\pi_1)^{a}{}_2 \times (\pi_2)^{b}{}_2 \times (\pi_3)^{c}{}_2 \times (\pi_4)^{d}{}_2 \times (\pi_5)^{e}{}_2 \times (\pi_6)^{f}{}_2 \times (\pi_7)^{g}{}_2 \times (\pi_8)^{h}{}_2 \times (\pi_9)^{i}{}_2 \times (\pi_{10})^{j}{}_2 \times (\pi_{11})^{k}{}_2]$$
$$Z_{3P} = K_3 \times [(\pi_1)^{a}{}_3 \times (\pi_2)^{b}{}_3 \times (\pi_3)^{c}{}_3 \times (\pi_4)^{d}{}_3 \times (\pi_5)^{e}{}_3 \times (\pi_6)^{f}{}_3 \times (\pi_7)^{g}{}_3 \times (\pi_8)^{h}{}_3 \times (\pi_9)^{i}{}_3 \times (\pi_{10})^{j}{}_3{}^{*} \times (\pi_{11})^{k}{}_3]$$

This is known as exponential form of model.

To determine the values of K_1, a_1, b_1, c_1, d_1, e_1, f_1, g_1, h_1, i_1, j_1, and k_1 and to arrive at the regression hyper-plane, the above equations are presented as follows.

Taking log on both sides of equations for overhauling time of crankshaft maintenance (z_1) we get 12 unknown terms in the equations:

$$\log z_{1P} = \log k_1 + a_1 \log \pi_1 + b_1 \log \pi_2 + c_1 \log \pi_3 + d_1 \log \pi_4 + e_1 \log \pi_5 + f_1 \log \pi_6 + g_1 \log \pi_7 + h_1 \log \pi_8 + i_1 \log \pi_9 + j_1 \log \pi_{10} + k_1 \log \pi_{11}$$

$$\text{Let, } Z_1 = \log z1, K_1 = \log k1, A = \log \pi_1, B = \log \pi_2, C = \log \pi_3, D = \log \pi_4, E = \log \pi_5, F = \log \pi_6, G = \log \pi_7, H = \log \pi_8, I = \log \pi_9, J = \log \pi_{10}, k_1 = \log \pi_{11}$$

Putting these values in log equation 4.11 the same can be written as:

$$Z_{1P} = K_1 + a_1A + b_1B + c_1C + d_1D + e_1E + f_1F + g_1G + h_1H + i_1I + j_1J + k_1K$$

This is a regression equation of Z on A, B, C, D, E, F, G, H, I, J, K in an n dimensional coordinate system this represents a regression hyper-plane as:

$$\sum z_{1P} = nK_1 + a_1 \sum A + b_1 \sum B + c_1 \sum C + d_1 \sum D + e_1 \sum E + f_1 \sum F + g_1 \sum G + h_1 \sum H + i_1 \sum I + j_1 \sum J + k_1 \sum K$$
$$\sum z_{1P}A = K_1 \sum A + a_1 \sum A^2 + b_1 \sum AB + c_1 \sum AC + d_1 \sum AD + e_1 \sum AE + f_1 \sum AF + g_1 \sum AG + h_1 \sum AH + i_1 \sum AI + j_1 \sum AJ + k_1 \sum AK$$
$$\sum z_{1P}B = K_1 \sum B + a_1 \sum AB + b_1 \sum B^2 + c_1 \sum BC + d_1 \sum BD + e_1 \sum BE + f_1 \sum BF + g_1 \sum BG + h_1 \sum BH + i_1 \sum BI + j_1 \sum BJ + k_1 \sum BK$$
$$\sum z_{1P}C = K_1 \sum C + a_1 \sum AC + b_1 \sum CB + c_1 \sum C^2 + d_1 \sum CD + e_1 \sum CE + f_1 \sum CF + g_1 \sum CG + h_1 \sum CH + i_1 \sum CI + j_1 \sum CJ + k_1 \sum CK$$
$$\sum z_{1P}D = K_1 \sum D + a_1 \sum AD + b_1 \sum BD + c_1 \sum CD + d_1 \sum D^2 + e_1 \sum DE + f_1 \sum DF + g_1 \sum DG + h_1 \sum DH + i_1 \sum DI + j_1 \sum DJ + k_1 \sum DK$$
$$\sum z_{1P}E = K_1 \sum E + a_1 \sum AE + b_1 \sum BE + c_1 \sum CE + d_1 \sum DE + e_1 \sum E^2 + f_1 \sum FE + g_1 \sum GE + h_1 \sum HE + i_1 \sum IE + j_1 \sum JE + k_1 \sum EK$$
$$\sum z_{1P}F = K_1 \sum F + a_1 \sum AF + b_1 \sum FB + c_1 \sum FC + d_1 \sum FD + e_1 \sum FE + f_1 \sum F^2 + g_1 \sum FG + h_1 \sum FH + i_1 \sum FI + j_1 \sum FJ + k_1 \sum FK$$

$$\Sigma z_{1P}G = K_1 \Sigma G + a_1 \Sigma AG + b_1 \Sigma GB + c_1 \Sigma GC + d_1 \Sigma GD + e_1 \Sigma GE + f_1 \Sigma GF + g_1 \Sigma G^2 + h_1 \Sigma GH + i_1 \Sigma GI + j_1 \Sigma GJ + k_1 \Sigma GK$$

$$\Sigma z_{1P}H = K_1 \Sigma H + a_1 \Sigma AH + b_1 \Sigma HB + c_1 \Sigma HC + d_1 \Sigma HD + e_1 \Sigma HE + f_1 \Sigma HF + g_1 \Sigma HG + h_1 \Sigma H^2 + i_1 \Sigma HI + j_1 \Sigma HJ + k_1 \Sigma HK$$

$$\Sigma z_{1P}I = K_1 \Sigma I + a_1 \Sigma AI + b_1 \Sigma IB + c_1 \Sigma IC + +d_1 \Sigma ID + +e_1 \Sigma IE + f_1 \Sigma IF + g_1 \Sigma IG + h_1 \Sigma IH + i_1 \Sigma I^2 + j_1 \Sigma IJ + k_1 \Sigma IK$$

$$\Sigma z_{1P}J = K_1 \Sigma J + a_1 \Sigma AJ + b_1 \Sigma JB + c_1 \Sigma JC + d_1 \Sigma JD + e_1 \Sigma JE + f_1 \Sigma JF + g_1 \Sigma JG + h_1 \Sigma JH + i_1 \Sigma JI + j_1 \Sigma J^2 + k_1 \Sigma JK$$

$$\Sigma z_{1P}K = K_1 \Sigma K + a_1 \Sigma AK + b_1 \Sigma KB + c_1 \Sigma KC + d_1 \Sigma KD + e_1 \Sigma KE + f_1 \Sigma KF + g_1 \Sigma KG + h_1 \Sigma KH + i_1 \Sigma KI + j_1 \Sigma KJ + k_1 \Sigma K^2$$

In the above set of equations, the values of the term Z on LHS and the multipliers of K_1, a_1, b_1, c_1, d_1, e_1, f_1, g_1, h_1, i_1, j_1, and k1 on RHS side are substituted to calculate the values of the unknowns (viz. K_1, a_1, b_1, c_1, d_1, e_1, f_1, g_1, h_1, i_1, j_1, and k_1).

After substituting these values in the equations, we will get a set of 12 equations, which are to be solved simultaneously to get values of unknowns (viz. K_1, a_1, b_1, c_1, d_1, e_1, f_1, g_1, h_1, i_1, j_1, and k_1). The above equation can be verified in the matrix form and values of unknown K_1, a_1, b_1, c_1, d_1, e_1, f_1, g_1, h_1, i_1, j_1, and k_1 can be obtained by using matrix method.

$$\mathbf{X = inv(W) \times Z}$$

The matrix obtained for solving these equations is:

$$Z_{1p} \times \begin{bmatrix} 1 \\ A \\ B \\ C \\ D \\ E \\ F \\ G \\ H \\ I \\ J \\ K \end{bmatrix} = \begin{bmatrix} n & A & B & C & D & E & F & G & H & I & J & K \\ A & A^2 & BA & CA & DA & EA & FA & GA & HA & IA & JA & KA \\ B & AB & B^2 & CB & DB & EB & FB & GB & HB & IB & JB & KB \\ C & AC & BC & C^2 & DC & EC & FC & GC & HC & IC & JC & KC \\ D & AD & BD & CD & D^2 & ED & FD & GD & HD & ID & JD & KD \\ E & AE & BE & CE & DE & E^2 & FE & GE & HE & IE & JE & KE \\ F & AF & BF & CF & DF & EF & F^2 & GF & HF & IF & JF & KF \\ G & AG & BG & CG & DG & EG & FG & G^2 & HG & IG & JG & KG \\ H & AH & BH & CH & DH & EH & FH & GH & H^2 & IH & JH & KH \\ I & AI & BI & CI & DI & EI & FI & GI & HI & I^2 & JI & KI \\ J & AJ & BJ & CJ & DJ & EJ & FJ & GJ & HJ & IJ & J^2 & KJ \\ K & AK & BK & CK & DK & EK & FK & GK & HK & IK & JK & K^2 \end{bmatrix} \times \begin{bmatrix} K_1 \\ a_1 \\ b_1 \\ c_1 \\ d_1 \\ e_1 \\ f_1 \\ g_1 \\ h_1 \\ i_1 \\ j_1 \\ k_1 \end{bmatrix}$$

A = 12 × 12 matrix of multipliers K_1, a_1, b_1, c_1, d_1, e_1, f_1, g_1, h_1, i_1, j_1, and k_1

B = 12 × 1 matrix of terms on LHS

C = 12 × 1 matrix of solutions of K_1, a_1, b_1, c_1, d_1, e_1, f_1, g_1, h_1, i_1, j_1, and k_1

Then, C = inverse (A) × B gives the unique values of K_1, a_1, b_1, c_1, d_1, e_1, f_1, g_1, h_1, i_1, j_1, and k_1. In the above equations, n is the number of sets of readings.

A, B, C, D, E, F, G, H, I, J, and K represent the independent pi terms π_1, π_2, π_3, π_4, π_5, π_6, π_7, π_8, π_9, π_{10}, and π_{11}, while z represents dependent pi term. Next, calculate the values of independent pi terms for corresponding dependent pi terms, helping to generate the equations in matrix form. Solving the above matrix by using MATLAB software makes model formulation precise, quick, and less cumbersome.

3.6 PROBLEM ASSOCIATED WITH FOSSIL FUELS

Because of the probable exhaustion and rising price of petroleum, as well as environmental problems generated by the burning of fossil fuels, the quest for alternative fuels has gotten a lot of attention. The alternative fuel not only avoids the petroleum crisis, but also reduces pollutant gases emitted by engines. The waste transformer oil is efficient for operating at high temperature stresses. They are highly viscous contains sludge with colour containing compound and oxidation product. The absorption of these impurity form waste transformer oil uses silica gel, and the resultant treated transformer oil is colourless. It demonstrates the same properties as diesel fuel. Hence various proportion blends such as 10% (that is 10% treated transformer oil and 90% diesel fuel), likewise 20%, 25%, 30%, and 40% are then made, as well as pure diesel fuel. The performance of the engine fuelled with these blends are evaluated thus:

1. Brake thermal efficiency (BTH).
2. Brake specific fuel consumption (BSFC).

Various factors are identified so as to optimize brake thermal efficiency and specific fuel consumption. There are many approaches to develop /upgrade engine performance such as method study and time study. The experimental data-based modelling approach is proposed to correlate fuel and engine characteristics with engine observations, such as brake thermal efficiency and brake specific fuel consumption.

The chemical stability of oil is influenced by three factors: heat, availability of oxygen, and the presence of a catalyst. Oil deterioration can be induced by the breakdown of hydrocarbons in oil at high temperatures. The oxygen concentration of insulating oil may cause an increase in acidity and the production of sludge. Catalysts such as iron and copper dissolve in oil while ageing, and may hasten the procedure.

3.6.1 DIESEL BLENDING

The hunt for alternative fuels has gained prominence following the probable shortage and rising prices of petroleum, as well as environmental problems generated by the burning of fossil fuels.

The diesel engine run was conducted with single cylinder diesel engine. The result obtained was fuelled with blends of treated transformer oil and diesel fuel varying proportion such as 10:90, 20:80, 25:75, 30:70, 40:60. The run were covered under varying load of 10 kg, 15 kg, 20 kg. The performance of engine was evaluated on the basis brake thermal efficiency (BTE), brake specific fuel consumption (BSFC).

The variables affecting the effectiveness of the phenomenon under consideration are blends of treated transformer oil with diesel and performance characteristics.

Dependent and independent variables for the diesel engine involved performance of diesel engine are present in Table 3.9.

TABLE 3.9
Independent and Dependent Variables: Diesel

Ser	Description	Variables	Symbol	Dimension
1	Load on engine	Independent	L	$[M^1L^0T^0]$
2	Blend	Independent	B	$[M^0L^0T^0]$
3	Flash point	Independent	Fp	$[M^0L^0T^0]$
4	Aniline point	Independent	Ap	$[M^0L^0T^0]$
5	Kinematic viscosity	Independent	Vi	$[M^0L^2T^{-1}]$
6	Density	Independent	D	$[M^1L^{-3}T^0]$
7	API gravity	Independent	Apig	$[M^0L^0T^0]$
8	Diesel index	Independent	Di	$[M^0L^0T^0]$
9	Cetane number	Independent	CN	$[M^0L^0T^0]$
10	Calorific value	Independent	Cv	$[M^0L^2T^{-2}]$
11	Time	Independent	T	$[M^0L^0T^1]$
12	Mass of fuel	Independent	Mf	$[M^1L^0T^{-1}]$
13	Bore diameter	Independent	Bd	$[M^0L^1T^0]$
14	Stroke length	Independent	Sl	$[M^0L^1T^0]$
15	Cubic capacity	Independent	Cc	$[M^0L^3T^0]$
16	Fuel tank capacity	Independent	Fc	$[M^0L^3T^0]$
17	Engine speed	Independent	N	$[M^0L^0T^{-1}]$
18	Brake thermal efficiency	Dependent	Bte	$[M^0L^0T^0]$
19	Brake specific flue consumption	Dependent	Bsfc	$[M^2L^2T^{-4}]$

3.6.2 Independent and Dependent π Term

$$\pi_1 = [(Cv)^a (Mf)^b (Bd)^c]L$$
$$[M^0L^0T^0] = [M^0L^2T^{-2}]^a[M^1L^0T^{-1}]^b[M^0L^1T^0]^c[M^1L^0T^0]^1$$
$$M = 0 = 0 + b + 0 + 1, b = -1$$
$$L = 0 = 2a + c,$$
$$T = 0 = -2a - b,$$
$$\therefore a = \tfrac{1}{2}; c = -1$$
$$\pi_1 = [\mathbf{Cv^{1/2}L/Mf\ Bd}]$$

$$\pi_2 = [(Cv)^a (Mf)^b (Bd)^c]B$$
$$[M^0L^0T^0] = [M^0L^2T^{-2}]^a[M^1L^0T^{-1}]^b[M^0L^1T^0]^c[M^0L^0T^0]^1$$
$$M = 0 = 0 + b + 0 + 0, b = 0$$
$$L = 0 = 2a + c,$$
$$T = 0 = -2a - b, a = 0$$
$$\therefore c = 0$$
$$\pi_2 = [\mathbf{B}]$$

$$\pi_3 = [(Cv)^a(Mf)^b(Bd)^c]Fp$$
$$[M^0L^0T^0] = [M^0L^2T^{-2}]^a[M^1L^0T^{-1}]^b[M^0L^1T^0]^c[M^0L^0T^0]^1$$
$$M = 0 = 0 + b + 0 + 0, b = 0$$
$$L = 0 = 2a + c,$$
$$T = 0 = -2a - b, a = 0$$
$$\therefore c = 0$$
$$\pi_3 = [\mathbf{Fp}]$$

$$\pi_4 = [(Cv)^a(Mf)^b(Bd)^c]Ap$$
$$[M^0L^0T^0] = [M^0L^2T^{-2}]^a[M^1L^0T^{-1}]^b[M^0L^1T^0]^c[M^0L^0T^0]^1$$
$$M = 0 = 0 + b + 0 + 0, b = 0$$
$$L = 0 = 2a + c,$$
$$T = 0 = -2a - b, a = 0$$
$$\therefore c = 0$$
$$\pi_4 = [\mathbf{Ap}]$$

$$\pi_5 = [(Cv)^a(Mf)^b(Bd)^c]Vi$$
$$[M^0L^0T^0] = [M^0L^2T^{-2}]^a[M^1L^0T^{-1}]^b[M^0L^1T^0]^c[M^0L^2T^0]^1$$
$$M = 0 = 0 + b + 0 + 0, b = 0$$
$$L = 0 = 2a + c + 2,$$
$$T = 0 = -2a - b - 1,$$
$$\therefore a = \tfrac{1}{2}; c = 0$$
$$\pi_5 = [\mathbf{Cv^{1/2}\ Vi/Bd}]$$

$$\pi_6 = [(Cv)^a(Mf)^b(Bd)^c]D$$
$$[M^0L^0T^0] = [M^0L^2T^{-2}]^a[M^1L^0T^{-1}]^b[M^0L^1T^0]^c[M^1L^{-3}T^0]^1$$
$$M = 0 = 0 + b + 0 + 1, b = -1$$
$$L = 0 = 2a + c - 3,$$
$$T = 0 = -2a - b,$$
$$\therefore a = \tfrac{1}{2}; c = 2$$
$$\pi_6 = [\mathbf{Cv^{1/2}\,Bd^2\,D/Mf}]$$

$$\pi_7 = [(Cv)^a(Mf)^b(Bd)^c]Apig$$
$$[M^0L^0T^0] = [M^0L^2T^{-2}]^a[M^1L^0T^{-1}]^b[M^0L^1T^0]^c[M^0L^0T^0]^1$$
$$M = 0 = 0 + b + 0 + 0, b = 0$$
$$L = 0 = 2a + c,$$
$$T = 0 = -2a - b, a = 0$$
$$\therefore c = 0$$
$$\pi_7 = [\mathbf{Apig}]$$

$$\pi_8 = [(Cv)^a (Mf)^b (Bd)^c]Di$$
$$[M^0L^0T^0] = [M^0L^2T^{-2}]^a [M^1L^0T^{-1}]^b [M^0L^1T^0]^c [M^0L^0T^0]^1$$
$$M = 0 = 0 + b + 0 + 0, b = 0$$
$$L = 0 = 2a + c,$$
$$T = 0 = -2a - b, a = 0$$
$$\therefore c = 0$$
$$\pi_8 = [\mathbf{Di}]$$

$$\pi_9 = [(Cv)^a (Mf)^b (Bd)^c]CN$$
$$[M^0L^0T^0] = [M^0L^2T^{-2}]^a [M^1L^0T^{-1}]^b [M^0L^1T^0]^c [M^0L^0T^0]^1$$
$$M = 0 = 0 + b + 0 + 0, b = 0$$
$$L = 0 = 2a + c,$$
$$T = 0 = -2a - b, a = 0$$
$$\therefore c = 0$$
$$\pi_9 = [\mathbf{CN}]$$

$$\pi_{10} = [(Cv)^a (Mf)^b (Bd)^c]T$$
$$[M^0L^0T^0] = [M^0L^2T^{-2}]^a [M^1L^0T^{-1}]^b [M^0L^1T^0]^c [M^0L^0T^1]^1$$
$$M = 0 = 0 + b + 0 + 0, b = 0$$
$$L = 0 = 2a + c,$$
$$T = 0 = -2a - b + 1, a = 1/2$$
$$\therefore c = -1$$
$$\pi_{10} = [\mathbf{Cv^{1/2}\,T/Bd}]$$

$$\pi_{11} = [(Cv)^a (Mf)^b (Bd)^c]Sl$$
$$[M^0L^0T^0] = [M^0L^2T^{-2}]^a [M^1L^0T^{-1}]^b [M^0L^1T^0]^c [M^0L^1T^0]^1$$
$$M = 0 = 0 + b + 0 + 0, b = 0$$
$$L = 0 = 2a + c + 1,$$
$$T = 0 = -2a - b, a = 1/2$$
$$\therefore c = -1$$
$$\pi_{11} = [\mathbf{Sl/Bd}]$$

$$\pi_{12} = [(Cv)^a (Mf)^b (Bd)^c]Cc$$
$$[M^0L^0T^0] = [M^0L^2T^{-2}]^a [M^1L^0T^{-1}]^b [M^0L^1T^0]^c [M^0L^3T^0]^1$$
$$M = 0 = 0 + b + 0 + 0, b = 0$$
$$L = 0 = 2a + c + 3,$$
$$T = 0 = -2a - b, a = 0$$
$$\therefore a = 0; c = -1$$
$$\pi_{12} = [\mathbf{Cc/Bd^3}]$$

$$\pi_{13} = [(Cv)^a(Mf)^b(Bd)^c]Fc$$
$$[M^0L^0T^0] = [M^0L^2T^{-2}]^a[M^1L^0T^{-1}]^b[M^0L^1T^0]^c[M^0L^3T^0]^1$$
$$M = 0 = 0 + b + 0 + 0, b = 0$$
$$L = 0 = 2a + c + 3,$$
$$T = 0 = -2a - b, a = 0$$
$$\therefore a = 0; c = -3$$
$$\pi_{13} = [\mathbf{Fc/Bd^3}]$$

$$\pi_{14} = [(Cv)^a(Mf)^b(Bd)^c]N$$
$$[M^0L^0T^0] = [M^0L^2T^{-2}]^a[M^1L^0T^{-1}]^b[M^0L^1T^0]^c[M^0L^0T^{-1}]^1$$
$$M = 0 = 0 + b + 0 + 0, b = 0$$
$$L = 0 = -2a + c,$$
$$T = 0 = -2a - b - 1, a = 1/2$$
$$\therefore a = \tfrac{1}{2}; c = -1$$
$$\pi_{14} = [\mathbf{NBd/Cv^{1/2}}]$$

$$\pi_{D1} = [(Cv)^a(Mf)^b(Bd)^c]Bte$$
$$[M^0L^0T^0] = [M^0L^2T^{-2}]^a[M^1L^0T^{-1}]^b[M^0L^1T^0]^c[M^0L^0T^{-1}]^1$$
$$M = 0 = 0 + b + 0 + 0, b = 0$$
$$L = 0 = -2a + c,$$
$$T = 0 = -2a - b, a = 0$$
$$\therefore c = 0$$
$$\pi_{D1} = [\mathbf{Bte}]$$

$$\pi_{D2} = [(Cv)^a(Mf)^b(Bd)^c]Bsfc$$
$$[M^0L^0T^0] = [M^0L^2T^{-2}]^a[M^1L^0T^{-1}]^b[M^0L^1T^0]^c[M^0L^0T^0]^1$$
$$M = 0 = 0 + b + 0 + 0, b = 0$$
$$L = 0 = 2a + c,$$
$$T = 0 = -2a - b, a = 0$$
$$\therefore c = 0$$
$$\pi_{D2} = [\mathbf{Bsfc}]$$

3.6.3 Establishment of Dimensionless Group of π Terms

These independent variables have been reduced into group of π terms. Lists of the Independent and Dependent π terms of the face drilling activity are shown in Tables 3.10 and 3.11.

3.6.4 Creation of Field Data-Based Model

Four independent π terms (π_1, π_2, π_3, π_4) and two dependent π terms (Z_1, Z_2) have been identified for model formulation in field study.

Each pi term is a function of the output terms (Shriwastawa n.d.; Dekker 1995),

TABLE 3.10

Independent Dimensionless π Terms Ser	Independent Dimensionless Ratios	Nature of Basic Physical Quantities
01	$\pi_1 = [([Cv^{1/2} Vi/Bd])([Cv^{1/2} T/Bd])([B])/([Fp])]$	Specification related to blend formation and time
02	$\pi_2 = [([Cv^{1/2} Bd^2 D/Mf])/([Cv^{1/2} L/Mf\ Bd])([Ap])]$	Specifications of fuel consumption and engine load
03	$\pi_3 = [([Apig])([Di])/([CN])]$	Specifications of fuel characteristic
04	$\pi_4 = [([Sl/Bd])([Cc/Bd^3])([Fc/Bd^3])([N\ Bd/Cv^{1/2}])]$	Engine specification

TABLE 3.11

Dependent Dimensionless π Terms Ser	Dependent Dimensionless Ratios or π Terms	Nature of Basic Physical Quantities
01	$Z_1 = [Bte]$	Brake thermal efficiency
02	$Z_2 = [Bsfc]$	Brake specific flue consumption

$$Z_1 = \text{function of } (\Pi_1, \Pi_2, \Pi_3, \Pi_4)$$
$$Z_2 = \text{function of } (\Pi_1, \Pi_2, \Pi_3, \Pi_4)$$

Where

$Z_1 = \Pi_{D1}$, First dependent π term = Bte

$Z_2 = \Pi_{D2}$, Second dependent π term = Bsfc

The most likely accurate mathematical form for the phenomenon's dimensions equations might be connections considered as being of exponentially nature.

$$(Z) = K \times [([Cv^{1/2} Vi/Bd])([Cv^{1/2} T/Bd])([B])/([Fp])]^a, [([Cv^{1/2} Bd^2\ D/Mf])/([Cv^{1/2} L/Mf\ Bd])([Ap])]^b, [([Apig])([Di])/([CN])]^c, [([Sl/Bd])([Cc/Bd^3])([Fc/Bd^3])([N\ Bd/Cv^{1/2}])]^d$$

3.6.5 Model Formulation by Identifying the Curve Fitting Constant and Various Indices of π Terms

By taking into account four independent variables and one dependent π term, multiple regression analysis assists in identifying the indices of the various π term in the model targeted at (Zhou and Xu 2014). Let the model be of the form:

$$(Z_1) = K_1 \times [(\pi_1)^{a1} \times (\pi_2)^{b1} \times (\pi_3)^{c1} \times (\pi_4)^{d1}]$$
$$(Z_2) = K_2 \times [(\pi_1)^{a2} \times (\pi_2)^{b2} \times (\pi_3)^{c2} \times (\pi_4)^{d2}]$$

To find the values of a_1, b_1, c_1, and d_1, the equation is:

$$\Sigma Z_1 = nK_1 + a_1 \times \Sigma A + b_1 \times \Sigma B + c_1 \times \Sigma C + d_1 \times \Sigma D$$
$$\Sigma Z_1 \times A = K_1 \times \Sigma A + a_1 \times \Sigma A \times A + b_1 \times \Sigma B \times A + c_1 \times \Sigma C \times A + d_1 \times \Sigma D \times A$$
$$\Sigma Z_1 \times B = K_1 \times \Sigma B + a_1 \times \Sigma A \times B + b_1 \times \Sigma B \times B + c_1 \times \Sigma C \times B + d_1 \times \Sigma D \times B$$
$$\Sigma Z_1 \times C = K_1 \times \Sigma C + a_1 \times \Sigma A \times C + b_1 \times \Sigma B \times C + c_1 \times \Sigma C \times C + d_1 \times \Sigma D \times C$$
$$\Sigma Z_1 \times D = K_1 \times \Sigma D + a_1 \times \Sigma A \times D + b_1 \times \Sigma B \times D + c_1 \times \Sigma C \times D + d_1 \times \Sigma D \times D$$

In the above set of equations, the values of K_1, a_1, b_1, c_1 and d_1 are substituted to compute the values of the unknowns. After substituting these values in the equations, one will get a set of five equations, which are to be solved simultaneously to get the values of K_1, a_1, b_1, $c_1$and d_1. The above equations can be transferer and used in the matrix form, and subsequently values of K_1, a_1, b_1, c_1, and d_1 can be obtained by adopting matrix analysis.

$$X_1 = \text{inv}(W) \times P_1$$

W = 5 x 5 matrix of the multipliers of K_1, a_1, b_1, c_1, and d_1
P_1 = 5 x 1 matrix on L H S and
X_1 = 5 x 1 matrix of solutions

Then, the matrix evaluated is given by:

Matrix

$$Z_1 \times \begin{bmatrix} 1 \\ A \\ B \\ C \\ D \end{bmatrix} = \begin{bmatrix} n & A & B & C & D \\ A & A^2 & BA & CA & DA \\ B & AB & B^2 & CB & DB \\ C & AC & BC & C^2 & DC \\ D & AD & BD & CD & D^2 \end{bmatrix} \times \begin{bmatrix} K_1 \\ a_1 \\ b_1 \\ c_1 \\ d_1 \end{bmatrix}$$

In the above equations, n is the number of sets of readings, A, B, C and D represent the independent π terms π_1, π_2, π_3 and π_4 while, Z represents dependent π term

3.7 PROBLEM ASSOCIATED WITH CONVENTIONAL POWER GENERATION

Many developing countries have large renewable energy resources, such as solar energy, wind power, biomass, and hydro energy. Developing countries are engaging with renewable energy due to environmental issues such as global warming and pollution, so they are implementing policies for development of renewable energy. The consumption of oil and natural gas can be optimized by proper use of renewable energy sources. Renewable energy is less expensive than fuel energy such as oil and gas. However, the supply of renewable energy depends on atmospheric conditions. In the remote areas the concept of solar chimney power plant is most effective as compared to the expensive distribution of electricity grid or alternative to the diesel generators. The capital investment of hydropower and tidal power is high in comparison to solar and wind energy. Solar power generation has the advantages of low maintenance, simple installation, silent operation, and long lifespan. The solar chimney power plant needs warm air to create the updraft to spin a turbine. The spinning of the turbine continues due to the collected warm air absorbed by the land when the sun is shining on it. This can be made more effective by covering the collector area with gravel. Solar power produce by the PV solar plant cannot provide power during night unless equipped with a storage system. The efficiency of the PV panel reduces if the panel becomes covered with the dust. Solar chimney power plants produce power at low running costs and high reliability. Few, however, have the ability to store sufficient energy during the day to maintain supply at night when solar radiation is negligible (Figure 3.1).

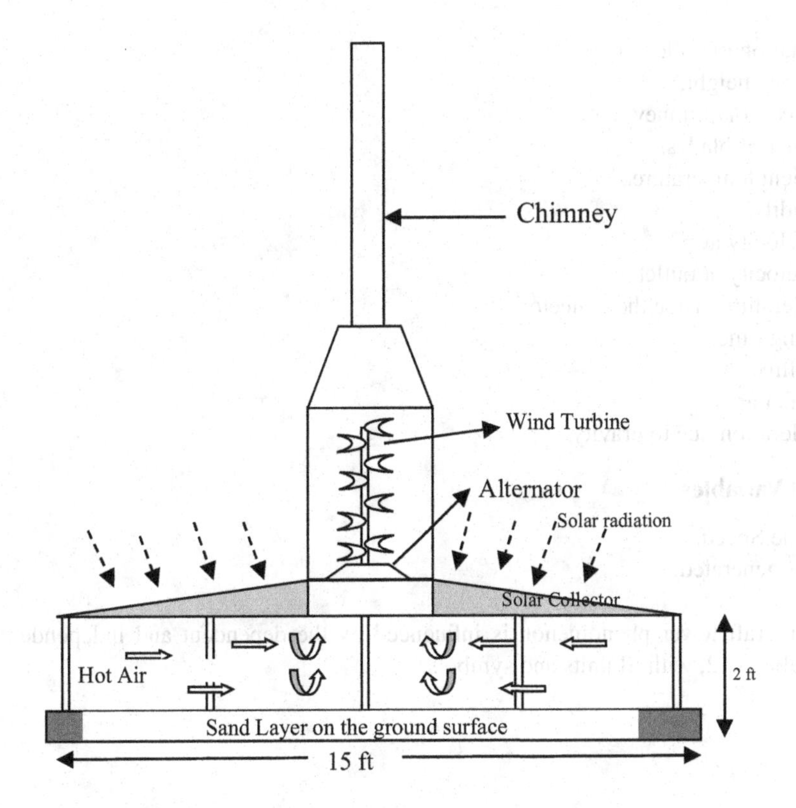

FIGURE 3.1 Solar updraft tower.

Diagram of solar updraft tower.

The mathematical model formed with all the variables involved in the design of the solar updraft tower. These independent variables are grouped such as variables related to collector, variables related to the chimney, variables related to atmospheric condition and variables related to heating condition. The indices of each grouped pie terms predict the performance of the dependent variables such speed and power produced by the turbine. Based on the mathematical model performance of the solar updraft is optimize.

Planning, execution and analysis of actual experimentation and obtaining reliable conclusions from the data generated by experimentation the research methodology suggested by Hilbert Schenk Jr. is very useful.

3.7.1 Identification of Variables Affecting the Phenomenon

Physical quantity that undergoes change is applied with the variables. If a physical quantity can be changed independently of the other quantities, then it is an independent variable. If the physical quantity changes in response to the variation of one or more number of variables, then it is termed as dependent or response variable. If the physical quantity that affects our test is changing in random and uncontrolled manner, then it is called an extraneous variable. The variables affecting the effectiveness of the phenomenon under consideration are collector materials, chimney height, chimney diameter, turbine blades, and solar radiation. The variables affecting in the context of solar up draft tower performance are:

Independent Variables

1. Diameter of collector.
2. Thermal conductivity of collector material.
3. Height of collector from ground level.
4. Thickness of covering collector material.

5. Inclination of collector.
6. Chimney height.
7. Diameter of chimney.
8. Number of blades.
9. Ambient temperature.
10. Humidity.
11. Air velocity at inlet.
12. Air velocity at outlet.
13. Temperature inside the collector.
14. Heating time.
15. Heat flux.
16. Air inlet area.
17. Acceleration due to gravity.

Dependent Variables

1. Turbine Speed.
2. Power generated.

The solar updraft tower phenomenon is influenced by the dependent and independent variables shown in Table 3.12, with SI units and symbols.

TABLE 3.12
Identification of Variables for Solar Updraft Tower

Ser	Description of Variables	Type of Variable	Symbol	Unit	Dimension
1	Diameter of collector	Independent	D_c	M	$M^0L^1T^0$
2	Thermal conductivity of collector Material	Independent	K	W/m⁰K	$M^1L^1T^{-3}\theta^{-1}$
3	Height of collector from ground level	Independent	Hgc	M	$M^0L^1T^0$
4	Thickness of covering collector material	Independent	Tcc	M	$M^0L^1T^0$
5	Inclination of collector	Independent	θ_c	rad	$M^0L^0T^0$
6	Chimney height	Independent	Hch	M	$M^0L^1T^0$
7	Diameter of chimney	Independent	Dch	M	$M^0L^1T^0$
8	Number of blades	Independent	N_b	–	$M^0L^0T^0$
9	Ambient temperature	Independent	T_a	°C	$M^0L^0T^0\theta^1$
10	Humidity	Independent	H_u	%	$M^0L^0T^0$
11	Air velocity at inlet	Independent	V_i	m/s	$M^0L^1T^{-1}$
12	Air velocity at outlet	Independent	V_o	m/s	$M^0L^1T^{-1}$
13	Temperature inside the collector	Independent	T_c	°C	$M^0L^0T^0\theta^1$
14	Heating time	Independent	T_h	sec	$M^0L^0T^1$
15	Heat Flux	Independent	Q	W/m²	$M^1L^0T^{-3}$
16	Air inlet area	Independent	Aoi	m²	$M^0L^2T^0$
17	Acceleration due to gravity	Independent	g	m/s²	$M^0L^1T^{-2}$
18	Turbine Speed	Dependent	N_T	rpm	$M^0L^0T^{-1}$
19	Power generated	Dependent	P_D	W	$M^1L^2T^{-3}$

M for mass, L for length, and T for time.

3.7.2 Formation of Pi (π) Terms for All Dependent and Independent Variables Affecting the Phenomenon

The Buckingham's π – Theorem is used for the dimensional analysis after identifying the dependant and independent variables. The method of dimensional analysis is followed step by step as explained below:

Total number of Variable = 19 Number of dependent variables = 02

Number of independent variables, n = 17 According to Buckingham's Π-Theorem,

No. of π terms = 19–3 = 16

$$\pi_{D1} = f_1(\pi_1, \pi_2, \pi_3, \pi_4, \pi_5, \pi_6, \pi_7, \pi_8, \pi_9, \pi_{10}, \pi_{11}, \pi_{12}, \pi_{13}, \pi_{14}, \pi_{15}, \pi_{16}) = 0$$

$$\pi_1 = [g^a \cdot k^b \cdot D^c c \cdot \theta^d]H_{gc}$$
$$\pi_1 = [M^0L^1T^{-2}]^a[MLT^{-3}\theta^{-1}]^b[M^0LT^0]^c[M^0L^0T^0]^d$$
$$\pi_1 = [M^0L^0T^0] = [M^0L^1T^{-2}]^a[MLT^{-3}\theta^{-1}]^b[M^0LT^0]^c[M^0L^0T^0]^d[M^0L^1T^0]$$
$$\pi_1 = [M^0L^0T^0] = M^bL^{a+b+c+1}T^{-2a-3b}\theta^{-1}$$

The values of a, b, and c are computed by equating the powers of M, L, and T on both sides as given in Table 3.13.

$$M \Rightarrow 0 = b$$
$$L \Rightarrow 0 = a + b + c + 1$$
$$T \Rightarrow 0 = -2a - 3b$$
$$\theta \Rightarrow 0 = -b$$

Solving, we get:

$$b = 0, a = 0, c = -1$$
$$\pi_1 = [g^0 \cdot k^0 \cdot Dc^{-1} \cdot \theta^0]H_{gc}$$
$$\therefore \pi_1 = \mathbf{(H_{gc})/D_c}$$

TABLE 3.13
The Dependent and Independent Dimensionless Ratios

Ser	Independent Dimensionless Ratios	Nature of Basic Physical Quantities
1	$\pi_1 = [(Hgc \times Tcc \times \theta_c)/Dc^2]$	Collector material
2	$\pi_2 = [(Hch \times Dch \times N_b)/Dc^2]$	Solar chimney
3	$\pi_3 = [Hu]$	Relative humidity
4	$\pi_4 = [(Ta \times To)(Vi\ VoDc^2/g^4)$	Ambient condition
5	$\pi_5 = [(g^{1/2}Th/Dc^{1/2})]\ [(Aoi/Dc^2)]$	Heating duration
Dependent Dimensionless Ratios or π Terms		
6	$\pi_6 = [(Dc\ Q/K)]$	Heat flux
7	$\pi D1 = [(Dc^{1/2}N_T/g^{1/2})]$	Turbine speed
8	$\pi D2 = [P_D/K\ Dc]$	Power developed

$$\pi_2 = [g^a \cdot k^b \cdot D^c c \cdot \theta^d]tcc$$
$$\pi_2 = [M^0L^1T^{-2}]^a[MLT^{-3}\theta^{-1}]^b[M^0 LT^0]^c[M^0L^0T^0]^d$$
$$\pi_2 = [M^0L^0T^0] = [M^0L^1T^{-2}]^a[MLT^{-3}\theta^{-1}]^b[M^0LT^0]^c[M^0L^0T^0]^d[M^0L^1T^0]$$
$$\pi_2 = [M^0L^0T^0] = M^bL^{a+b+c+1}T^{-2a-3b}\theta^{-1}$$
$$M \Rightarrow 0 = b$$
$$L \Rightarrow 0 = a + b + c + 1$$
$$T \Rightarrow 0 = -2a - 3b$$
$$\theta \Rightarrow 0 = -b$$
$$\pi_2 = [g^0 \cdot k^0 \cdot Dc^{-1} \cdot \theta^0](Tcc)$$
$$\therefore \boldsymbol{\pi_2} = \mathbf{(Tcc)/Dc}$$

$$\pi_3 = [M^0L^1T^{-2}]^a[MLT^{-3}\theta^{-1}]^b[M^0LT^0]^c[M^0L^0T^0]^d[M^0L^0T^0\theta^{+1}]$$
$$M \Rightarrow 0 = b$$
$$L \Rightarrow 0 = a + b + c$$
$$T \Rightarrow 0 = -2a - 3b$$
$$\theta \Rightarrow 0 = -b + 1$$
$$b = 0, a = 0, c = 0$$
$$\therefore \boldsymbol{\pi_6} = [\boldsymbol{\theta_c}]$$

$$\pi_4 = [g^a \cdot k^b \cdot D^c c \cdot \theta^d](Hch)$$
$$\pi_4 = [M^0L^1T^{-2}]^a[MLT^{-3}\theta^{-1}]^b[M^0LT^0]^c[M^0L^0T^0]^d$$
$$\pi_4 = [M^0L^0T^0] = [M^0L^1T^{-2}]^a[M LT^{-3}\theta^{-1}]^b[M^0LT^0]^c[M^0L^0T^0]^d[M^0L^1T^0]$$
$$\pi_4 = [M^0L^0T^0] = M^bL^{a+b+c+1}T^{-2a-3b}\theta^{-1}$$
$$M \Rightarrow 0 = b$$
$$L \Rightarrow 0 = a + b + c + 1$$
$$T \Rightarrow 0 = -2a - 3b$$
$$\theta \Rightarrow 0 = -b,\ b = 0\ ,\ a = 0\ ,\ c = -1$$
$$\pi_4 = [g^0 \cdot k^0 \cdot Dc^{-1} \cdot \theta^0](Hch)$$
$$\therefore \boldsymbol{\pi_4} = [\mathbf{(Hch)/Dc}]$$

$$\pi_5 = [g^a \cdot k^b \cdot D^c c \cdot \theta^d](Dch)$$
$$\pi_5 = [M^0L^1T^{-2}]^a[MLT^{-3}\theta^{-1}]^b[M^0LT^0]^c[M^0L^0T^0]^d$$
$$\pi_5 = [M^0L^0T^0] = [M^0L^1T^{-2}]^a[M LT^{-3}\theta^{-1}]^b[M^0LT^0]^c[M^0L^0T^0]^d[M^0L^1T^0]$$
$$\pi_5 = [M^0L^0T^0] = M^bL^{a+b+c+1}T^{-2a-3b}\theta^{-1}$$
$$M \Rightarrow 0 = b$$
$$L \Rightarrow 0 = a + b + c + 1$$
$$T \Rightarrow 0 = -2a - 3b$$
$$\theta \Rightarrow 0 = -b,\ b = 0\ ,\ a = 0\ ,\ c = -1$$
$$\pi_5 = [g^0 \cdot k^0 \cdot Dc^{-1} \cdot \theta^0](Dch)$$
$$\therefore \boldsymbol{\pi_5} = [\mathbf{(Dch)/Dc}]$$

$$\pi_6 = [M^0L^0T^0] = [M^0L^1T^{-2}]^a[MLT^{-3}\theta^{-1}]^b[M^0LT^0]^c[M^0L^0T^0]^d[Nb]$$

$M \Rightarrow 0 = b$

$L \Rightarrow 0 = a + b + c$

$T \Rightarrow 0 = -2a - 3b$

$\theta \Rightarrow 0 = -b$

$b = 0, a = 0, c = 0$

$\therefore \pi_6 = [\mathbf{Nb}]$

$$\pi_7 = [(To)] = [M^0L^0T^0] = [M^0L^1T^{-2}]^a[MLT^{-3}\theta^{-1}]^b[M^0LT^0]^c[M^0L^0T^0]^d[M^0L^0T^0]$$

$M \Rightarrow 0 = b$

$L \Rightarrow 0 = a + b + c$

$T \Rightarrow 0 = -2a - 3b$

$\theta \Rightarrow 0 = -b$

$b = 0, a = 0, c = 0$

$\therefore \pi_7 = [(\mathbf{Ta})]$

$$\pi_8 = [(To)] = [M^0L^0T^0] = [M^0L^1T^{-2}]^a[MLT^{-3}\theta^{-1}]^b[M^0LT^0]^c[M^0L^0T^0]^d[M^0L^0T^0]$$

$M \Rightarrow 0 = b$

$L \Rightarrow 0 = a + b + c$

$T \Rightarrow 0 = -2a - 3b$

$\theta \Rightarrow 0 = -b$

$b = 0, a = 0, c = 0$

$\therefore \pi_8 = [(\mathbf{To})]$

$$\pi_9 = [M^0L^0T^0] = [M^0L^1T^{-2}]^a[MLT^{-3}\theta^{-1}]^b[M^0LT^0]^c[M^0L^0T^0]^d[M^0L^1T^{-1}]$$

$M \Rightarrow 0 = b$

$L \Rightarrow 0 = a + b + c + 1$

$T \Rightarrow 0 = -2a - 3b - 1$

$\theta \Rightarrow 0 = -b$

$b = 0, a = -2, c = 1$

$\therefore \pi_9 = [(\mathbf{ViDc/g^2})]$

$$\pi_{10} = [M^0L^0T^0] = [M^0L^1T^{-2}]^a[MLT^{-3}\theta^{-1}]^b[M^0LT^0]^c[M^0L^0T^0]^d[M^0L^1T^{-1}]$$

$M \Rightarrow 0 = b$

$L \Rightarrow 0 = a + b + c + 1$

$T \Rightarrow 0 = -2a - 3b - 1$

$\theta \Rightarrow 0 = -b$

$b = 0, a = -2, c = 1$

$\therefore \pi_{10} = [(\mathbf{VoDc/g^2})]$

$$\pi_{11} = [M^0L^0T^0] = [M^0L^1T^{-2}]^a[MLT^{-3}\theta^{-1}]^b[M^0LT^0]^c[M^0L^0T^0]^d[(Hu)]$$
$$M \Rightarrow 0 = b$$
$$L \Rightarrow 0 = a + b + c$$
$$T \Rightarrow 0 = -2a - 3b$$
$$\Theta \Rightarrow 0 = -b$$
$$b = 0, a = 0, c = 0$$
$$\therefore \pi_{11} = [\mathbf{Hu}]$$

$$\pi_{12} = [M^0L^0T^0] = [M^0L^1T^{-2}]^a[MLT^{-3}\theta^{-1}]^b[M^0LT^0]^c[M^0L^0T^0]^d[M^0L^0T^1]$$
$$M \Rightarrow 0 = b$$
$$L \Rightarrow 0 = a + b + c + 1$$
$$T \Rightarrow 0 = -2a - 3b - 1$$
$$\theta \Rightarrow 0 = -b$$
$$b = 0, a = 1/2, c = -1/2$$
$$\therefore \pi_{12} = [(\mathbf{g^{1/2}Th/Dc^{1/2}})]$$
$$\therefore \pi_{13} = [(\mathbf{Aoi/Dc^2})]$$

$$\pi_{14} = [M^0L^0T^0] = [M^0L^1T^{-2}]^a[MLT^{-3}\theta^{-1}]^b[M^0LT^0]^c[M^0L^0T^0]^d[M^1L^0T^{-3}]$$
$$M \Rightarrow 0 = b + 1$$
$$L \Rightarrow 0 = a + b + c$$
$$T \Rightarrow 0 = -2a - 3b - 3$$
$$\theta \Rightarrow 0 = -b$$
$$b = -1, a = 0, c = 1$$
$$\therefore \pi_{14} = [(\mathbf{DcQ/K})]$$
$$\therefore \pi_{15D1} = [(\mathbf{Dc^{1/2}N_T/g^{1/2}})]$$

$$\pi_{16}D2 = [M^0L^0T^0] = [M^0L^1T^{-2}]^a[MLT^{-3}\theta^{-1}]^b[M^0LT^0]^c[M^0L^0T^0]^d[M^1L^2T^{-3}]$$
$$M \Rightarrow 0 = b + 1$$
$$L \Rightarrow 0 = a + b + c + 2$$
$$T \Rightarrow 0 = -2a - 3b - 3$$
$$\theta \Rightarrow 0 = -b$$
$$b = -1, a = 0, c = -1$$
$$\therefore \pi_{16D2} = [\mathbf{P_D/KDc}]$$

3.7.3 Formulation of Experimental Data Base Model for Solar Updraft Tower

Six independent π terms (π_1, π_2, π_3, π_4, π_5, π_6) and two dependent π terms (π_{D1}, π_{D2}) have been identified for model formulation.

Each dependent π term is a function of the available independent π terms:

$$Z_1 = f(\pi_1, \pi_2, \pi_3, \pi_4, \pi_5, \pi_6)$$
$$Z_2 = f(\pi_1, \pi_2, \pi_3, \pi_4, \pi_5, \pi_6)$$

Here, f stands for "function of", where

$Z_1 = \pi_{D1}$, First dependent π term = $[(Dc^{1/2}\ N_T/g^{1/2})]$
$Z_2 = \pi_{D2}$, Second dependent π term = $[P_D/K\ D_c]$

To evolve the probable mathematical form of dimensional equation of the phenomenon, we can assume that the relationship amongst the dependent and independent π terms to be of exponential form. A general equation can be written as follows:

$$\pi_D = k \times [(\pi_1)^a \times (\pi_2)^b \times (\pi_3)^c \times (\pi_4)^d \times (\pi_5)^e \times (\pi_6)^f]$$

This equation, in fact, represents the following independent π terms involved in the solar updraft tower phenomena:

$$(Z) = K\{[(H_{gc}T_{cc}\theta_c)/Dc^2]^a, [H_{ch}D_{ch}N_b/Dc^2]^b, [Hu]^c, [(T_aT_o)(V_iD_c/g^2)(V_oD_c/g^2)]^d,$$
$$[(g^{1/2}Th/Dc^{1/2})(A_{oi}/Dc^2)]^e, [(D_cQ/K)]^f\}$$

The equation contains seven unknowns, the Curve fitting constant K, and the indices a, b, c, d, e, and f. To get the values of these unknowns we need a minimum of seven sets of values of π_1, π_2, π_3, π_4, π_5, π_6 and π_D.

As per the plan in the design of experimentation, 80 sets of such values as given. If any arbitrary seven sets are chosen from this table and the values of the unknowns, K, a, b, c, d, e, and f are computed. To be very specific one can select ${}^nC_r$ number combinations of r number of sets chosen out of n number of sets of values or ${}^{80}C_7$ in our case. Solving these many sets and finding their solutions is a herculean task. Hence, employ the Curve Fitting Technique (Eastman Kodak Co. Ltd. 1983a). To implement this method to experiment, it is imperative to have the equation in the form:

$$Y = a + bV + cW + dX.... \tag{3.1}$$

By taking log on both sides of Equation 3.1, the above form of equation can be obtained thus:

$$\log\pi_D = \log k + a \times \log(\pi_1) + b \times \log(\pi_2) + c \times \log(\pi_3) + d \times \log(\pi_4) + e \times \log(\pi_5) + f \times \log(\pi_6) \tag{3.2}$$

Let us denote log π_D = Z log k = K log (π_1) = A log (π_2) = B log (π_3) = C log (π_4) = D log (π_5) = E log (π_6) = F

Now, Equation 3.2 can be written as:

$$Z = K + a \times A + b \times B + c \times C + d \times D + e \times E = f \times F + g \times G \tag{3.3}$$

Equation 3.3 is a regression equation of Z on A, B, C, D, E, F, and G in an N-dimensional coordinate system. This represents a regression hyper-plane. The curve fitting method, in order to include every set of readings, instead of any eight randomly chosen sets of readings, instructs us to add all these equations. This results into the following equation:

$$\Sigma Z = nK + a \times \Sigma A + b \times \Sigma B + c \times \Sigma C + d \times \Sigma D + e \times \Sigma E + f \times \Sigma F$$

Number of readings, n, is 80.

As mentioned earlier, there are seven unknowns in this equation, K, a, b, c, d, e, and f, hence requiring seven different equations to solve. In order to achieve these equations, multiply equation six with A, B C, D, E, and F to get a set of seven different equations as shown below:

$$\Sigma Z \times A = K \times \Sigma A + a \times \Sigma A \times A + b \times \Sigma B \times A + c \times \Sigma C \times A + d \times \Sigma D \times A + e \times \Sigma E \times A$$
$$+ f \times \Sigma F \times A \Sigma Z \times B = K \times \Sigma B + a \times \Sigma A \times B + b \times \Sigma B \times B + c \times \Sigma C \times B + d \times \Sigma D \times B$$
$$+ e \times \Sigma E \times B + f \times \Sigma F \times B \Sigma Z \times C = K \times \Sigma C + a \times \Sigma A \times C + b \times \Sigma B \times C + c \times \Sigma C \times C$$
$$+ d \times \Sigma D \times C + e \times \Sigma E \times C + f \times \Sigma F \times C \Sigma Z \times D = K \times \Sigma D + a \times \Sigma A \times D + b \times \Sigma B \times D$$
$$+ c \times \Sigma C \times D + d \times \Sigma D \times D + e \times \Sigma E \times D + f \times \Sigma F \times D \Sigma Z \times E = K \times \Sigma E + a \times \Sigma A \times E$$
$$+ b \times \Sigma B \times E + c \times \Sigma C \times E + d \times \Sigma D \times E + e \times \Sigma E \times E + f \times \Sigma F \times E \Sigma Z \times F = K \times \Sigma F$$
$$+ a \times \Sigma A \times F + b \times \Sigma B \times F + c \times \Sigma C \times F + d \times \Sigma D \times F + e \times \Sigma E \times F + f \times \Sigma F \times F$$

A set of seven equations for the seven unknowns. To solve these equations, use Matrix analysis. The Matrix format of an equation is of the form.

$$Z = W \times X$$

In the above equation, X is the unknown matrix. This equation is solved to determine the value of X by taking an inverse of matrix W. The final equation solved by using MATLAB is as follows:

$$X = \mathrm{inv}(W) \times Z$$

Solving this equation for every dependent π term, the model is formed.

4 Experimentation

4.1 INTRODUCTION

The steps involved in formulating field or experimental data-based model suggested by H. Schenck Jr., have been proposed in this investigation. It involves the collection of the relevant data by direct measurement with the help of required instruments and the data collected from the field or experimentation. The identified independent variables have been reduced into a group of pi terms, which are discussed. The developed models at the end of this chapter show the relationships amongst the dimensionless pi terms of the phenomenon.

4.2 INSTRUMENTATION AND DATA COLLECTION

4.2.1 Instrumentation for Crankshaft Maintenance Activity

The crankshaft maintenance activity is influenced by the following variables and the instrumentation used for their measurement types are presented (Table 4.1).

TABLE 4.1
Independent and Dependent Variables and Instrumentation-Crankshaft Maintenance Activity

Ser	Variable Description	Variable Type	Symbol	Dimension	Dimension Measurement
1	Overhauling time of crankshaft maintenance activity	Dependent	T_{cs}	$[M^0L^0T^1]$	Digital watch/calculated
2	Human energy consumed in crankshaft maintenance activity	Dependent	HE_{cs}	$[M^1L^2T^{-2}]$	Calculated from pulse rate
3	Productivity of crank shaft overhauling	Dependent	Pd_{cs}	$[M^0L^0T^{-1}]$	Calculated
4	Age of worker	Independent	Ag_s	$[M^0L^0T^1]$	Questionnaire
5	Experience of worker	Independent	Ex_s	$[M^0L^0T^1]$	Questionnaire
6	Skill of worker	Independent	Sk_s	$[M^0L^0T^0]$	Questionnaire/grading scale
7	Enthusiasm of worker	Independent	En_s	$[M^0L^0T^0]$	Questionnaire/grading scale
8	Habits of worker	Independent	Hb_s	$[M^0L^0T^0]$	Questionnaire/grading scale
9	Health of worker	Independent	hl_s	$[M^0L^0T^0]$	Questionnaire/grading scale
10	Anthropometric data of worker	Independent	Ant_s	$[M^0L^1T^0]$	Measuring tape
11	Temperature of workstation	Independent	Tmp_s	$[ML^2T^{-2}]$	DBT on mobile
12	Humidity of workstation	Independent	Hmd_s	$[M^0L^0T^0]$	hygrometer
13	Noise of workstation	Independent	Noi_s	$[M^0L^0T^0]$	Noise meter on mobile
14	Illumination of workstation	Independent	ilm_s	$[M^1L^0T^{-3}]$	Lux meter on mobile
15	Diameter of split pin	Independent	Dsp	$[M^0L^1T^0]$	Vernier calliper
16	Length of split pin	Independent	Lsp	$[M^0L^1T^0]$	Measuring tape
17	Diameter of saddle nut	Independent	Dsn	$[M^0L^1T^0]$	Measuring tape
18	Length of saddle nut	Independent	Lsn	$[M^0L^1T^0]$	Measuring tape
19	Diameter of saddle bolt	Independent	Dsb	$[M^0L^1T^0]$	Measuring tape

(Continued)

DOI: 10.1201/9781003318699-4

TABLE 4.1 (Continued)

Ser	Variable Description	Variable Type	Symbol	Dimension	Dimension Measurement
20	Length of saddle bolt	Independent	Lsb	$[M^0L^1T^0]$	Measuring tape
21	Diameter of small Allen bolt	Independent	Dsmab	$[M^0L^1T^0]$	Measuring tape
22	Length of small Allen bolt	Independent	Lsmab	$[M^0L^1T^0]$	Measuring tape
23	Diameter of side Allen bolt	Independent	Dsiab	$[M^0L^1T^0]$	Measuring tape
24	Length of side Allen bolt	Independent	Lsiab	$[M^0L^1T^0]$	Measuring tape
25	Diameter of saddle cap	Independent	Dsc	$[M^0L^1T^0]$	Measuring tape
26	Length of saddle cap	Independent	Lsc	$[M^0L^1T^0]$	Vernier calliper
27	Diameter of main journal of crankshaft	Independent	Dmj	$[M^0L^1T^0]$	Measuring tape
28	Length of main journal of crankshaft	Independent	Lmj	$[M^0L^1T^0]$	Measuring tape
29	Diameter of CP journal of crankshaft	Independent	Dcpj	$[M^0L^1T^0]$	Measuring tape
30	Length of CP journal of crankshaft	Independent	Lcpj	$[M^0L^1T^0]$	Measuring tape
31	Diameter of bend ring	Independent	Dbr	$[M^0L^1T^0]$	Vernier calliper
32	Length of bend ring	Independent	Lbr	$[M^0L^1T^0]$	Measuring tape
33	Diameter of small Allen spanner	Independent	Dsmsp	$[M^0L^1T^0]$	Vernier calliper
34	Length of small Allen spanner	Independent	Lsmsp	$[M^0L^1T^0]$	Vernier calliper
35	Diameter of side Allen spanner	Independent	Dsisp	$[M^0L^1T^0]$	Vernier calliper
36	Length of side Allen spanner	Independent	Lsisp	$[M^0L^1T^0]$	Measuring tape
37	Diameter of barring rod	Independent	Dbrr	$[M^0L^1T^0]$	Vernier calliper
38	Length of barring rod	Independent	Lbrr	$[M^0L^1T^0]$	Measuring tape
39	Diameter of main journal bearing	Independent	Dmjb	$[M^0L^1T^0]$	Measuring tape
40	Thickness of main journal bearing	Independent	Tmjb	$[M^0L^1T^0]$	Vernier calliper
41	Diameter of crank pin journal bearing	Independent	Dcpjb	$[M^0L^1T^0]$	Measuring tape
42	Thickness of crank pin journal bearing	Independent	Tcpjb	$[M^0L^1T^0]$	Vernier calliper
43	Length of frame	Independent	Lfrm	$[M^0L^1T^0]$	Measuring tape
44	Width of frame	Independent	Wfrm	$[M^0L^1T^0]$	Measuring tape
45	Height of frame	Independent	Hfrm	$[M^0L^1T^0]$	Measuring tape
46	Kerosene (solvent)	Independent	Ker	$[M^1L^0T^0]$	Container of 5 L
47	Lube oil	Independent	Loil	$[M^1L^0T^0]$	Container of 5 L
48	Emery belt	Independent	Eb	$[M^1L^0T^0]$	Weighing machine
49	Compressed air	Independent	Cmar	$[M^1L^0T^0]$	Compressor air flow meter
50	Axial clearance of main journal	Independent	Ax mj	$[M^0L^1T^0]$	Vernier calliper
51	Axial clearance of crank pin	Independent	Axcp	$[M^0L^1T^0]$	Vernier calliper
52	Saddle bolt elongation	Independent	Elsb	$[M^0L^1T^0]$	Vernier calliper

The field data has been collected from diesel loco shed. The readings have been collected at two workstations with a team of three workers at each location in each work shift. The anthropometric data is collected at two workstations. The weight of the individual worker has been measured with digital scales. The measurements include (a) stature; (b) shoulder height; (c) elbow height; (d) eye height; (e) fingertip length; (f) shoulder breadth; (g) hip breath; and (h) hand breadth across thumb, using measuring tape. The anthropometric data of workers (π_1) has been arrived at by calculation. Thirty loco shed workers have been enquired through questionnaire for the data such as age of worker, experience of performing work, general health status, and habits are rated on a scale of 1–10. Three supervisors also have been questioned for data such as skills in performing work, enthusiasm in performing work of the individual worker to rate them on a similar 1–10 scale.

Posture means the geometry of outline of body adopted by the operator/worker. Postures have been arrived at by observing the activity at the workstation and are numerically rated as 1 and 2 for standing and bending positions of workers respectively. The data of workers (π_2) has been arrived at by calculation.

The following scales are used for grading of workers data of loco shed.

Considerations	Poor	Average	Good
Skill of worker	0–4	4.1–7	7.1–10
Habits of worker	0–4	4.1–7	7.1–10
Health of worker	0–4	4.1–7	7.1–10
Enthusiasm of worker	0–4	4.1–7	7.1–10

The parameters such as length of saddle nut and bolt, length of small Allen bolt, length of saddle cap, diameter of liner, and length of bend ring, length of spanner, and length of barring rod, axial clearance of crankpin, saddle bolt elongation, thickness of main journal bearing, and dimensions of frames, have been measured with a 5 m measuring tape and vernier calliper. Then the values of (π_3, π_4, π_6, and π_7) have been arrived at by calculation.

The parameters such as weight of compressed air and emery belt have been measured by compressed air flow meter and by weighting machine respectively. The parameters such as kerosene and lube oil have been measured by container of 5 L. Then, the value of (π_5) has been arrived at by calculation.

The extraneous variables are atmospheric temperature, humidity, illumination, and surrounding noise level. The temperature at the workplace and relative humidity have been measured with the dry bulb temperature and relative humidity measuring digital meter. The illumination and noise at the workplace have been measured with a light meter and noise meter, an application on android smart phone, to measure intensity of light and noise. These readings have been used for calculating π_8, π_9, π_{10}, and π_{11}.

The overhauling time of crankshaft maintenance activity and liner piston maintenance activity has been measured with digital watch. During these maintenance activities, the operators (M1, M2, and M3) are using their stored human energy (HE) which can be estimated in pulse using stopwatch. The field human energy consumed by the person performing the task is calculated in Nm. It is calculated by using the formula illustrated in last section. For this, the pulse rate of the worker is measured before the maintenance activity begins, and after its completion.

4.2.2 Data Collection f rom Field for Crankshaft Maintenance Activity

The field data has been collected from diesel locoshed. The readings have been collected from liner piston workstation for duration of 15 months (Tables 4.2–4.8).

4.2.3 Instrumentation Used for Liner Piston Maintenance Activity

The liner piston maintenance activity is influenced by the following variables and the instrumentation used for their measurement (Table 4.9).

M, L, and T are the symbols for mass, length, and time respectively. The same instruments as used in crankshaft maintenance activity have been used in liner piston maintenance activity and the dependent variables have been calculated.

TABLE 4.2
Anthropometric Data of Loco Shed Worker of Crankshaft Maintenance Activity

Ser	Weight (kg)	Stature (mm) (a)	Shoulder HT (mm) (b)	Elbow HT (mm) (c)	Eye HT (mm) (d)	Fingertip Length (mm) (e)	Shoulder Breadth (mm) (f)	Hip Breadth (mm) (g)	Hand Breadth Across Thumb (mm) (h)	$\pi_1 = [(a \times c \times e \times g)/(b \times d \times f \times h)]$
1	65	1676	1372	1036.30	1524.00	670.56	487.70	128.06	128.00	1.143135
2	66	1676	1341	1036.30	1424.00	640.08	518.20	121.90	121.10	1.131234
3	70	1646	1372	1127.80	1524.00	701.04	457.20	118.87	118.80	1.362632
4	68	1585	1311	944.88	1493.50	335.28	487.70	125.92	124.70	0.531259
5	72	1737	1433	1097.30	1554.50	670.56	457.20	128.01	128.20	1.253709
6	71	1707	1433	1127.80	1585.00	670.56	457.20	131.06	131.10	1.243331
7	73	1798	1433	1097.30	1676.40	853.44	518.20	134.11	134.10	1.353348
8	65	1615	1341	1005.80	1524.00	609.56	457.20	128.02	128.00	1.060013
9	74	1646	1372	1036.30	1524.00	670.56	487.70	128.11	128.00	1.122789
10	70	1707	1372	1036.30	1615.40	670.56	457.20	124.96	125.00	1.170874
11	62	1646	1311	1005.80	1554.50	335.28	487.70	131.06	131.10	0.558635
12	63	1707	1372	975.36	1615.40	731.52	518.20	128.01	128.00	1.060747
13	58	1707	1402	1127.80	1585.00	670.56	457.20	134.11	134.10	1.270381
14	78	1768	1463	1097.30	1645.90	640.08	518.20	124.96	125.00	0.994939
15	62	1707	1372	1036.30	1585.00	640.08	548.60	131.06	131.60	0.945105
16	60	1615	1341	1036.30	1524.00	609.60	457.20	128.02	128.10	1.091746
17	61	1646	1372	1066.30	1524.00	640.08	457.20	121.92	121.90	1.175664
18	69	1707	1402	914.40	1554.50	640.08	487.70	128.02	128.10	0.939648
19	57	1676	1402	914.40	1554.50	609.60	487.70	128.02	128.10	0.878923
20	70	1676	1433	944.88	1554.50	609.60	487.70	124.97	125.00	0.889097
21	71	1645	1372	1127.80	1524.00	701.04	457.20	121.92	121.90	1.360616
22	60	1737	1341	1097.30	1524.00	762.00	426.70	131.06	131.60	1.658750
23	66	1737	1433	1097.30	1585.00	670.56	457.20	131.07	131.10	1.231518
24	67	1676	1463	944.88	1554.50	609.60	487.70	128.02	128.10	0.870337
25	68	1676	1463	944.88	1585.00	609.60	487.70	128.03	128.10	0.853733
26	65	1737	1341	1097.30	1524.00	762.00	426.70	131.06	131.60	1.658750
27	70	1737	1433	1097.30	1585.00	670.56	457.20	131.07	131.10	1.231518
28	66	1678	1463	944.88	1554.50	609.60	487.70	129.02	128.10	0.878182
29	64	1677	1463	944.88	1585.00	612.60	487.70	128.03	128.10	0.858446
30	68	1676	1461	948.88	1585.00	609.60	487.70	128.03	128.10	0.858521

TABLE 4.3
Loco Shed Workers Data of Crankshaft Maintenance Activity

		Average Age	Average Experience	Average Skill	Enthusiasm	Health	Habits
Ser	Weight	Ags	Exs	Sks	Ens	hls	hbs
1	65	40.00	16.33	8.33	8.67	8.23	6.77
2	66	39.66	18.66	7.90	7.30	8.27	7.25
3	70	40.33	18.00	8.33	7.66	8.93	6.80
4	68	44.33	15.33	7.00	6.20	7.77	7.16
5	72	43.00	20.33	8.50	6.53	8.40	6.90
6	71	40.33	16.33	8.50	8.23	8.47	7.26
7	73	39.66	17.00	7.83	8.20	8.10	7.20
8	65	43.60	18.33	7.83	8.10	8.57	7.00
9	74	44.60	18.00	7.83	6.87	6.17	6.66
10	70	40.33	16.33	8.33	7.27	8.03	6.90
11	62	40.33	16.33	8.33	7.20	8.66	6.90
12	63	45.00	22.66	8.17	6.53	7.97	7.80
13	58	40.33	16.33	8.37	7.33	8.00	7.33
14	78	40.33	16.33	8.37	7.73	7.13	7.46
15	62	39.33	14.00	7.50	7.66	8.50	7.50
16	60	39.66	13.67	8.05	8.23	7.83	7.50
17	61	40.00	16.33	8.50	6.50	7.00	5.83
18	69	41.00	18.67	7.90	6.53	7.33	6.93
19	57	43.66	18.33	8.10	6.50	7.70	7.07
20	70	43.66	18.33	7.93	6.13	7.50	6.50
21	71	44.33	17.67	7.66	7.00	5.33	7.00
22	60	40.00	16.33	9.33	9.33	8.33	6.33
23	66	40.33	16.33	8.60	8.33	6.33	7.00
24	67	39.66	17.00	8.10	7.66	7.00	6.66
25	68	45.00	19.67	8.16	7.16	8.17	7.10
26	65	40.00	15.67	8.00	8.56	8.50	7.90
27	70	42.00	19.67	8.16	7.50	7.33	7.66
28	66	40.00	16.33	8.00	8.33	8.50	7.66
29	64	40.33	16.33	8.50	8.10	8.30	7.17
30	68	45.00	22.00	8.33	7.20	8.66	6.90

TABLE 4.4
Data Related to Specification of Crankshaft of Crankshaft Maintenance Activity

	Split Pin Dia	L (Split Pin)	D (Saddle Nut)	l (Saddle Nut)	D (Saddle Bolt)	L (Saddle Bolt)	D (Small Alen Bolt)	L (Small Alen Bolt)	D (Side Alen Bolt)	L (Side Alen Bolt)	D (Saddle Cap)	L (Saddle Cap)	d (Main Journal)	L (Main Journal)	D (Crankpin)	L (Crankpin)	D (J. Bearing)	T (J. Bearing)	D (C. Bearing)	T (C. Bearing)
	mm	mm	mm	mm	mm	mm	mm	mm	mm	mm	mm	mm	mm	mm	mm	mm	mm	mm	mm	Mm
Sr	Dsp	Lsp	Dsn	Lsn	Dsb	Lsbs	Dsmab	Lsmab	Dsiab	Lsiab	Dsc	Lsc	Dmj	Lmj	Dcpj	Lcpj	Dmjb	Tmib	Dcpjb	Tcpjb
1	5	70	36	28	30	385	16	27	24	200	192	72	180	80.4	170	80	180.2	6	170.2	6
2	5	70	36	28	30	385	16	27	24	200	192	72	180	80.4	170	80	180.2	6	170.2	6
3	5	70	36	28	30	385	16	27	24	200	192	72	180	80.4	170	80	180.2	6	170.2	6
4	5	70	36	28	30	385	16	27	24	200	192	72	180	80.4	170	80	180.2	6	170.2	6
5	5	70	36	28	30	385	16	27	24	200	192	72	180	80.4	170	80	180.2	6	170.2	6
6	5	70	36	28	30	385	16	27	24	200	192	72	180	80.4	170	80	180.2	6	170.2	6
7	5	70	36	28	30	385	16	27	24	200	192	72	180	80.4	170	80	180.2	6	170.2	6
8	5	70	36	28	30	385	16	27	24	200	192	72	180	80.4	170	80	180.2	6	170.2	6
9	5	70	36	28	30	385	16	27	24	200	192	72	180	80.4	170	80	180.2	6	170.2	6
10	5	70	36	28	30	385	16	27	24	200	192	72	180	80.4	170	80	180.2	6	170.2	6
11	5	70	36	28	30	385	16	27	24	200	192	72	180	80.4	170	80	180.2	6	170.2	6
12	5	70	36	28	30	385	16	27	24	200	192	72	180	80.4	170	80	180.2	6	170.2	6
13	5	70	36	28	30	385	16	27	24	200	192	72	180	80.4	170	80	180.2	6	170.2	6
14	5	70	36	28	30	385	16	27	24	200	192	72	180	80.4	170	80	180.2	6	170.2	6
15	5	70	36	28	30	385	16	27	24	200	192	72	180	80.4	170	80	180.2	6	170.2	6
16	5	70	36	28	30	385	16	27	24	200	192	72	180	80.4	170	80	180.2	6	170.2	6
17	5	70	36	28	30	385	16	27	24	200	192	72	180	80.4	170	80	180.2	6	170.2	6
18	5	70	36	28	30	385	16	27	24	200	192	72	180	80.4	170	80	180.2	6	170.2	6
19	5	70	36	28	30	385	16	27	24	200	192	72	180	80.4	170	80	180.2	6	170.2	6
20	5	70	36	28	30	385	16	27	24	200	192	72	180	80.4	170	80	180.2	6	170.2	6

21	5	70	36	28	30	385	16	27	24	200	192	72	180	80.4	170	80	180.2	6	170.2	6
22	5	70	36	28	30	385	16	27	24	200	192	72	180	80.4	170	80	180.2	6	170.2	6
23	5	70	36	28	30	385	16	27	24	200	192	72	180	80.4	170	80	180.2	6	170.2	6
24	5	70	36	28	30	385	16	27	24	200	192	72	180	80.4	170	80	180.2	6	170.2	6
25	5	70	36	28	30	385	16	27	24	200	192	72	180	80.4	170	80	180.2	6	170.2	6
26	5	70	36	28	30	385	16	27	24	200	192	72	180	80.4	170	80	180.2	6	170.2	6
27	5	70	36	28	30	385	16	27	24	200	192	72	180	80.4	170	80	180.2	6	170.2	6
28	5	70	36	28	30	385	16	27	24	200	192	72	180	80.4	170	80	180.2	6	170.2	6
29	5	70	36	28	30	385	16	27	24	200	192	72	180	80.4	170	80	180.2	6	170.2	6
30	5	70	36	28	30	385	16	27	24	200	192	72	180	80.4	170	80	180.2	6	170.2	6

TABLE 4.5
Data Related to Specification of Tools of Crankshaft Maintenance Activity

	D(Bend Ring)	L(Bend Ring)	D(Spanner Small)	L(Spanner Small)	D(Spanner Side)	L(Spanner Side)	D(Barring Rod)	L(Barring Rod)
	mm	mm	mm	mm	mm	mm	mm	Mm
Ser	Dbr	Lbr	Dsmsp	Lsmsp	Dsisp	Lsisp	Dbrr	Lbrr
1	55	570	12	135	19	190	25	1000
2	55	570	12	135	19	190	25	1000
3	55	570	12	135	19	190	25	1000
4	55	570	12	135	19	190	25	1000
5	55	570	12	135	19	190	25	1000
6	55	570	12	135	19	190	25	1000
7	55	570	12	135	19	190	25	1000
8	55	570	12	135	19	190	25	1000
9	55	570	12	135	19	190	25	1000
10	55	570	12	135	19	190	25	1000
11	55	570	12	135	19	190	25	1000
12	55	570	12	135	19	190	25	1000
13	55	570	12	135	19	190	25	1000
14	55	570	12	135	19	190	25	1000
15	55	570	12	135	19	190	25	1000
16	55	570	12	135	19	190	25	1000
17	55	570	12	135	19	190	25	1000
18	55	570	12	135	19	190	25	1000
19	55	570	12	135	19	190	25	1000
20	55	570	12	135	19	190	25	1000
21	55	570	12	135	19	190	25	1000
22	55	570	12	135	19	190	25	1000
23	55	570	12	135	19	190	25	1000
24	55	570	12	135	19	190	25	1000
25	55	570	12	135	19	190	25	1000
26	55	570	12	135	19	190	25	1000
27	55	570	12	135	19	190	25	1000
28	55	570	12	135	19	190	25	1000
29	55	570	12	135	19	190	25	1000
30	55	570	12	135	19	190	25	1000

4.2.4 Data Collection from the Field for Liner Piston Maintenance Activity

The field data has been collected from diesel locoshed, Nagpur. The readings have been collected from liner piston workstation for duration of 15 months (Tables 4.10–4.16).

4.2.5 Basis for Arriving at Number of Observations

The number of observations for crankshaft maintenance activity and liner piston maintenance activity is considered as 30 based on following formula (Table 4.17).

TABLE 4.6
Data Related to Specification of Solvents and Axial Clearance of Crankshaft Maintenance Activity

	Kerosene	Lube Oil	Compress Air	Emery Belt	Crankpin Axial Clearance	MJ Axial Clearance	Bolt Elongation
	Lit	Lit	kg	kg	mm	mm	Mm
Ser	Ker	Loil	Cmar	Eb	Axcp	Axmj	Elsb
1	2	0.5	7	0.21	0.21	0.18	0.54
2	2.5	0.5	8	0.17	0.21	0.18	0.53
3	2	0.5	6	0.19	0.22	0.18	0.54
4	2	0.4	8	0.23	0.18	0.2	0.54
5	2	0.8	7	0.24	0.91	0.18	0.54
6	2	0.8	7	0.28	0.21	0.18	0.53
7	2.5	0.6	8	0.21	0.21	0.21	0.54
8	2	0.5	8	0.19	0.21	0.18	0.54
9	2	0.6	7	0.24	0.2	0.18	0.54
10	2	0.6	7	0.21	0.21	0.25	0.56
11	2.5	0.8	6	0.23	0.21	0.18	0.54
12	2	0.4	9	0.24	0.21	0.18	0.54
13	2	0.7	7	0.21	0.26	0.17	0.54
14	2.5	0.5	7	0.26	0.21	0.18	0.54
15	2	0.5	8	0.21	0.21	0.19	0.54
16	2	0.5	8	0.23	0.27	0.18	0.55
17	2.5	0.8	9	0.21	0.21	0.18	0.54
18	2	0.5	7	0.23	0.21	0.28	0.54
19	2.5	0.7	9	0.24	0.28	0.18	0.55
20	2	0.5	8	0.28	0.21	0.22	0.54
21	2.5	0.9	8	0.24	0.21	0.18	0.54
22	2	0.6	9	0.21	0.21	0.19	0.53
23	2.5	0.5	9	0.24	0.22	0.18	0.54
24	2	0.8	7	0.21	0.21	0.18	0.54
25	2	0.4	7	0.24	0.24	0.23	0.55
26	2.5	0.6	7	0.21	0.21	0.21	0.54
27	2	0.5	7	0.21	0.25	0.21	0.56
28	2	0.5	8	0.24	0.21	0.18	0.54
29	2	0.8	7	0.21	0.26	0.18	0.53
30	2.5	0.5	8	0.24	0.23	0.18	0.54

$$N = \left(\frac{\sqrt{\frac{B^2}{A^2}(n\sum(fX^2) - (\sum fx)^2)}}{\sum fX} \right)^2$$

N = No. readings

B = 2 for 95%

A = 0.05 for ± 5% desired precision

TABLE 4.7

Data Related to – Workstation, Temperature, Humidity, Illumination, and Noise at Workplace of Crankshaft Maintenance Activity

	Length of Frame	Width of Frame	Height of Frame	Temperature	Humidity	Illumination	Noise
	mm	mm	mm	°C	%	Lux	Db
Ser	Lfrm	Wfrm	Hfrm	temp	hum	lum	Noise
1	1740	1041	650	31.2	31	215	62.75
2	1740	1041	650	26	31	217	72
3	1740	1041	650	32.8	27	220	70
4	1740	1041	650	34	27	222	85
5	1740	1041	650	36	30	230	80
6	1740	1041	650	38.1	30	235	84
7	1740	1041	650	43	24	255	79
8	1740	1041	650	35.3	24	255	73
9	1740	1041	650	39.3	27	255	76
10	1740	1041	650	39.8	27	255	77
11	1740	1041	650	41.2	55	220	74
12	1740	1041	650	45.5	55	219	72.4
13	1740	1041	650	30	77	227	75
14	1740	1041	650	29.7	77	232	78
15	1740	1041	650	31.2	80	237	83
16	1740	1041	650	27.5	80	240	82
17	1740	1041	650	33.6	74	245	83
18	1740	1041	650	30.8	74	250	81
19	1740	1041	650	30.6	61	242	80
20	1740	1041	650	29.4	61	239	82
21	1740	1041	650	31	55	233	81
22	1740	1041	650	29.7	55	198	70
23	1740	1041	650	29.2	56	228	78
24	1740	1041	650	28.7	56	217	80
25	1740	1041	650	28.6	53	234	70
26	1740	1041	650	28.4	53	219	75
27	1740	1041	650	33.2	42	230	72
28	1740	1041	650	31.2	42	244	80
29	1740	1041	650	27.5	29	221	82
30	1740	1041	650	29.4	29	222	83

Substituting these values in formula we get,

$$N = \left(\frac{\frac{\sqrt{2^2}}{0.05^2}(30 * 56528270) - (40800)}{40800} \right)^2$$

$$N = (5.47)^2$$

$$N = 29.99 \sim .30$$

No. readings, $N = 30$

TABLE 4.8
Data Related to pi Term of Dependent Variables of Crankshaft Maintenance – Overhauling Time (z_{1C}), Human Energy Consumed in Crankshaft Maintenance Activity (z_{2C}) and Productivity of Crankshaft Overhauling (z_{3C})

	Field Overhauling Time of Crankshaft	Accn Due to Gravity	Length of Saddle Bolt	Conversion Factor, 1w = 1.341E-3 HP	Field HE Consumed in Crankshaft Maintenance	Total Engine Maintenance Time	Productivity of Crankshaft Maintenance
	Mins	m/s²	meter		Nm	Min	
Ser	Tm	g	Llin		HE		Pd
1	1408	9.81	0.385	1.341E−3	6239.21209	21600	0.065185185
2	1453	9.81	0.385	1.341E−3	5145.55945	21600	0.067268519
3	1598	9.81	0.385	1.341E−3	4866.26765	21600	0.073981481
4	1151	9.81	0.385	1.341E−3	4927.61255	21600	0.053287037
5	1892	9.81	0.385	1.341E−3	4458.05000	21600	0.088981481
6	1887	9.81	0.385	1.341E−3	4462.13800	21600	0.087731481
7	1243	9.81	0.385	1.341E−3	4883.43391	21600	0.057546296
8	1296	9.81	0.385	1.341E−3	4001.92236	21600	0.060000000
9	1280	9.81	0.385	1.341E−3	4341.45170	21600	0.059259259
10	1180	9.81	0.385	1.341E−3	3795.88180	21600	0.054629630
11	1239	9.81	0.385	1.341E−3	3252.17384	21600	0.057361111
12	999	9.81	0.385	1.341E−3	4306.95400	21600	0.041435185
13	1200	9.81	0.385	1.341E−3	5180.93474	21600	0.055555556
14	1312	9.81	0.385	1.341E−3	3516.79124	21600	0.060740741
15	1247	9.81	0.385	1.341E−3	2948.34915	21600	0.057731481
16	1173	9.81	0.385	1.341E−3	5043.92500	21600	0.053842593
17	1407	9.81	0.385	1.341E−3	4573.41286	21600	0.065138889
18	1377	9.81	0.385	1.341E−3	5145.32250	21600	0.063750000
19	1325	9.81	0.385	1.341E−3	6028.71095	21600	0.061342593
20	1440	9.81	0.385	1.341E−3	3942.82564	21600	0.066666667
21	1442	9.81	0.385	1.341E−3	6010.02340	21600	0.066759259
22	1225	9.81	0.385	1.341E−3	6610.83126	21600	0.056712963
23	1205	9.81	0.385	1.341E−3	4319.92666	21600	0.055787037
24	1450	9.81	0.385	1.341E−3	3879.49652	21600	0.067129630
25	1470	9.81	0.385	1.341E−3	3354.10736	21600	0.068055556
26	1362	9.81	0.385	1.341E−3	3796.26056	21600	0.063055556
27	1388	9.81	0.385	1.341E−3	4436.31134	21600	0.064259259
28	1422	9.81	0.385	1.341E−3	3908.52308	21600	0.065833333
29	1305	9.81	0.385	1.341E−3	4118.29925	21600	0.060416667
30	1424	9.81	0.385	1.341E−3	3541.89292	21600	0.065925926

Thus, number of observations of crankshaft maintenance activity is considered as 30. A similar method is adopted to calculate number of readings for liner piston maintenance activity.

4.2.6 Calculation of Field Human Energy Consumed in Maintenance Activity

Crankshaft maintenance activity of locoshed is performed by three workers, viz. M_1, M_2, and M_3. So, field human energy is calculated for each worker after completion of their maintenance task, and total field human energy is taken as average of these workers' human energy. The chronicle sequence and the task of each worker during crankshaft maintenance activity is:

a. Taking out split pin (14 Nos) – In this process, M_1 holds one end of split pin by left hand and pulls it by cutting plier by right hand. M_2 stands on other side of crankshaft and holds

TABLE 4.9
Instrumentation Used for Measurement-Liner Piston Maintenance Activity

Ser	Description of Variables	Type of Variable	Symbol	Dimension	Dimension Measurement
1	Overhauling time for liner piston maintenance activity	Dependent	To	$[M^0L^0T^1]$	Digital watch/calculated
2	Human energy consumed in liner piston activity	Dependent	H E	$[M^1L^2T^{-2}]$	Calculated from pulse rate
3	Productivity of liner piston maintenance activity	Dependent	Pd	$[M^0L^0T^{-1}]$	Calculated
4	Age of worker	Independent	Aw	$[M^0L^0T^1]$	Questionnaire
5	Experience of worker	Independent	Exw	$[M^0L^0T^1]$	Questionnaire
6	Skill of worker	Independent	Skw	$[M^0L^0T^0]$	Questionnaire/grading scale
7	Enthusiasm of worker	Independent	Ew	$[M^0L^0T^0]$	Questionnaire/grading scale
8	Habits of worker	Independent	Hw	$[M^0L^0T^0]$	Questionnaire/grading scale
9	Health of worker	Independent	HlW	$[M^0L^0T^0]$	Questionnaire/grading scale
10	Anthropometric data of worker	Independent	Ad	$[M^0L^1T^0]$	Measuring tape
11	Temperature of workstation	Independent	Tws	$[ML^2T^{-2}]$	DBT on mobile
12	Humidity of workstation	Independent	Hws	$[M^0L^0T^0]$	Hygrometer
13	Noise of workstation	Independent	Nws	$[M^0L^0T^0]$	Noise meter on mob
14	Illumination of workstation	Independent	IlWs	$[M^1L^0T^{-3}]$	Lux meter on mob
15	Diameter of liner	Independent	Dlin	$[M^0L^1T^0]$	Measuring tape
16	Length of liner	Independent	Llin	$[M^0L^1T^0]$	Measuring tape
17	Diameter of piston	Independent	Dpis	$[M^0L^1T^0]$	Measuring tape
18	Length of piston	Independent	Lprng	$[M^0L^1T^0]$	Measuring tape
19	Diameter of piston rings	Independent	Dprng	$[M^0L^1T^0]$	Measuring tape
20	Thickness of piston rings	Independent	Tprng	$[M^0L^1T^0]$	Vernier calliper
21	Length of piston stroke	Independent	Lpstrk	$[M^0L^1T^0]$	Calculated
22	Length of piston pin	Independent	Lppin	$[M^0L^1T^0]$	Measuring tape
23	Diameter of piston pin	Independent	Dppin	$[M^0L^1T^0]$	Measuring tape
24	Length of connecting rod	Independent	Lcr	$[M^0L^1T^0]$	Measuring tape
25	Diameter of big end bearing	Independent	Dbgbr	$[M^0L^1T^0]$	Measuring tape
26	thickness of big end bearing	Independent	Tbgbr	$[M^0L^1T^0]$	Vernier calliper

27	Diameter of little end bearing	Independent	Dltbr	$[M^0L^1T^0]$	Measuring tape
28	Length of little end bearing	Independent	Lltbr	$[M^0L^1T^0]$	Measuring tape
29	Diameter of big end nut	Independent	Dbnt	$[M^0L^1T^0]$	Measuring tape
30	Length of big end nut	Independent	Lbnt	$[M^0L^1T^0]$	Measuring tape
31	Diameter of big end bolt	Independent	Dbbl	$[M^0L^1T^0]$	Measuring tape
32	Length of big end bolt	Independent	Lbbl	$[M^0L^1T^0]$	Measuring tape
33	small end bearing axial clearance	Independent	Scl	$[M^0L^1T^0]$	Vernier calliper
34	big end bearing axial clearance	Independent	Bcl	$[M^0L^1T^0]$	Vernier calliper
35	bolt elongation	Independent	Bel	$[M^0L^1T^0]$	Vernier calliper
36	Diameter of bracket compressor to set piston rings	Independent	Dbrcm	$[M^0L^1T^0]$	Measuring tape
37	Length of bracket compressor to set piston rings	Independent	Lbrcm	$[M^0L^1T^0]$	Measuring tape
38	Diameter of bracket expander to remove piston and liner ring	Independent	Dbrex	$[M^0L^1T^0]$	Measuring tape
39	Length of Bracket expander to remove piston and liner ring	Independent	Lbrex	$[M^0L^1T^0]$	Measuring tape
40	Diameter of socket for big end nut bolt	Independent	Dst	$[M^0L^1T^0]$	Measuring tape
41	Length of socket for big end nut bolt	Independent	Lst	$[M^0L^1T^0]$	Measuring tape
42	Length of hammer	Independent	Lhmr	$[M^0L^1T^0]$	Measuring tape
43	Diameter of hammer	Independent	Dhmr	$[M^0L^1T^0]$	Vernier calliper
44	Length of nose plier	Independent	Lnpl	$[M^0L^1T^0]$	Measuring tape
45	Length of liner puller	Independent	Llipl	$[M^0L^1T^0]$	Measuring tape
46	Diameter of liner puller	Independent	Dlipl	$[M^0L^1T^0]$	Vernier calliper
47	Length of bush puller	Independent	Lbpl	$[M^0L^1T^0]$	Measuring tape
48	Diameter of bush puller	Independent	Dbpl	$[M^0L^1T^0]$	Vernier calliper
49	Kerosene (solvent)	Independent	Ke	$[M^0L^1T^0]$	Container of 5 lit
50	Emery belt	Independent	Eb	$[M^1L^0T^0]$	Weighing machine
51	Lube oil	Independent	Loil	$[M^1L^0T^0]$	Container of 5 lit
52	Orient 11 solvent	Independent	O11	$[M^1L^0T^0]$	Container of 5 lit
53	Height of connecting rod workstation	Independent	Hcrt	$[M^0L^1T^0]$	Measuring tape
54	Length connecting rod workstation	Independent	Lcrt	$[M^0L^1T^0]$	Measuring tape
55	Breath of connecting rod workstation	Independent	Bcrt	$[M^0L^1T^0]$	Measuring tape

TABLE 4.10
Anthropometric Data of Locoshed Workers of Liner Piston Maintenance Activity

	Age	Weight	Stature	Shoulder ht	Elbow ht	Eye ht	Finger Tip ht	Shoulder Breadth	Hip Breadth	Hand Breadth Across Thumb	π_1 = [(a*c*e*g)/(b*d*f*h)]
			(a)	(b)	(c)	(d)	(e)	(f)	(g)	(h)	
Ser	Year	kg	mm	mm	mm	mm	mm	mm	mm	mm	
1	43	65	1676	1372	1036.3	1524	670.56	487.7	128.06	128	1.143135
2	45	66	1676	1341	1036.3	1424	640.08	518.2	121.9	121.1	1.131234
3	45	70	1646	1372	1127.8	1524	701.04	457.2	118.87	118.8	1.362632
4	46	68	1585	1311	944.88	1493.5	335.28	487.7	125.92	124.7	0.531259
5	47	72	1737	1433	1097.3	1554.5	670.56	457.2	128.01	128.2	1.253709
6	42	71	1707	1433	1127.8	1585	670.56	457.2	131.06	131.1	1.243331
7	32	73	1798	1433	1097.3	1676.4	853.44	518.2	134.11	134.1	1.353348
8	43	65	1615	1341	1005.8	1524	609.56	457.2	128.02	128	1.060013
9	43	74	1646	1372	1036.3	1524	670.56	487.7	128.11	128	1.122789
10	42	70	1707	1372	1036.3	1615.4	670.56	457.2	124.96	125	1.170874
11	45	62	1646	1311	1005.8	1554.5	335.28	487.7	131.06	131.1	0.558635
12	46	63	1707	1372	975.36	1615.4	731.52	518.2	128.01	128	1.060747
13	44	58	1707	1402	1127.8	1585	670.56	457.2	134.11	134.1	1.270381
14	44	78	1768	1463	1097.3	1645.9	640.08	518.2	124.96	125	0.994939
15	44	62	1707	1372	1036.3	1585	640.08	548.6	131.06	131.6	0.945105
16	44	60	1615	1341	1036.3	1524	609.6	457.2	128.01	128.1	1.091746
17	44	61	1646	1372	1066.3	1524	640.08	457.2	121.92	121.9	1.175664
18	48	69	1707	1402	914.4	1554.5	640.08	487.7	128.01	128.1	0.939648
19	30	57	1676	1402	914.4	1554.5	609.6	487.7	128.01	128.1	0.878923
20	40	70	1676	1433	944.88	1554.5	609.6	487.7	124.96	125	0.889097
21	45	71	1645	1372	1127.8	1524	701.04	457.2	121.92	121.9	1.360616
22	30	60	1737	1341	1097.3	1524	762	426.7	131.06	131.6	1.65875
23	35	66	1737	1433	1097.3	1585	670.56	457.2	131.07	131.1	1.231518
24	37	67	1676	1463	944.88	1554.5	609.6	487.7	128.02	128.1	0.870337
25	36	68	1676	1463	944.88	1585	609.6	487.7	128.03	128.1	0.853733
26	45	65	1737	1341	1097.3	1524	762	426.7	131.06	131.6	1.65875
27	35	70	1737	1433	1097.3	1585	670.56	457.2	131.07	131.1	1.231518
28	35	66	1678	1463	944.88	1554.5	609.6	487.7	129.02	128.1	0.878182
29	35	64	1677	1463	944.88	1585	612.6	487.7	128.03	128.1	0.858446
30	31	68	1676	1461	948.88	1585	609.6	487.7	128.03	128.1	0.858521

one end split pin by left hand and pulls it by cutting plier by right hand. M_3 rotate flywheel to get suitable position to take out split pin.

b. Taking outside Allen bolt (14 Nos) – In this, M_1 fix 19 mm spanner on Allen bolts and rotates it in anticlockwise direction for loosening Allen bolt. M_2 helps to M_1 to rotate spanner in anticlockwise direction. M_3 rotate flywheel to get suitable position to take out split pin.
c. Loosening of saddle nut (14 Nos) and bolt (14 Nos) of 7 saddle. – In this, M_1 fix 55 mm bent ring on saddle nut and rotate it in anticlockwise direction for loosening Allen bolt. M_2 Helps to M_1 to rotate spanner in anticlockwise direction. M_3 rotate flywheel to get suitable position to take out split pin.
d. Keeping engine block in horizontal position by crane. – In this, M_1 helps to crane man to guide for keeping engine in required position. M_2 locks the hooks of belt in angles and

TABLE 4.11
Loco Shed Workers Data of Liner Piston Maintenance Activity

Ser	Average Age	Average Experience	Average Skill	Enthusiasm	Health	Habit
1	38.00	13.33	7.80	7.2667	7.46667	8.63333
2	43.00	15.33	7.97	6.1666	8.80000	6.16667
3	39.67	16.00	7.50	7.2300	7.30000	7.00000
4	39.33	12.00	8.43	7.1000	8.33333	7.46667
5	41.33	15.00	7.97	6.8000	9.63333	8.16667
6	41.33	15.00	7.63	7.2000	7.26667	8.66667
7	34.67	16.00	7.73	8.0000	7.00000	8.66667
8	41.33	15.00	8.50	7.0000	7.40000	7.33333
9	40.00	17.00	7.67	7.9667	8.66670	7.16667
10	43.00	15.33	8.10	6.0000	8.33333	7.23333
11	39.33	14.00	8.00	8.4333	7.26667	8.76667
12	43.67	20.67	7.67	8.5667	7.50000	7.90000
13	43.00	15.33	7.97	7.7000	8.50000	7.76667
14	45.67	19.67	7.00	7.3000	8.16667	7.86667
15	39.67	16.00	8.00	7.6667	8.33333	7.50000
16	43.00	17.00	7.33	8.0000	7.96667	8.93333
17	43.00	15.33	7.80	7.0667	6.00000	6.40000
18	39.33	14.00	7.53	7.0000	6.83333	6.16667
19	44.33	19.67	8.33	7.0000	8.00000	7.00000
20	37.00	11.67	7.67	6.6667	7.00000	7.00000
21	43.00	15.00	7.40	7.9667	8.56667	8.20000
22	43.00	15.00	7.90	8.0667	8.20000	8.00000
23	45.00	17.33	8.10	7.7333	8.43333	6.40000
24	44.67	22.00	8.50	8.0000	7.00000	7.73300
25	39.67	13.00	7.93	7.0000	6.33330	7.66667
26	44.66	20.67	8.50	10.0557	5.85807	6.51856
27	43.00	15.33	8.30	8.2100	5.01556	5.83667
28	39.66	15.33	8.10	8.1433	4.96370	5.74778
29	43.66	20.66	8.76	10.1400	5.92444	6.63333
30	43.00	15.00	7.46	7.8200	4.74889	5.42667

helps to crane man to guide for keeping engine block in required position. M_3 helps to M_1 and M_2.

e. Taking out loosed saddle nut – In this, M_1, and M_2 take out loosed saddle nut by hand by rotating nuts in anticlockwise direction. M_3 rotate flywheel to get suitable position to take out saddle nut.
f. Loosening of small Allen bolt – In this, M_1 fix 12 mm spanner on Allen bolt and rotate the spanner in anticlockwise direction for loosening Allen bolt. M_2 helps to M_1 to rotate spanner in anticlockwise direction while M_3 rotate flywheel to get suitable position to take out small Allen bolt.
g. Taking out 2, 3, 5, and 6 no's saddle with its bolts (8 Nos). – In this, M_1 and M_2 pulls saddle by hand and keeps on ground while M_3 rotate flywheel by tummy bar to get suitable position to take out saddle.
h. Cleaning 2, 3, 5, and 6 no's main journal (MJ) and saddle. – In this, M_1, M_2 cleans MJ by kerosene and M_3 cleans saddle by kerosene.

TABLE 4.12
Data Related to Specification of Liner Piston of Liner Piston Maintenance Activity

	Diameter of Liner	Length of Liner	Diameter of Piston	Length of Piston	Diameter of Piston Ring	Thickness of Piston Ring	Piston Stroke Length	Diameter of Piston Pin	Length of Piston Pin	Length of C/Rod	Diameter of Big End Bearing	Thickness of Big End Bearing	Diameter of Little End Bearing	Thickness of Little End Bearing	Diameter of Big End Nut	Length of Big End Nut	Diameter of Big End Bolt	Length of Big End Bolt
	mm	mm	mm	mm	mm	mm	mm	mm	mm	mm	mm	mm	mm	mm	mm	mm	mm	mm
Sr	Dlin	Llin	Dpis	Lpis	Dprng	Tprng	Lpstrk	Dppin	Lppin	Lcr	Dbgbr	Tbgbr	Dltbr	Tltbr	Dbnt	Lbnt	Dbbl	Lbbl
1	240	627	238	356	230	5	280	100	200	800	170.2	6	100	6	24	265	26	24
2	240	627	238	356	230	5	280	100	200	800	170.2	6	100	6	24	265	26	24
3	240	627	238	356	230	5	280	100	200	800	170.2	6	100	6	24	265	26	24
4	240	627	238	356	230	5	280	100	200	800	170.2	6	100	6	24	265	26	24
5	240	627	238	356	230	5	280	100	200	800	170.2	6	100	6	24	265	26	24
6	240	627	238	356	230	5	280	100	200	800	170.2	6	100	6	24	265	26	24
7	240	627	238	356	230	5	280	100	200	800	170.2	6	100	6	24	265	26	24
8	240	627	238	356	230	5	280	100	200	800	170.2	6	100	6	24	265	26	24
9	240	627	238	356	230	5	280	100	200	800	170.2	6	100	6	24	265	26	24
10	240	627	238	356	230	5	280	100	200	800	170.2	6	100	6	24	265	26	24
11	240	627	238	356	230	5	280	100	200	800	170.2	6	100	6	24	265	26	24
12	240	627	238	356	230	5	280	100	200	800	170.2	6	100	6	24	265	26	24
13	240	627	238	356	230	5	280	100	200	800	170.2	6	100	6	24	265	26	24
14	240	627	238	356	230	5	280	100	200	800	170.2	6	100	6	24	265	26	24
15	240	627	238	356	230	5	280	100	200	800	170.2	6	100	6	24	265	26	24
16	240	627	238	356	230	5	280	100	200	800	170.2	6	100	6	24	265	26	24
17	240	627	238	356	230	5	280	100	200	800	170.2	6	100	6	24	265	26	24
18	240	627	238	356	230	5	280	100	200	800	170.2	6	100	6	24	265	26	24
19	240	627	238	356	230	5	280	100	200	800	170.2	6	100	6	24	265	26	24

20	240	627	238	356	230	5	280	100	200	800	170.2	6	100	6	24	265	26	24
21	240	627	238	356	230	5	280	100	200	800	170.2	6	100	6	24	265	26	24
22	240	627	238	356	230	5	280	100	200	800	170.2	6	100	6	24	265	26	24
23	240	627	238	356	230	5	280	100	200	800	170.2	6	100	6	24	265	26	24
24	240	627	238	356	230	5	280	100	200	800	170.2	6	100	6	24	265	26	24
25	240	627	238	356	230	5	280	100	200	800	170.2	6	100	6	24	265	26	24
26	240	627	238	356	230	5	280	100	200	800	170.2	6	100	6	24	265	26	24
27	240	627	238	356	230	5	280	100	200	800	170.2	6	100	6	24	265	26	24
28	240	627	238	356	230	5	280	100	200	800	170.2	6	100	6	24	265	26	24
29	240	627	238	356	230	5	280	100	200	800	170.2	6	100	6	24	265	26	24
30	240	627	238	356	230	5	280	100	200	800	170.2	6	100	6	24	265	26	24

TABLE 4.13
Data Related to Specification of Tools of Liner Piston Maintenance Activity

	Diameter of Bracket Compressor	Length of Bracket Compressor	Diameter of Bracket Expander	Length of Bracket Expander	Diameter of Socket Tool	Length of Socket Tool	Diameter of Hammer	Length of Hammer	Length of Nose Plier	Diameter of Liner Puller	Length of Liner Puller	Diameter of Bush Puller	Length of Bush Puller
	mm	mm	mm	mm	mm	mm	mm	mm	mm	mm	mm	mm	mm
Ser	Dbr.cm	lbr.cm	dbr.ex	lbr.ex	d.st	L.st	D.hmr	L.hmr	l.npl	D.li.pl	L.li.pl	Db.pl	Lb.pl
1	240	80	260	5	41	280	35	370	85	24	142	24	98
2	240	80	260	5	41	280	35	370	85	24	142	24	98
3	240	80	260	5	41	280	35	370	85	24	142	24	98
4	240	80	260	5	41	280	35	370	85	24	142	24	98
5	240	80	260	5	41	280	35	370	85	24	142	24	98
6	240	80	260	5	41	280	35	370	85	24	142	24	98
7	240	80	260	5	41	280	35	370	85	24	142	24	98
8	240	80	260	5	41	280	35	370	85	24	142	24	98
9	240	80	260	5	41	280	35	370	85	24	142	24	98
10	240	80	260	5	41	280	35	370	85	24	142	24	98
11	240	80	260	5	41	280	35	370	85	24	142	24	98
12	240	80	260	5	41	280	35	370	85	24	142	24	98
13	240	80	260	5	41	280	35	370	85	24	142	24	98
14	240	80	260	5	41	280	35	370	85	24	142	24	98
15	240	80	260	5	41	280	35	370	85	24	142	24	98
16	240	80	260	5	41	280	35	370	85	24	142	24	98
17	240	80	260	5	41	280	35	370	85	24	142	24	98
18	240	80	260	5	41	280	35	370	85	24	142	24	98
19	240	80	260	5	41	280	35	370	85	24	142	24	98
20	240	80	260	5	41	280	35	370	85	24	142	24	98

21	240	80	260	5	41	280	35	370	85	24	142	24	98
22	240	80	260	5	41	280	35	370	85	24	142	24	98
23	240	80	260	5	41	280	35	370	85	24	142	24	98
24	240	80	260	5	41	280	35	370	85	24	142	24	98
25	240	80	260	5	41	280	35	370	85	24	142	24	98
26	240	80	260	5	41	280	35	370	85	24	142	24	98
27	240	80	260	5	41	280	35	370	85	24	142	24	98
28	240	80	260	5	41	280	35	370	85	24	142	24	98
29	240	80	260	5	41	280	35	370	85	24	142	24	98
30	240	80	260	5	41	280	35	370	85	24	142	24	98

TABLE 4.14

Data Related to Specification of Solvents and Axial Clearance of Liner Piston Maintenance Activity

Variable	Kerosene	Emery Belt	Orient 11	Lube Oil	Big End Clearance	Small End Clearance	Bolt Elongation
Symbol	ker2	Eb2	o11	Loil2	b.cl	s.cl	Bel
unit	kg	kg	ltr	lit	mm	mm	mm
1	1.2255	0.21	5.0	0.5	0.21	0.18	0.54
2	2.0425	0.18	4.5	0.5	0.21	0.18	0.53
3	1.6340	0.19	6.0	0.5	0.22	0.18	0.54
4	1.6340	0.23	8.0	1.0	0.18	0.20	0.53
5	1.6340	0.18	7.0	0.8	0.19	0.18	0.54
6	1.6340	0.18	5.0	0.5	0.21	0.18	0.53
7	2.0425	0.18	5.0	1.0	0.21	0.21	0.54
8	1.6340	0.19	5.0	0.5	0.21	0.18	0.54
9	1.6340	0.18	5.5	0.6	0.20	0.18	0.54
10	1.6340	0.21	4.5	0.6	0.21	0.23	0.56
11	2.0425	0.23	5.0	1.0	0.20	0.18	0.54
12	1.6340	0.18	5.0	0.5	0.21	0.18	0.54
13	1.6340	0.21	5.0	1.0	0.26	0.17	0.54
14	2.0425	0.18	5.0	0.5	0.21	0.18	0.54
15	1.6340	0.21	5.0	0.5	0.21	0.19	0.54
16	1.6340	0.23	5.0	0.5	0.23	0.18	0.55
17	2.0425	0.21	4.5	1.0	0.21	0.18	0.54
18	1.6340	0.23	5.0	0.5	0.21	0.23	0.54
19	2.0425	0.18	5.0	0.7	0.28	0.18	0.55
20	1.6340	0.18	5.5	0.5	0.21	0.22	0.54
21	2.0425	0.18	5.0	1.0	0.21	0.18	0.54
22	1.6340	0.21	5.0	0.6	0.21	0.19	0.53
23	2.0425	0.18	5.0	0.5	0.22	0.18	0.54
24	1.6340	0.21	5.0	1.0	0.21	0.18	0.54
25	1.6340	0.18	5.0	0.4	0.24	0.23	0.55
26	2.0425	0.21	5.5	0.6	0.21	0.21	0.54
27	1.6340	0.21	5.0	0.5	0.25	0.21	0.56
28	1.6340	0.18	4.5	0.5	0.21	0.18	0.54
29	1.6340	0.21	5.0	1.0	0.23	0.18	0.53
30	1.2250	0.18	5.0	0.5	0.23	0.18	0.54

i. Polishing of saddle and main journal – No 2, 3, 5, and 6 of crank shaft – In this, M_1 and M_2 Sits on wooden block and polishes MJ by emery belt while M_3 cleans and polishes saddle by emery paper in sitting posture.
j. Cleaning of MJ and shaft crankpin – Here, M_1 brings compressor, and apply compressed air on shaft by holding pipe by right hand. M_2 rotate flywheel to get suitable position to clean by compressed air and M_3 cleans and polishes saddle by emery paper.
k. Refitting of 2, 3, 5, and 6 no's saddle – Here, M_1, and M_2 apply lube oil on MJ and crank pin and keeps bearing on MJ, remove saddle nut from saddle, fits bearing inside the saddle, keep saddle on crankpin, and tighten nut by hands. M_3 rotate flywheel to get suitable position and helps to give bearings, saddles, and helps to tighten saddle nut.

TABLE 4.15
Data Related to – Workstation, Temperature, Humidity, Illumination, and Noise at Workplace of Liner Piston Maintenance

	Length of Connecting Rod Table	Breath of Connecting Rod Table	Height of Connecting Rod Table	Temperature	Humidity	Illumination	Noise
	mm	mm	mm	°C	%	lux	Db
Ser	l.crt	b.crt	h.crt	temp	hum	ill	Noise
1	1670	710	840	31.6	54	215	73
2	1670	710	840	26	54	217	72
3	1670	710	840	30.8	43	220	70
4	1670	710	840	32.2	43	222	85
5	1670	710	840	36	30	230	80
6	1670	710	840	39.8	30	235	84
7	1670	710	840	42	24	255	79
8	1670	710	840	40.4	24	255	73
9	1670	710	840	45.2	27	255	76
10	1670	710	840	45.8	27	255	77
11	1670	710	840	41.7	55	220	74
12	1670	710	840	30.6	55	219	72.4
13	1670	710	840	29.6	77	227	75
14	1670	710	840	29.6	77	232	78
15	1670	710	840	30.6	80	237	83
16	1670	710	840	29.8	80	240	82
17	1670	710	840	34.2	74	245	83
18	1670	710	840	32.2	74	250	81
19	1670	710	840	31.2	61	242	80
20	1670	710	840	30.8	61	239	82
21	1670	710	840	29.2	55	233	81
22	1670	710	840	30	55	198	77
23	1670	710	840	29.6	56	228	88
24	1670	710	840	29.1	56	217	78
25	1670	710	840	27.4	53	234	75
26	1670	710	840	29	53	219	89
27	1670	710	840	30.4	42	230	78
28	1670	710	840	27.7	42	244	80
29	1670	710	840	33.25	29	221	84
30	1670	710	840	39.25	29	222	83

l. Taking out 1, 4, and 7 no's saddle – In this, M1 and M2 pulls saddle by bent ring and keeps it on ground while M3 rotate flywheel to get suitable position to take out saddle.
m. Cleaning1, 4, and 7 no's main journal and saddle – Here, M_1, and M_2 cleans MJ by kerosene while M_3 cleans saddle by kerosene in sitting position.
n. Cleaning and polishing of main journal – 1, 4 and 7 crankshafts – In this, M_1 and M_2 sits on wooden block and polishes MJ by emery belt while M_3 cleans and polishes saddle by emery paper in sitting posture.
o. Refitting1, 4, and 7 no's saddle – Here, M_1, and M_2 apply lube oil on MJ and crank pin and keeps bearing on MJ, remove saddle nut from saddle, fits bearing inside the saddle, keep

TABLE 4.16

Data Related to Dependent Variables of Liner Maintenance – Overhauling Time of Liner Piston Maintenance Activity (z_{1P}), Human Energy Consumed in Liner Piston Maintenance Activity (z_{2P}) and Productivity of Liner Piston Maintenance Activity (z_{3P})

	Field Maintenance Time of Liner	Accn Due to Gravity	Length of Liner		Field Human Energy Consumed in Liner	Total Engine Maintenance Time	Productivity of Liner Maintenance, Z3 = (Tm//14400)
	Mins	m/s²	meter	Conversion Factor 1 w =	Nm	Min	
Ser	Tm	g	Llin	1.341E–3 HP	HE		pd
1	1261	9.81	0.627	1.341E–3	3748.7606	14400	0.087569444
2	1294	9.81	0.627	1.341E–3	4862.5418	14400	0.089861111
3	1245	9.81	0.627	1.341E–3	4057.8981	14400	0.086458333
4	1275	9.81	0.627	1.341E–3	3678.0328	14400	0.088541667
5	1243	9.81	0.627	1.341E–3	4808.2231	14400	0.086319444
6	1277	9.81	0.627	1.341E–3	4429.2952	14400	0.084513889
7	1371	9.81	0.627	1.341E–3	3253.2378	14400	0.088055556
8	1264	9.81	0.627	1.341E–3	4854.9091	14400	0.087777778
9	1229	9.81	0.627	1.341E–3	2693.7386	14400	0.091388889
10	1303	9.81	0.627	1.341E–3	3753.1199	14400	0.090486111
11	1279	9.81	0.627	1.341E–3	1276.7862	14400	0.088819444
12	1282	9.81	0.627	1.341E–3	3003.3246	14400	0.089027778
13	1467	9.81	0.627	1.341E–3	3991.3103	14400	0.081041667
14	1263	9.81	0.627	1.341E–3	2899.1201	14400	0.084236111
15	1257	9.81	0.627	1.341E–3	3829.6065	14400	0.085069444
16	1296	9.81	0.627	1.341E–3	3460.6366	14400	0.083055556
17	1298	9.81	0.627	1.341E–3	3921.166	14400	0.083194444
18	1258	9.81	0.627	1.341E–3	2530.981	14400	0.083888889
19	1255	9.81	0.627	1.341E–3	3608.7828	14400	0.085208333
20	1457	9.81	0.627	1.341E–3	4186.0923	14400	0.081041667
21	1276	9.81	0.627	1.341E–3	3452.4871	14400	0.081666667
22	1249	9.81	0.627	1.341E–3	3578.6017	14400	0.084652778
23	1461	9.81	0.627	1.341E–3	3505.4559	14400	0.080625000
24	1410	9.81	0.627	1.341E–3	3462.0396	14400	0.090972222
25	1870	9.81	0.627	1.341E–3	2059.5107	14400	0.137500000
26	1898	9.81	0.627	1.341E–3	2747.2011	14400	0.141944444
27	1262	9.81	0.627	1.341E–3	3484.4112	14400	0.087916667
28	1883	9.81	0.627	1.341E–3	2067.5503	14400	0.136666667
29	1295	9.81	0.627	1.341E–3	3002.9846	14400	0.089236111
30	1325	9.81	0.627	1.341E–3	3571.2078	14400	0.084722222

saddle on crankpin, and tighten nut by hands. M_3 rotate flywheel to get suitable position and helps to give bearings, saddles, and helps to tighten saddle nut.

p. Measuring of clearance of fitted bearing – In this, M_1 keeps magnetic dial gauge on saddle and take reading for bearing no 3 and 4 as 0.017 mm, for bearing no 1 and 6 as 0.018 mm and for bearing no 2 and 5 as 0.020 mm clearance. M_2 helps to give left right movement of crankshaft by tummy bar get required clearance and M_3 rotate flywheel to get suitable position of crankshaft for required clearance.

TABLE 4.17
Data Related to Field Overhauling Time of Crankshaft Maintenance Activity

Ser	Overhauling Time of Crankshaft in Minutes (x)	f	f × x	f × x²
1	1408	1	1408	1982464
2	1453	1	1453	2111209
3	1598	1	1598	2553604
4	1151	1	1151	1324801
5	1892	1	1892	3579664
6	1887	1	1887	3560769
7	1243	1	1243	1545049
8	1296	1	1296	1679616
9	1280	1	1280	1638400
10	1180	1	1180	1392400
11	1239	1	1239	1535121
12	0999	1	0999	0998001
13	1200	1	1200	1440000
14	1312	1	1312	1721344
15	1247	1	1247	1555009
16	1173	1	1173	1375929
17	1407	1	1407	1979649
18	1377	1	1377	1896129
19	1325	1	1325	1755625
20	1440	1	1440	2073600
21	1442	1	1442	2079364
22	1225	1	1225	1500625
23	1205	1	1205	1452025
24	1450	1	1450	2102500
25	1470	1	1470	2160900
26	1362	1	1362	1855044
27	1388	1	1388	1926544
28	1422	1	1422	2022084
29	1305	1	1305	1703025
30	1424	1	1424	2027776
Σ	**40800**	**30**	**40800**	**56528270**

q. Fitting small Allen bolt (14 Nos) – M_1 fix 12 mm spanner on Allen bolt and rotate it in clockwise direction to tight Allen bolt. M_2 helps to M_1 to rotate spanner in clockwise direction and M_3 rotate flywheel to get suitable position to tight small Allen bolt.

r. Keeping engine block in vertical position by crane – M_1 helps to crane man to guide for keeping engine in required position. M_2 and M_3 locks the hooks of belt in angles and helps to crane man to guide for keeping engine in required position

s. Checking all bearing alignments – M_1 sets all bearing alignment properly by fixing bent ring on saddle nut and rotating in clockwise direction and M_2 helps to M_1 to fix bent ring and helps to rotate it in clockwise direction. M_3 rotate flywheel to get suitable position of crank shaft to set bearing alignment properly.

t. Tightening of all saddle nut bolts with bolt elongation up to 0.53 mm – 0.56 mm – Here, M_1 fix 55 mm bent ring on saddle nut and rotate it in clockwise direction to tight Allen bolt and

M_2 helps to M_1 to rotate bent ring in clockwise direction. M_3 rotate flywheel to get suitable position to take out split pin.

u. Tightening of side bolt by spanner – Here, M_1 fix 19 mm spanner on Allen bolt and rotate it in clockwise direction to tight Allen bolt and M_2 helps to M_1 to rotate spanner in anticlockwise direction. M_3 rotate flywheel to get suitable position to take out split pin

v. Fitting of new split pin – In this, M_1 holds one end split pin by left hand and push it by cutting plier by right hand. M_2 stands other side of crankshaft and holds one end of split pin by left hand and push it to get fix and fix it by cutting pliers by right hand. M_3 rotate flywheel to get suitable position to take out split pin.

4.2.7 Calculation of Human Energy Consumed in Crankshaft Maintenance Activity

The following formula is used to calculate field human energy consumed in crankshaft maintenance activity. As the crankshaft maintenance operation is performed by three workers. So the field human energy is calculated of each worker and total field human energy is estimated by taking average of this human energy.

$$HE_1 = w1gh \times \left(\frac{(p2m1 - p1m1)Tm}{(p2s1 - p1s1)T1s} \right),$$

$$HE_2 = w2gh \times \left(\frac{(p2m2 - p1m2)Tm}{(p2s2 - p1s2)T2s} \right),$$

$$HE_3 = w3gh \times \left(\frac{(p2m3 - p1m3)Tm}{(p2s3 - p1s3)T3s} \right)$$

$$HE = \left(\frac{HE_1 + HE_2 + HE_3}{3} \right)$$

Where:

HE_1 = human energy consumed of first worker
HE_2 = human energy consumed of second worker
HE_3 = human energy consumed of third worker
HE = total human energy consumed in crankshaft maintenance activity
g = Acceleration due to gravity
h = height of staircase climbed by workers
p1s1 = pulse rate of first worker before climbing staircase
p2s1 = pulse rate of first worker after climbing staircase
p1s2 = pulse rate of second worker before climbing staircase
p2s2 = pulse rate of second worker after climbing staircase
p1s3 = pulse rate of third worker before climbing staircase
p2s3 = pulse rate of third worker after climbing staircase
p1c1 = pulse rate of first worker before beginning of crankshaft maintenance activity
p1c2 = pulse rate of second worker before beginning of crankshaft maintenance activity
p1c3 = pulse rate of third worker before beginning of crankshaft maintenance activity
p2c1 = pulse rate of first worker after completion of crankshaft maintenance activity
p2c2 = pulse rate of second worker after completion of crankshaft maintenance activity

p2c3 = pulse rate of third worker after completion of crankshaft maintenance activity
T1s = time of climbing of staircase of first worker
T2s = time of climbing of staircase of second worker
T3s = time of climbing of staircase of third worker
Tm = time required to complete crankshaft maintenance operation
(a), (b)...(v) = chronicle sequence of crankshaft maintenance activity
w1 = weight of first worker = 66 kg
w2 = weight of second worker = 73 kg
w3 = weight of third worker = 67 kg
$g = 9.81\ m/s^2$, h = 21.8 m, w1 = 66 kg, w2 = 73 kg, w3 = 67 kg

The values of the above terms are calculated and shown in Table 4.18 for each sequence of crankshaft maintenance activity. Thus, for the first observation of crankshaft maintenance activity, we get the human energy consumed of each worker as given below:

$$HE_1 = 3114.0\text{Nm}, \quad HE_2 = 5683.7\text{Nm}, \quad HE_3 = 9919.7\text{Nm}$$

Thus the values of HE_1, HE_2, and HE_3 are substituted in total human energy formula to calculate the total field human energy consumed in crankshaft maintenance activity as shown below:

$$HE = \left(\frac{3114.0 + 5683.7 + 9919.7}{3}\right) = 6239.212\,\text{Nm}$$

Similarly, field human energy consumed in crankshaft maintenance activity and liner piston maintenance activity is calculated for all other observations (Table 4.18).

4.2.8 Instrumentation and Collection of Data for Solar Updraft Tower

The solar chimney uses warm air to generate power. The solar chimney power plant system consists of four major components: collector, chimney, energy storage layer, and turbo generators at the base. Solar radiation is absorbed by the collector plates, and warms up the air under the solar collector, which is then sucked towards the centre of the vertical turbine shell from the base. Thus, the draft produce drives the turbine and generate the solar power. The wind turbine is located at the bottom of the chimney (Figure 4.1).

The variables affecting the phenomenon under consideration are collector materials, chimney height, chimney diameter, turbine blades, and solar radiation. All dependent and independent parameters are converted into dimensionless terms. Table 4.20 shows dependent and independent variables with units and symbols. Experimental setup is designed and fabricated to execute the experimentation according to experimentation plan. The experimental setup basically consists of concrete base with angle structural frame to locate the chimney and collector sheets as well as to locate turbine at the bottom of chimney. The site is selected on the basis of availability of wind velocity and proximity to open sky so that sun rays are available throughout the daytime. The base of around 4500 mm diameter is covered with brick joints of three bricks layers with concrete. This circular base of bricks is filled with sand, which act as a heat reservoir. On the periphery of the brick circle, eight angles of 600 mm height are erected, to locate the frame of the collector and chimney.

The MS angle frame is fabricated of 4500 mm diameter on which the collector sheets are mounted and there is a 600 mm diameter circular hole at the centre for mounting the turbine and frame to hold the chimney pipe. This frame is mounted on the angles erected at the periphery of the bricks joint.

TABLE 4.18
Calculation of Field Human Energy Consumed in Crankshaft Maintenance Activity

Ser	Tm	p1c1	p2c1	p1s1	p2s1	T1s	HE_1	p1c2	p2c2	p1s1	p2s2	T2s	HE_2	p1c3	p2c3	p1s3	p2s3	T3s	HE_3
(a)	30	30	70	72	106	0.816	90.62	72	89	71	95	0.81	301.9	70	79	72	90	0.76	183.5
(b)	35	35	70	72	106	0.816	174.7	72	102	71	95	0.81	456.7	72	115	72	90	0.76	751.5
(c)	172	172	70	72	106	0.816	88.91	70	118	71	95	0.81	148.7	70	119	72	90	0.76	174.3
(d)	40	40	70	72	106	0.816	8.495	71	73	71	95	0.81	26.64	72	72	72	90	0.76	0.0
(e)	22	22	70	72	106	0.816	0	71	73	71	95	0.81	48.44	73	105	72	90	0.76	889.8
(f)	18	18	72	72	106	0.816	151.0	72	91	71	95	0.81	562.4	72	95	72	90	0.76	781.6
(g)	100	100	71	72	106	0.816	23.78	72	73	71	95	0.81	5.328	71	100	72	90	0.76	177.4
(h)	135	135	72	72	106	0.816	15.10	72	98	71	95	0.81	102.6	72	95	72	90	0.76	104.2
(i)	273	273	71	72	106	0.816	36.10	72	101	71	95	0.81	56.60	72	102	72	90	0.76	67.2
(j)	22	22	72	72	106	0.816	30.89	72	84	71	95	0.81	290.6	72	77	72	90	0.76	139.0
(k)	25	25	73	72	106	0.816	95.15	73	81	71	95	0.81	170.5	72	103	72	90	0.76	758.5
(l)	41	41	72	72	106	0.816	66.31	74	76	71	95	0.81	25.99	72	116	72	90	0.76	656.5
(m)	48	48	72	72	106	0.816	63.71	75	83	71	95	0.81	88.81	74	117	72	90	0.76	548.0
(n)	88	88	72	72	106	0.816	166.0	72	120	71	95	0.81	290.6	72	123	72	90	0.76	354.5
(o)	33	33	71	72	106	0.816	288.3	72	118	71	95	0.81	742.8	72	110	72	90	0.76	704.4
(p)	42	42	72	72	106	0.816	186.1	75	113	71	95	0.81	482.1	75	121	72	90	0.76	670.0
(q)	49	49	72	72	106	0.816	367.5	72	101	71	95	0.81	315.3	72	125	72	90	0.76	661.7
(r)	25	25	73	72	106	0.816	13.5	74	77	71	95	0.81	63.94	74	77	72	90	0.76	73.4
(s)	40	40	72	72	106	0.816	314	72	102	71	95	0.81	399.66	72	118	72	90	0.76	703.5
(t)	45	45	72	72	106	0.816	385	75	112	71	95	0.81	438.1	73	121	72	90	0.76	652.5
(u)	80	80	72	72	106	0.816	208	72	99	71	95	0.81	179.8	72	118	72	90	0.76	351.7
(v)	45	45	72	72	106	0.816	339	72	113	71	95	0.81	485.5	72	110	72	90	0.76	516.6
1408							**3112.2**						**5683**						**9919.8**

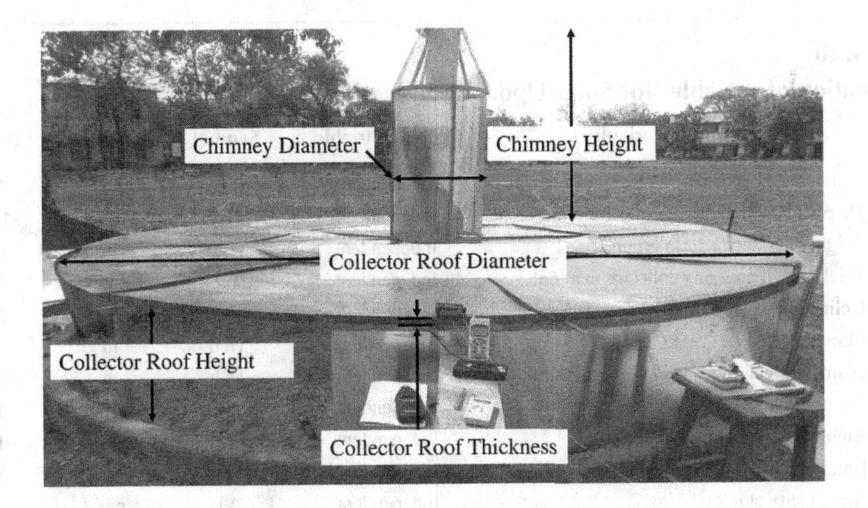

FIGURE 4.1 Experimental set up with instrumentation.

Diagram of solar updraft tower.

The different types of collector sheets were examined each time on the experimental setup to determine the effects of updraft. Chimney pipes of diameter 150 mm and heights 3600 mm and 4800 mm are placed one at a time, and subsequently changed during the experiment to predict the performance of the turbine. The lower base of the solar tower or the pipe is fitted with a vertical wind turbine equipped with blades. Table 4.20 shows the description of all the independent and dependent variables of the experimental setup of solar updraft tower.

The correlation for the independent variables and dependent variables such turbine power output and turbine speed are formulated by the mathematical model. Since the numbers of the independent variables are more thus these variables are grouped to predicate the performance and to reduce the complexity and to obtain the simplicity in the behaviour of the event, the pi terms are reduced as suggested by Schenk Jr. The pi terms related to the independent variables collector material, solar chimney, relative humidity, ambient condition, and solar radiation are reduced to form a single new pi term. Table 4.20 shows the new pi terms of independent variables in reduced form. Thus, the total of seventeen pi terms of independent variables are reduced to six new pi terms as shown in table Tables 4.19–4.22.

TABLE 4.19
Selection of Instruments

Ser	Instruments	Parameters
01	Digital tachometer	RPM of the turbine
02	Digital thermocouple with data loggers	Temperature of air
03	Spectral pyranometer	Solar radiation
04	Digital anemometer	Air velocity

TABLE 4.20
Identification of Variables for Solar Updraft Tower

Ser	Description of Variables	Type of Variable	Symbol	Unit	Dimension
01	Diameter of collector	Independent	Dc	m	$M^0L^1T^0$
02	Thermal conductivity of collector material	Independent	K	W/mK	$M^1L^1T^{-3}\theta^{-1}$
03	Height of collector from ground level	Independent	Hgc	m	$M^0L^1T^0$
04	Thickness of covering collector material	Independent	Tcc	m	$M^0L^1T^0$
05	Inclination of collector	Independent	θc	rad	$M^0L^0T^0$
06	Chimney height	Independent	Hch	m	$M^0L^1T^0$
07	Diameter of chimney	Independent	Dch	m	$M^0L^1T^0$
08	No. of blades	Independent	Nb	–	$M^0L^0T^0$
09	Ambient temperature	Independent	Ta	0C	$M^0L^0T^0\theta^1$
10	Humidity	Independent	Hu	%	$M^0L^0T^0$
11	Air velocity at inlet	Independent	Vi	m/s	$M^0L^1T^{-1}$
12	Air velocity at outlet	Independent	Vo	m/s	$M^0L^1T^{-1}$
13	Temperature inside the collector	Independent	Tc	0C	$M^0L^0T^0\theta^1$
14	Heating time	Independent	Th	sec	$M^0L^0T^1$
15	Heat flux	Independent	Q	W/m^2	$M^1L^0T^{-3}$
16	Air inlet area	Independent	Aoi	m^2	$M^0L^2T^0$
17	Acceleration due to gravity	Independent	g	m/s^2	$M^0L^1T^{-2}$
18	Turbine speed	Dependent	Ts	rpm	$M^0L^0T^{-1}$
19	Power generated	Dependent	Pd	W	$M^1L^2T^{-3}$

TABLE 4.21
Grouped Independent Pi Terms

Ser	Independent Dimensionless Ratios	Nature of Physical Quantities
01	π_1 = [(Hgc Tcc θc)/Dc2]	Collector material
02	π_2 = [Hch Dch Nb/Dc2]	Solar chimney
03	π_3 = [Hu]	Relative humidity
04	π_4 = [(Ta To)(ViDc/g2) (VoDc/g2)]	Ambient condition
05	π_5 = [(g1/2Th/Dc1/2)(Aoi/Dc2)]	Heating duration
06	π_6 = [(DcQ/K)]	Heat flux
Dependent dimensionless ratios or π terms		
01	πD1 = [(Dc1/2 N/g1/2)]	Turbine speed
02	πD2 = [Po/KDc]	Power developed

4.2.9 Instrumentation and Collection of Data for the Engine Performance by Using the Alternative Fuels

The variables affecting the effectiveness of the phenomenon under consideration are blends of treated transformer oil with diesel and performance characteristics. The dependent (or response) variables are:

- Brake thermal efficiency (BTH).
- Brake specific fuel consumption (BSFC).

TABLE 4.22
Calculations of Pi Terms

	Independent Pi Terms					Dependent Pi Terms		
Ser	π_1	π_2	π_3	π_4	π_5	π_6	π_{01}	π_{02}
1	2.48E–05	0.053	0.16	21.11	3.543	2981.25	152.4	1.5
2	2.48E–05	0.053	0.17	21.37	3.543	3026.25	157.14	1.6667
3	2.48E–05	0.053	0.18	25.27	3.543	3093.75	161.2	1.8333
4	2.48E–05	0.053	0.16	24.12	3.543	3065.63	159.17	1.75
5	2.48E–05	0.053	0.19	28.03	3.543	3121.88	162.56	1.9167
6	2.48E–05	0.053	0.17	26.65	3.543	3105	161.2	1.8333
7	2.48E–05	0.053	0.18	23.63	3.543	3037.5	159.17	1.75
8	2.48E–05	0.053	0.19	24.12	3.543	3065.63	159.17	1.75
9	2.48E–05	0.053	0.20	29.54	3.543	3150	169.33	2.0833
10	2.48E–05	0.053	0.20	28.03	3.543	3121.88	162.56	1.9167
11	1.24E–05	0.053	0.16	18.61	3.543	11497.5	142.24	4.1667
12	1.24E–05	0.053	0.17	17.4	3.543	11362.5	138.85	3.8889
13	1.24E–05	0.053	0.18	20.83	3.543	11700	149.01	5.6667
14	1.24E–05	0.053	0.16	18.61	3.543	11497.5	142.24	4.1667
15	1.24E–05	0.053	0.19	20.83	3.543	11700	149.01	5.6667
16	1.24E–05	0.053	0.17	18.61	3.543	11497.5	142.24	4.1667
17	1.24E–05	0.053	0.18	21.23	3.543	11362.5	140.88	4.1667
18	1.24E–05	0.053	0.19	17.4	3.543	11475	138.85	3.8889
19	1.24E–05	0.053	0.20	25.57	3.543	11925	159.17	6.3333
20	1.24E–05	0.053	0.20	24.21	3.543	11812.5	155.78	6
21	1.24E–05	0.053	0.16	17.81	3.543	11842.1	142.24	4.386
22	1.24E–05	0.053	0.16	17.81	3.543	11842.1	142.24	4.386
23	1.24E–05	0.053	0.18	19.95	3.543	11889.5	149.01	5.9649
24	1.24E–05	0.053	0.17	17.59	3.543	11723.7	142.24	4.386
25	1.24E–05	0.053	0.20	23.23	3.543	12031.6	155.78	6.6667
26	1.24E–05	0.053	0.19	17.59	3.543	11723.7	142.24	4.386
27	1.24E–05	0.053	0.20	23.43	3.543	12031.6	155.78	6.6667
28	1.24E–05	0.053	0.17	17.81	3.543	11842.1	142.24	4.386
29	1.24E–05	0.053	0.18	17.44	3.543	11723.7	142.24	4.386
30	1.24E–05	0.053	0.19	19.95	3.543	11889.5	149.01	5.9649
31	1.24E–05	0.053	0.16	17.03	3.543	13500	132.08	3.8889
32	1.24E–05	0.053	0.17	16.64	3.543	13162.5	130.04	3.7778
33	1.24E–05	0.053	0.17	17	3.543	13500	132.08	3.8889
34	1.24E–05	0.053	0.16	17	3.543	13218.8	132.08	3.8889
35	1.24E–05	0.053	0.18	16.5	3.543	13078.1	130.04	3.7778
36	1.24E–05	0.053	0.18	18.85	3.543	13640.6	135.46	4.1667
37	1.24E–05	0.053	0.19	18.85	3.543	13640.6	135.46	4.1667
38	1.24E–05	0.053	0.20	26	3.543	13781.3	142.24	4.4444
39	1.24E–05	0.053	0.19	16.64	3.543	13162.5	130.04	3.7778
40	1.24E–05	0.053	0.20	26.22	3.543	13781.3	142.24	4.4444
41	2.48E–05	0.071	0.19	30.27	3.543	3121.88	179.49	2.625
42	2.48E–05	0.071	0.18	25.69	3.543	3037.5	162.56	1.9167
43	2.48E–05	0.071	0.18	27.38	3.543	3093.75	169.33	2.0833

(Continued)

TABLE 4.22 (Continued)

Ser	Independent Pi Terms					Dependent Pi Terms		
	π_1	π_2	π_3	π_4	π_5	π_6	π_{01}	π_{02}
44	2.48E–05	0.071	0.19	26.22	3.543	3065.63	162.56	1.9167
45	2.48E–05	0.071	0.16	23.03	3.543	2981.25	161.2	1.8333
46	2.48E–05	0.071	0.2	31.4	3.543	3121.88	186.26	2.8194
47	2.48E–05	0.071	0.17	23.32	3.543	3026.25	161.2	1.8333
48	2.48E–05	0.071	0.16	26.22	3.543	3065.63	162.56	1.9167
49	2.48E–05	0.071	0.2	32.95	3.543	3150	189.65	3.1111
50	2.48E–05	0.071	0.17	28.87	3.543	3105	169.33	2.0833
51	1.24E–05	0.071	0.16	19.54	3.543	11497.5	149.01	5.6667
52	1.24E–05	0.071	0.19	18.31	3.543	11475	142.24	4.1667
53	1.24E–05	0.071	0.18	21.83	3.543	11700	152.4	6
54	1.24E–05	0.071	0.18	22.24	3.543	11362.5	152.4	6
55	1.24E–05	0.071	0.19	21.83	3.543	11700	152.4	5.6667
56	1.24E–05	0.071	0.2	26.63	3.543	11925	162.56	6.6667
57	1.24E–05	0.071	0.16	19.54	3.543	11497.5	149.01	5.6667
58	1.24E–05	0.071	0.17	18.31	3.543	11362.5	142.24	4.1667
59	1.24E–05	0.071	0.17	19.54	3.543	11497.5	142.24	4.1667
60	1.24E–05	0.071	0.2	25.26	3.543	11812.5	159.17	6.3333
61	1.24E–05	0.071	0.18	17.44	3.543	11723.7	142.24	4.386
62	1.24E–05	0.071	0.17	17.81	3.543	11842.1	142.24	4.386
63	1.24E–05	0.071	0.2	23.43	3.543	12031.6	155.78	6.6667
64	1.24E–05	0.071	0.2	23.23	3.543	12031.6	155.78	6.6667
65	1.24E–05	0.071	0.19	19.95	3.543	11889.5	149.01	5.9649
66	1.24E–05	0.071	0.17	17.59	3.543	11723.7	142.24	4.386
67	1.24E–05	0.071	0.16	17.81	3.543	11842.1	142.24	4.386
68	1.24E–05	0.071	0.19	17.59	3.543	11723.7	142.24	4.386
69	1.24E–05	0.071	0.18	19.95	3.543	11889.5	149.01	5.9649
70	1.24E–05	0.071	0.16	17.81	3.543	11842.1	142.24	4.386
71	1.24E–05	0.071	0.2	26.22	3.543	13781.3	142.24	4.4444
72	1.24E–05	0.071	0.17	16.64	3.543	13162.5	130.04	3.7778
73	1.24E–05	0.071	0.2	26	3.543	13781.3	142.24	4.4444
74	1.24E–05	0.071	0.17	17	3.543	13500	132.08	3.8889
75	1.24E–05	0.071	0.19	16.64	3.543	13162.5	130.04	3.7778
76	1.24E–05	0.071	0.16	17.03	3.543	13500	132.08	3.8889
77	1.24E–05	0.071	0.18	16.5	3.543	13078.1	130.04	3.7778
78	1.24E–05	0.071	0.19	18.85	3.543	13640.6	135.46	4.1667
79	1.24E–05	0.071	0.16	17	3.543	13218.8	132.08	3.8889
80	1.24E–05	0.071	0.18	18.85	3.543	13640.6	135.46	4.1667

4.2.10 Establishment of Dimensionless Group of π Terms

These independent variables have been reduced into group of π terms. List of the Independent and Dependent π terms of the face drilling activity are shown in Tables 4.23 and 4.24.

4.2.11 Creation of Field Data-Based Model

Four independent π terms (π_1, π_2, π_3, π_4) and two dependent π terms (Z_1, Z_2) have been identified for model formulation of field study.

TABLE 4.23
Independent Dimensionless π Terms

Ser	Independent Dimensionless Ratios	Nature of Basic Physical Quantities
01	$\pi_1 = [([Cv^{1/2}\ Vi/Bd])([Cv^{1/2}\ T/Bd])([B])/([Fp])]$	Specification related to blend formation and time
02	$\pi_2 = [([Cv^{1/2}\ Bd^2\ D/Mf])/([Cv^{1/2}\ L/Mf\ Bd])([Ap])]$	Specifications of fuel consumption and engine load
03	$\pi_3 = [([Apig])([Di])/([CN])]$	Specifications of fuel characteristic
04	$\pi_4 = [([Sl/Bd])([Cc/Bd^3])([Fc/Bd^3])([N\ Bd/Cv^{1/2}])]$	Engine specification

TABLE 4.24
Dependent Dimensionless π Terms

Ser	Dependent Dimensionless Ratios or π Terms	Nature of Basic Physical Quantities
01	$Z_1 = [Bte]$	Brake thermal efficiency
02	$Z_2 = [Bsfc]$	Brake specific flue consumption

Each pi term is a function of the output terms (Shriwastawa n.d.; Dekker 1995),

$$Z_1 = \text{function of}\,(\Pi_1, \Pi_2, \Pi_3, \Pi_4)$$
$$Z_2 = \text{function of}\,(\Pi_1, \Pi_2, \Pi_3, \Pi_4)$$

Where:

$Z_1 = \Pi_{D1}$, First dependent π term = Bte
$Z_2 = \Pi_{D2}$, Second dependent π term = Bsfc

The most likely accurate mathematical form for the phenomenon's dimensions equations might be connections considered as being exponential.

$$(Z) = K \times [([Cv^{1/2}Vi/Bd])([Cv^{1/2}T/Bd])]([B])/([Fp])]^a, [([Cv^{1/2}Bd^2D/Mf])/([Cv^{1/2}L/Mf\,Bd])([Ap])]^b, [([Apig])([Di])/([CN])]^c, [([Sl/Bd])([Cc/Bd^3])([Fc/Bd^3])([N\,Bd/Cv^{1/2}])]^d$$

4.2.12 Model Formulation by Identifying the Curve Fitting Constant and Various Indices of π Terms

By taking into account four independent variables and one dependent π term, multiple regression analysis assists in identifying the indices of the various π term in the model targeted at (Zhou and Xu 2014). Let the model aimed at be of the form:

$$(Z_1) = K_1 \times [(\pi_1)^{a1} \times (\pi_2)^{b2} \times (\pi_3)^{c1} \times (\pi_4)^{d_1}]$$
$$(Z_2) = K_2 \times [(\pi_1)^{a2} \times (\pi_2)^{b2} \times (\pi_3)^{c2} \times (\pi_4)^{d_2}]$$

To find the values of a_1, b_1, c_1, and d_1, the equation is presented as follows:

$$\Sigma Z_1 = nK_1 + a_1 \times \Sigma A + b_1 \times \Sigma B + c_1 \times \Sigma C + d_1 \times \Sigma D$$
$$\Sigma Z_1 A = K_1 \times \Sigma A + a_1 \times \Sigma A \times A + b_1 \times \Sigma B \times A + c_1 \times \Sigma C \times A + d_1 \times \Sigma D \times A$$
$$\Sigma Z_1 B = K_1 \times \Sigma B + a_1 \times \Sigma A \times B + b_1 \times \Sigma B \times B + c_1 \times \Sigma C \times B + d_1 \times \Sigma D \times B$$
$$\Sigma Z_1 C = K_1 \times \Sigma C + a_1 \times \Sigma A \times C + b_1 \times \Sigma B \times C + c_1 \times \Sigma C \times C + d_1 \times \Sigma D \times C$$
$$\Sigma Z_1 D = K_1 \times \Sigma D + a_1 \times \Sigma A \times D + b_1 \times \Sigma B \times D + c_1 \times \Sigma C \times D + d_1 \times \Sigma D \times D$$

In the above set of equations, the values of K_1, a_1, b_1, c_1, and d_1 are substituted to compute the values of the unknowns. After substituting these values in the equations, one will get a set of five equations, which are to be solved simultaneously to get the values of K_1, a_1, b_1, c_1 and d_1. The above equations can be transfer used in the matrix form and subsequently values of K_1, a_1, b_1, c_1, and d_1 can be obtained by adopting matrix analysis.

$$X_1 = \text{inv}(W) \times P_1$$

W = 5 × 5 matrix of the multipliers of K_1, a_1, b_1, c_1, and d1
P_1 = 5 × 1 matrix on L H S and
X_1 = 5 × 1 matrix of solutions

Then, the matrix evaluated is given by,

Matrix

$$Z_1 \times \begin{bmatrix} 1 \\ A \\ B \\ C \\ D \end{bmatrix} = \begin{bmatrix} n & A & B & C & D \\ A & A^2 & BA & CA & DA \\ B & AB & B^2 & CB & DB \\ C & AC & BC & C^2 & DC \\ D & AD & BD & CD & D^2 \end{bmatrix} \times \begin{bmatrix} K_1 \\ a_1 \\ b_1 \\ c_1 \\ d_1 \end{bmatrix}$$

In the above equations, n is the number of sets of readings, A, B, C, and D represent the independent π terms π_1, π_2, π_3, and π_4, while Z represents a dependent π term.

4.2.13 Basis for Arriving at Number of Observations

The numbers of observation taken is 15, based on the probability concept of degree of uncertainty. The formula for calculating number of readings is:

$$\sqrt{N} = [\{x/Zc\} - \mu]\sigma$$

Where:

x = mean
μ = median
σ = standard deviation
N = number of readings

For N ≥15, Zc = 2.58 for Certainty (Confidence level 99%)

Zc = 1.96 for Certainty (Confidence level 95%)
Zc = 1.645 for Certainty (Confidence level 90%)

Selecting Zc = 2.58 for Certainty with the Confidence level 99%, satisfied the number of readings

$$N = [\{x/Tc\} - \mu]^2 \sigma$$

For N < 30, Ţc = 2.48 for Certainty (Confidence level 99%)

Ţc = 1.71 for Certainty (Confidence level 95%)
Ţc = 1.32 for Certainty (Confidence level 90%)

Selecting Ţc = 2.48 for Certainty with the Confidence level 99%, satisfied the no. of readings (Table 4.25–4.29).

TABLE 4.25
Data Related to Dimensions Term π_1, π_2, π_3, and π_4

Load (Lo) (Kg)	CV (cm²/s²)	Mf (kg/s)	Bore dia (cm)	P1	P2	P3	P4
5	4.61481E+11	0.000139276	8.75	2787167569	10	38	70
5	4.60912E+11	0.000123	8.75	3154029973	20	38.5	71.11
5	4.61481E+11	0.000109162	8.75	3556041381	25	39	72.22
5	4.61481E+11	0.000103564	8.75	3748259834	30	44.5	75
5	4.60912E+11	0.000101475	8.75	3823066634	40	45	76.11
10	4.61481E+11	0.000175609	8.75	4421024419	10	38	70
10	4.60912E+11	0.00016236	8.75	4778833293	20	38.5	71.11
10	4.61481E+11	0.00014425	8.75	5382116684	25	39	72.22
10	4.61481E+11	0.000134633	8.75	5766553590	30	44.5	75
10	4.60912E+11	0.000123	8.75	6308059946	40	45	76.11
15	4.61481E+11	0.000186129	8.75	6256710645	10	38	70
15	4.60912E+11	0.000176478	8.75	6594789944	20	38.5	71.11
15	4.61481E+11	0.000168292	8.75	6919864308	25	39	72.22
15	4.61481E+11	0.000155346	8.75	7496519667	30	44.5	75
15	4.60912E+11	0.000144964	8.75	8028439932	40	45	76.11

TABLE 4.26
Data Related to Dimensions Term π_5 and π_6

Vi (cm²/s)	CV (cm²/s²)	Bore Dia (cm)	P5	Mf (kg/s)	P6
3.372	4.61481E+11	8.75	5.67287E–07	0.000139276	3.02E+11
3.772	4.60912E+11	8.75	6.34972E–07	0.000123	3.43E+11
3.85	4.61481E+11	8.75	6.47703E–07	0.000109162	3.85E+11
4.33	4.61481E+11	8.75	7.28455E–07	0.000103564	4.06E+11
5	4.60912E+11	8.75	8.41692E–07	0.000101475	4.16E+11
3.372	4.61481E+11	8.75	5.67287E–07	0.000175609	2.39E+11
3.772	4.60912E+11	8.75	6.34972E–07	0.00016236	2.6E+11
3.85	4.61481E+11	8.75	6.47703E–07	0.00014425	2.91E+11
4.33	4.61481E+11	8.75	7.28455E–07	0.000134633	3.12E+11
5	4.60912E+11	8.75	8.41692E–07	0.000123	3.43E+11
3.372	4.61481E+11	8.75	5.67287E–07	0.000186129	2.26E+11
3.772	4.60912E+11	8.75	6.34972E–07	0.000176478	2.39E+11
3.85	4.61481E+11	8.75	6.47703E–07	0.000168292	2.5E+11
4.33	4.61481E+11	8.75	7.28455E–07	0.000155346	2.7E+11
5	4.60912E+11	8.75	8.41692E–07	0.000144964	2.91E+11

TABLE 4.27
Data Related to Dimensions Term π_7, π_8, π_9, π_{10}, and π_{11}

P7	P8	P9	CV (cm2/s2)	Time (s)	Bore dia (cm)	P10	Stroke (cm)	P11
70	61.91704	54.5803	4.61481E+11	290	8.75	22514740	11	1.257143
71.11	61.3888	54.1999	4.60912E+11	330	8.75	25604415	11	1.257143
72.22	63.48456	55.7089	4.61481E+11	370	8.75	28725702	11	1.257143
75	65.44396	57.11965	4.61481E+11	390	8.75	30278443	11	1.257143
76.11	70.642	60.86224	4.60912E+11	400	8.75	31035655	11	1.257143
70	61.91704	54.5803	4.61481E+11	230	8.75	17856518	11	1.257143
71.11	61.3888	54.1999	4.60912E+11	250	8.75	19397284	11	1.257143
72.22	63.48456	55.7089	4.61481E+11	280	8.75	21738369	11	1.257143
75	65.44396	57.11965	4.61481E+11	300	8.75	23291110	11	1.257143
76.11	70.642	60.86224	4.60912E+11	330	8.75	25604415	11	1.257143
70	61.91704	54.5803	4.61481E+11	217	8.75	16847236	11	1.257143
71.11	61.3888	54.1999	4.60912E+11	230	8.75	17845502	11	1.257143
72.22	63.48456	55.7089	4.61481E+11	240	8.75	18632888	11	1.257143
75	65.44396	57.11965	4.61481E+11	260	8.75	20185629	11	1.257143
76.11	70.642	60.86224	4.60912E+11	280	8.75	21724958	11	1.257143

TABLE 4.28
Data Related to Dimensions Term π_{12} and π_{13}

Bore Dia (cm)	Cubic Cap (cm3)	P12	Fuel Tank Cap (cm3)	P13
8.75	1323	1.974857	11000	16.41983
8.75	1323	1.974857	11000	16.41983
8.75	1323	1.974857	11000	16.41983
8.75	1323	1.974857	11000	16.41983
8.75	1323	1.974857	11000	16.41983
8.75	1323	1.974857	11000	16.41983
8.75	1323	1.974857	11000	16.41983
8.75	1323	1.974857	11000	16.41983
8.75	1323	1.974857	11000	16.41983
8.75	1323	1.974857	11000	16.41983
8.75	1323	1.974857	11000	16.41983
8.75	1323	1.974857	11000	16.41983
8.75	1323	1.974857	11000	16.41983
8.75	1323	1.974857	11000	16.41983
8.75	1323	1.974857	11000	16.41983

TABLE 4.29
Data Related to Dimensions Term π_{14} and Term πD_1 and πD_2

Speed (Rps)	Bore Dia (cm)	CV (cm^2/s^2)	P14	BTE	BSFC
25	8.75	4.61481E+11	0.000322	25.82726	100
25	8.75	4.60912E+11	0.000322	29.281	110
25	8.75	4.61481E+11	0.000322	32.95202	134
25	8.75	4.61481E+11	0.000322	34.73321	156
25	8.75	4.60912E+11	0.000322	35.49212	120
25	8.75	4.61481E+11	0.000322	40.84398	95
25	8.75	4.60912E+11	0.000322	44.23152	120
25	8.75	4.61481E+11	0.000322	49.72311	136
25	8.75	4.61481E+11	0.000322	53.27476	154
25	8.75	4.60912E+11	0.000322	58.38561	171
25	8.75	4.61481E+11	0.000322	57.86132	124
25	8.75	4.60912E+11	0.000322	61.10097	138
25	8.75	4.61481E+11	0.000322	63.99409	160
25	8.75	4.61481E+11	0.000322	69.41268	174
25	8.75	4.60912E+11	0.000322	74.2919	184

5 Formulation of Mathematical Model

5.1 FORMULATION OF FIELD DATA-BASED MODEL FOR CRANKSHAFT MAINTENANCE OPERATION

For crankshaft maintenance operation, eleven independent pi terms (π_1, π_2, π_3, π_4, π_5, π_6, π_7, π_8, π_9, π_{10}, π_{11}) and three dependent pi terms (π_{D1}, π_{D2}, π_{D3}) have been identified from field for the model formulation.

$$z_{1C} = f(\pi_{D1}) = f(\pi_1, \pi_2, \pi_3, \pi_4, \pi_5, \pi_6, \pi_7, \pi_8, \pi_9, \pi_{10}, \pi_{11})$$

$$z_{2C} = f(\pi_{D2}) = f(\pi_1, \pi_2, \pi_3, \pi_4, \pi_5, \pi_6, \pi_7, \pi_8, \pi_9, \pi_{10}, \pi_{11})$$

$$z_{3C} = f(\pi_{D3}) = f(\pi_1, \pi_2, \pi_3, \pi_4, \pi_5, \pi_6, \pi_7, \pi_8, \pi_9, \pi_{10}, \pi_{11})$$

f stands for "function of"

Where:

$z_{1C} = \pi_{D1}$, first dependent pi term, overhauling time of crankshaft maintenance activity

$z_{2C} = \pi_{D2}$, second dependent pi term, human energy consumed in crankshaft maintenance activity

$z_{3C} = \pi_{D3}$, third dependent pi term, productivity of crankshaft maintenance activity

To evolve the probable exact mathematical form of the dimensional equation of the phenomenon, we can assume that the relationship between dependent and independent pi terms to be of exponential form. We can write the general equation as follows:

$$z_C = k \times (\pi_1)^a \times (\pi_2)^b \times (\pi_3)^c \times (\pi_4)^d \times (\pi_5)^e \times (\pi_6)^f \times (\pi_7)^g \times (\pi_8)^h \times (\pi_9)^i \times (\pi_{10})^j \times (\pi_{11})^k \quad (5.1)$$

This equation represents the following independent pi terms involved in the crankshaft maintenance activity.

Dependent pi term:

$$\begin{aligned}
z_C = \pi_D = K \times \{ & [(a \times c \times e \times g)/(b \times d \times f \times h)]^a, \\
& [(Ags/Exs), (sks/Ens), (hls/Hbs)]^b, \quad \text{--- \textbf{anthropometric data}} \\
& [(Dsps/Dsns) \times (Lsps/Dsns) \times (Dsmab/Dsns) \times (Lsns/Dsns) \times (Dsbs/Dsns) \\
& \times (Lsbs/Dsns) \times (Lsmab/Dsn) \times (Dsiab/Dsns) \times (Lsiab/Dsns) \\
& \times (Dsc/Dsns) \times (Lsc/Dsns) \times (Dmj/Dsns) \times (Lmj/Dsns) \times *(Dcpj/Dsns) \\
& \times (Lcpj/Dsns) \times (tmjb/Dsns) \times (Dmjb/Dsns) \times (Dcpjb/Dsns) \qquad (5.2) \\
& \times (tcpjb/Dsns)]^c, \quad \text{--- \textbf{workers data}} \\
& [(Dsmsp/Dbr), (Lbr/Dbr), (Lsmsp/Dbr), (Dsisp/Dbr), (Lsisp/Dbr), \\
& (Dbrr/Dbr), (Lbrr/Dbr)]^d, \quad \text{--- \textbf{specification of tools}} \\
& [(ker /l.oil), (Ca/E.belt)]^e, \quad \text{--- \textbf{specification of solvent}} \\
& [(Axcp/Elsb), (Axmj/Elsb)]^f, \quad \text{--- \textbf{axial clearance}} \\
& [(Lfrm/hfrm), (Wfrm/hfrm)]^g, \quad \text{--- \textbf{workstation data}}
\end{aligned}$$

DOI: 10.1201/9781003318699-5

$[temp/100]^h$, --- **temperature at workplace**
$[(Humd)]^i$, --- **humidity at workplace**
$[((illmus \times Ags)^3)/Wt]^j$, --- **Illumination at workplace**
$[(Noise)]^k$ --- **noise at workplace**
}

Equation 5.2 contains 12 unknowns, the curve fitting constant K and indices a, b, c, d, e, f, g, h, i, j, and k. To get the values of these unknowns, we need minimum a 12 set of values of (π_1, π_2, π_3, π_4, π_5, π_6, π_7, π_8, π_9, π_{10}, π_{11}, π_D).

As per the plan in the design of experimentation, we have 30 sets of such values. If any arbitrary 12 sets from table of (pi terms of crankshaft) are selected then the values of unknown K and indices a, b, c, d, e, f, g, h, i, j, and k can be computed, but we may not have unique solution which represents a best curve fit for the remaining sets of values. To be specific one can find out ${}^nC_r$ combinations of r number of sets chosen out of n number of sets of values ${}^{30}C_{12}$ in our case. Solving these many sets and finding their solutions is a Herculean task. Hence, we employ the curve fitting technique. To implement this method to our field data, it is imperative to have the equation in the form:

$$Z = a + bX + cY + dZ + \cdots\cdots \quad (5.3)$$

Equation (5.1) can be rendered in the form of equation (5.3) by taking log on both sides of equation (5.1).

$$\begin{aligned} \log\pi_D = {} & \log k + a \times \log\pi_1 + b \times \log\pi_2 + c \times \log\pi_3 + d \times \log\pi_4 + e \times \log\pi_5 + f \times \log\pi_6 \\ & + g \times \log\pi_7 + h \times \log\pi_8 + i \times \log\pi_9 + j \times \log\pi_{10} + k \times \log\pi_{11} \end{aligned} \quad (5.4)$$

Let us denote:

$$\begin{array}{llll} z_C = \log\pi_{D1} & B = \log\pi_2 & E = \log\pi_5 & H = \log\pi_8 \\ K_1 = \log k & C = \log\pi_3 & F = \log\pi_6 & I = \log\pi_9 \\ A = \log\pi_1 & D = \log\pi_4 & G = \log\pi_7 & J = \log\pi_{10} \\ & & & K = \log\pi_{11} \end{array}$$

Putting these values in equations 6.1.4, the same can be written as:

$$z_C = K_1 + a \times A + b \times B + c \times C + d \times D + e \times E + f \times F + g \times G + h \times H + i \times I + j \times J + k \times K \quad (5.5)$$

Equation 5.5 is a regression equation of Z on A, B, C, D, E, F, G, H, I, J, and K in an n-dimensional coordinate system. This represents a regression hyperplane. Substituting the data collected from field, we will get one such equation for 30 such equations. The curve fitting method, in order to include every set of readings, instead of any 12 randomly chosen sets of readings, instructs us to add all these equations. This results into the following equation:

$$\sum z_C = nK + a\sum A + b\sum B + c\sum C + d\sum D + e\sum E + f\sum F + g\sum G + h\sum H + i\sum I + j\sum J + k\sum K \quad (5.6)$$

Where number of readings, n = 30.

As mentioned earlier, there are 11 unknowns in this equation, K and indices a, b, c, d, e, f, g, h, i, j. Hence, we will require 11 different equations to solve. In order to achieve these equations, we multiply equation 5.6 with A, B, C, D, E, F, G, H, I, J, and K to get a set of 11 different equations as shown below:

$$\sum z_C A = K\sum A + a\sum A^2 + b\sum AB + c\times\sum AC + d\sum AD + e\sum AE + f\sum AF + g\sum AG + h\sum AH$$
$$+ i\sum AI + j\sum AJ + k\sum AK$$

$$\sum z_C B = K\sum B + a\sum AB + b\sum B^2 + c\times\sum BC + d\sum BD + e\sum BE + f\sum BF + g\sum BG + h\sum BH$$
$$+ i\sum BI + j\sum BJ + k\sum BK$$

$$\sum z_C C = K\sum C + a\sum AC + b\sum CB + c\times\sum C^2 + d\sum CD + e\sum CE + f\sum CF + g\sum CG + h\sum CH$$
$$+ i\sum CI + j\sum CJ + k\sum CK$$

$$\sum z_C D = K\sum D + a\sum AD + b\sum BD + c\times\sum CD + d\sum D^2 + e\sum DE + f\sum DF + g\sum DG + h\sum DH$$
$$+ i\sum DI + j\sum DJ + k\sum DK$$

$$\sum z_C E = K\sum E + a\sum AE + b\sum BE + c\times\sum CE + d\sum DE + e\sum E^2 + f\sum FE + g\sum GE + h\sum HE$$
$$+ i\sum IE + j\sum JE + k\sum EK$$

$$\sum z_C F = K\sum F + a\sum AF + b\sum FB + c\times\sum FC + d\sum FD + e\sum FE + f\sum F^2 + g\sum FG + h\sum FH$$
$$+ i\sum FI + j\sum FJ + k\sum FK$$

$$\sum z_C G = K\sum G + a\sum AG + b\sum GB + c\times\sum GC + d\sum GD + e\sum GE + f\sum GF + g\sum G^2 + h\sum GH$$
$$+ i\sum GI + j\sum GJ + k\sum GK$$

$$\sum z_C H = K\sum H + a\sum AH + b\sum HB + c\times\sum HC + d\sum HD + e\sum HE + f\sum HF + g\sum HG + h\sum H^2$$
$$+ i\sum HI + j\sum HJ + k\sum HK$$

$$\sum z_C I = K\sum I + a\sum AI + b\sum IB + c\times\sum IC + d\sum ID + e\sum IE + f\sum IF + g\sum IG + h\sum IH$$
$$+ i\sum I^{2+} j\sum IJ + k\sum IK$$

$$\sum z_C J = K\sum J + a\sum AJ + b\sum JB + c\times\sum JC + d\sum JD + e\sum JE + f\sum JF + g\sum JG + h\sum JH$$
$$+ i\sum JI + j\sum J^2 + k\sum JK$$

$$\sum z_C K = K\sum K + a\sum AK + b\sum KB + c\times\sum KC + d\sum KD + e\sum KE + f\sum KF + g\sum KG + h\sum KH$$
$$+ i\sum KI + j\sum KJ + k\sum K^2$$

Including equation 5.6, we have a set of 11 equations for the 11 unknowns. To solve these equations, we shall use matrix analysis. The matrix format of an equation is of the form:

$$\mathbf{z_C} = \mathbf{W} \times \mathbf{X}$$

In the above equation X is the unknown matrix. We solve this equation to determine the value of X by taking an inverse of matrix W. The final equation solved by using MATLAB is as follows.

$$\mathbf{X} = \mathbf{inv(W)} \times \mathbf{z_C}$$

Solving this equation for every dependent pi term, the model is formed.

5.1.1 Model Formulation for π_{D1}, Overhauling Time of Crankshaft Maintenance Activity by Identifying the Curve Fitting Constant and Various Indices of pi Terms

The multiple regression analysis helps to identify the indices of different pi terms in the model aimed at, by considering eleven independent pi terms and one dependent pi term. Let the model be of the form for dependent pi term (π_{D1}) overhauling time of crankshaft maintenance activity:

$$Z_{1C} = \pi_{D1} = k_1 \times (\pi_1)^{a1} \times (\pi_2)^{b1} \times (\pi_3)^{c1} \times (\pi_4)^{d1} \times (\pi_5)^{e1} \times (\pi_6)^{f1} \times (\pi_7)^{g1} \times (\pi_8)^{h1} \times (\pi_9)^{i1} \times (\pi_{10})^{j1} \times (\pi_{11})^{k1} \quad (5.7)$$

To determine the values of k_1, a_1, b_1, c_1, d_1, e_1, f_1, g_1, h_1, i_1, j_1, k_1 and to arrive at the regression hyperplane, the above equations are presented as follows:

$$\begin{aligned}
\Sigma z_{1C} &= nK_1 + a_1 \Sigma A + b_1 \Sigma B + c_1 \Sigma C + d_1 \Sigma D + e_1 \Sigma E + f_1 \Sigma F + g_1 \Sigma G \\
&\quad + h_1 \Sigma H + i_1 \Sigma I + j_1 \Sigma J + k_1 \Sigma K \\
\Sigma z_{1C}A &= K_1 \Sigma A + a_1 \Sigma A^2 + b_1 \Sigma AB + c_1 \Sigma AC + d_1 \Sigma AD + e_1 \Sigma AE \\
&\quad + f_1 \Sigma AF + g_1 \Sigma AG + h_1 \Sigma AH + i_1 \Sigma AI + j_1 \Sigma AJ + k_1 \Sigma AK \\
\Sigma z_{1C}B &= K_1 \Sigma B + a_1 \Sigma AB + b_1 \Sigma B^2 + c_1 \Sigma BC + d_1 \Sigma BD + e_1 \Sigma BE \\
&\quad + f_1 \Sigma BF + g_1 \Sigma BG + h_1 \Sigma BH + i_1 \Sigma BI + j_1 \Sigma BJ + k_1 \Sigma BK \\
\Sigma z_{1C}C &= K_1 \Sigma C + a_1 \Sigma AC + b_1 \Sigma CB + c_1 \Sigma C^2 + d_1 \Sigma CD + e_1 \Sigma CE \\
&\quad + f_1 \Sigma CF + g_1 \Sigma CG + h_1 \Sigma CH + i_1 \Sigma CI + j_1 \Sigma CJ + k_1 \Sigma CK \\
\Sigma z_{1C}D &= K_1 \Sigma D + a_1 \Sigma AD + b_1 \Sigma BD + c_1 \Sigma CD + d_1 \Sigma D^2 + e_1 \Sigma DE \\
&\quad + f_1 \Sigma DF + g_1 \Sigma DG + h_1 \Sigma DH + i_1 \Sigma DI + j_1 \Sigma DJ + k_1 \Sigma DK \\
\Sigma z_{1C}E &= K_1 \Sigma E + a_1 \Sigma AE + b_1 \Sigma BE + c_1 \Sigma CE + d_1 \Sigma DE + e_1 \Sigma E^2 \\
&\quad + f_1 \Sigma FE + g_1 \Sigma GE + h_1 \Sigma HE + i_1 \Sigma IE + j_1 \Sigma JE + k_1 \Sigma EK \\
\Sigma z_{1C}F &= K_1 \Sigma F + a_1 \Sigma AF + b_1 \Sigma FB + c_1 \Sigma FC + d_1 \Sigma FD + e_1 \Sigma FE \\
&\quad + f_1 \Sigma F^2 + g_1 \Sigma FG + h_1 \Sigma FH + i_1 \Sigma FI + j_1 \Sigma FJ + k_1 \Sigma FK \\
\Sigma z_{1C}G &= K_1 \Sigma G + a_1 \Sigma AG + b_1 \Sigma GB + c_1 \Sigma GC + d_1 \Sigma GD + e_1 \Sigma GE \\
&\quad + f_1 \Sigma GF + g_1 \Sigma G^2 + h_1 \Sigma GH + i_1 \Sigma GI + j_1 \Sigma GJ + k_1 \Sigma GK \\
\Sigma z_{1C}H &= K_1 \Sigma H + a_1 \Sigma AH + b_1 \Sigma HB + c_1 \Sigma HC + d_1 \Sigma HD + e_1 \Sigma HE \\
&\quad + f_1 \Sigma HF + g_1 \Sigma HG + h_1 \Sigma H^2 + i_1 \Sigma HI + j_1 \Sigma HJ + k_1 \Sigma HK \\
\Sigma z_{1C}I &= K_1 \Sigma I + a_1 \Sigma AI + b_1 \Sigma IB + c_1 \Sigma IC + d_1 \Sigma ID + e_1 \Sigma IE \\
&\quad + f_1 \Sigma IF + g_1 \Sigma IG + h_1 \Sigma IH + i_1 \Sigma I^2 + j_1 \Sigma IJ + k_1 \Sigma IK \\
\Sigma z_{1C}J &= K_1 \Sigma J + a_1 \Sigma AJ + b_1 \Sigma JB + c_1 \Sigma JC + d_1 \Sigma JD + e_1 \Sigma JE \\
&\quad + f_1 \Sigma JF + g_1 \Sigma JG + h_1 \Sigma JH + i_1 \Sigma JI + j_1 \Sigma J^2 + k_1 \Sigma JK \\
\Sigma z_{1C}K &= K_1 \Sigma K + a_1 \Sigma AK + b_1 \Sigma KB + c_1 \Sigma KC + d_1 \Sigma KD + e_1 \Sigma KE \\
&\quad + f_1 \Sigma KF + g_1 \Sigma KG + h_1 \Sigma KH + i_1 \Sigma KI + j_1 \Sigma KJ + k_1 \Sigma K^2
\end{aligned} \quad (5.8)$$

Where n is the number of sets of the values.

In the above set of equations, the values of the multipliers K_1, a_1, b_1, c_1, d_1, e_1, f_1, g_1 h_1, i_1, j_1, and k_1 are substituted to compute the values of the unknowns (viz. K_1, a_1, b_1, c_1, d_1, e_1, f_1, g_1 h_1, i_1, j_1, and k_1). The values of the terms on LHS and the multipliers of a_1, b_1, c_1, d_1, e_1, f_1, g_1 h_1, i_1, j_1, and k_1 in the set of equation are calculated simultaneously to get the values of K_1, a_1, b_1, c_1, d_1, e_1, f_1, g_1 h_1, i_1, j_1, and k_1. The above equations can be verified in the matrix form and further values of K_1, a_1, b_1, c_1, d_1, e_1, f_1, g_1 h_1, i_1, j_1, and k1 can be obtained by using matrix analysis.

The matrix obtained is given by:

$$\mathbf{z_{1C} = W_1 \times X_1}$$

$$Z_{1C} \times \begin{bmatrix} 1 \\ A \\ B \\ C \\ D \\ E \\ F \\ G \\ H \\ I \\ J \\ K \end{bmatrix} = \begin{bmatrix} n & A & B & C & D & E & F & G & H & I & J & K \\ A & A^2 & BA & CA & DA & EA & FA & GA & HA & IA & JA & KA \\ B & AB & B^2 & CB & DB & EB & FB & GB & HB & IB & JB & KB \\ C & AC & BC & C^2 & DC & EC & FC & GC & HC & IC & JC & KC \\ D & AD & BD & CD & D^2 & ED & FD & GD & HD & ID & JD & KD \\ E & AE & BE & CE & DE & E^2 & FE & GE & HE & IE & JE & KE \\ F & AF & BF & CF & DF & EF & F^2 & GF & HF & IF & JF & KF \\ G & AG & BG & CG & DG & EG & FG & G^2 & HG & IG & JG & KG \\ H & AH & BH & CH & DH & EH & FH & GH & H^2 & IH & JH & KH \\ I & AI & BI & CI & DI & EI & FI & GI & HI & I^2 & JI & KI \\ J & AJ & BJ & CJ & DJ & EJ & FJ & GJ & HJ & IJ & J^2 & KJ \\ K & AK & BK & CK & DK & EK & FK & GK & HK & IK & JK & K^2 \end{bmatrix} \times \begin{bmatrix} K_1 \\ a_1 \\ b_1 \\ c_1 \\ d_1 \\ e_1 \\ f_1 \\ g_1 \\ h_1 \\ i_1 \\ j_1 \\ k_1 \end{bmatrix}$$

W_1 = 12 × 12 matrix of the multipliers of K_1, a_1, b_1, c_1, d_1, e_1, f_1, g_1 h_1, i_1, j_1, and k_1
X_1 = 12 × 1 matrix of solutions of values of K_1, a_1, b_1, c_1, d_1, e_1, f_1, g_1 h_1, i_1, j_1, and k_1
Z_1 = 12 × 1 matrix of the terms on LHS

In the above matrix, n is the number of readings A, B, C, D, E, F, G, H, I, J, and K represent the independent pi terms π_1, π_2, π_3, π_4, π_5, π_6, π_7, π_8, π_9, π_{10}, and π_{11} while z_{1C} represents dependent pi term (overhauling time of crankshaft maintenance activity). Next, calculate the values of independent pi terms for corresponding dependent pi terms which helps to form the equation in matrix form. It is recommended to use MATLAB software for this purpose for making this process of model formulation quickest and least cumbersome.

$$\mathbf{z_{1C} = W_1 \times X_1}$$

$$\begin{bmatrix} z_1 \\ 337.18 \\ 7.90 \\ 165.89 \\ 1105.5 \\ 586.12 \\ 708.16 \\ -277.4 \\ 213.16 \\ -165.1 \\ 555.02 \\ 1819.6 \\ 636.39 \end{bmatrix} = \begin{bmatrix} & & & & & w_1 & & & & & & \\ 30 & 0.7 & 14.75 & 98.35 & 52.14 & 63.01 & -24.6 & 18.96 & -14.68 & 49.38 & 161.89 & 56.62 \\ 0.7 & 0.38 & 0.331 & 2.29 & 1.22 & 1.47 & -0.52 & 0.44 & -0.34 & 1.13 & 3.71 & 1.29 \\ 14.75 & 0.33 & 8.249 & 48.34 & 25.6 & 30.71 & -11.5 & 9.3 & -7.16 & 24.14 & 79.61 & 27.85 \\ 98.36 & 2.3 & 48.35 & 322.4 & 171 & 206.6 & -80.9 & 62.1 & -48.14 & 161.9 & 530.77 & 185.6 \\ 52.15 & 1.22 & 25.63 & 170.9 & 90.7 & 109.5 & -42.91 & 32.9 & -25.52 & 85.85 & 281.42 & 98.42 \\ 36.02 & 1.47 & 30.72 & 206.6 & 110 & 132.87 & -52 & 39.3 & -30.8 & 103.7 & 340 & 118.9 \\ -24.6 & -.52 & -11.7 & -80.93 & -42.9 & -52 & 20.73 & -15.6 & 12.08 & -40.68 & -133.1 & -46.5 \\ 18.97 & 0.44 & 9.3 & 62.1 & 33.97 & 39.83 & -15.60 & 11.9 & -9.28 & 31.2 & 102.3 & 35.79 \\ -14.7 & -0.35 & -7.17 & -48.1 & -25.5 & -30.8 & 12 & -9.28 & 7.30 & 24.29 & -79.2 & -27.7 \\ 49.39 & 1.14 & 24.14 & 161.9 & 85.9 & 103.7 & -40.6 & 31.2 & -24.29 & 82.2 & 266.3 & 93.2 \\ 161.9 & 3.71 & 79.62 & 530.7 & 281 & 340 & -133.2 & 102.3 & -79.2 & 266.5 & 873.8 & 305.5 \\ 56.62 & 1.29 & 27.86 & 185.6 & 98.4 & 118.9 & -46.57 & 35.7 & -27.7 & 93.24 & 305.5 & 106.8 \end{bmatrix} \times \begin{bmatrix} x_1 \\ K_1 \\ a_1 \\ b_1 \\ c_1 \\ d_1 \\ e_1 \\ f_1 \\ g_1 \\ h_1 \\ i_1 \\ j_1 \\ k_1 \end{bmatrix}$$

After multiplying inverse of matrix W_1 with Z_1 we get X_1 matrix with result values of K_1 and indices a_1, b_1, c_1, d_1, e_1, f_1, g_1 h_1, i_1, j_1, k_1

$$X_1 = inv(W_1) \times z_{1C}$$

But K_1 is log value so convert into normal value take antilog of K_1
Thus, antilog (0.5421) = 1.7198

$K_1 = 1.7198$	$c_1 = 2.8137$	$f_1 = -0.0025$	$i_1 = -0.125$
$a_1 = 0.0873$	$d_1 = 0.2115$	$g_1 = 3.5409$	$j_1 = -0.0.609$
$b_1 = 0.1026$	$e_1 = -0.2101$	$h_1 = -0.4451$	$k_1 = 0.1131$

Hence the model for dependent term π_{D1}, overhauling time of crankshaft maintenance operation is given as:

$$\begin{aligned} \pi_{D1} &= k_1 \times (\pi_1)^{a1} \times (\pi_2)^{b1} \times (\pi_3)^{c1} \times (\pi_4)^{d1} \times (\pi_5)^{e1} \times (\pi_6)^{f1} \times (\pi_7)^{g1} \times (\pi_8)^{h1} \times (\pi_9)^{i1} \times (\pi_{10})^{j1} \times (\pi_{11})^{k1} \\ z_{1C} &= \pi_{D1} = 1.7198(\pi_1)^{0.0873}(\pi_2)^{0.1026}(\pi_3)^{2.8137}(\pi_4)^{0.2115}(\pi_5)^{-0.2101}(\pi_6)^{-0.0025}(\pi_7)^{3.5409} \\ &(\pi_8)^{-0.4451}(\pi_9)^{-0.125}(\pi_{10})^{-0.0609}(\pi_{11})^{0.1131} \end{aligned} \tag{5.9}$$

Overhauling time of crankshaft maintenance activity, dependent pi term:

$$\begin{aligned} z_{1C} = K \times \{&[(a \times c \times e \times g)/(b \times d \times f \times h)]^{0.0873}, \\ &[(Ags/Exs), (sks/Ens), (hls/Hbs)]^{0.1026}, \\ &[(Dsps/Dsns) \times (Lsps/Dsns) \times (Dsmab/Dsns) \times (Lsns/Dsns) \times (Dsbs/Dsns) \\ &\times (Lsbs/Dsns) \times (Lsmab/Dsns) \times (Dsiab/Dsns) \times (Lsiab/Dsns) \times (Dsc/Dsns) \\ &\times (Lsc/Dsns) \times (Dmj/Dsns) \times (Lmj/Dsns) \times (Dcpj/Dsns) \times (Lcpj/Dsns) \\ &\times (tmjb/Dsns) \times (Dmjb/Dsns) \times (Dcpjb/Dsns) \times (tcpjb/Dsns)]^{2.8137}, \\ &[(Dsmsp/Dbr), (Lbr/Dbr), (Lsmsp/Dbr), (Dsisp/Dbr), (Lsisp/Dbr), \\ &(Dbrr/Dbr), (Lbrr/Dbr)]^{0.2115}, \\ &[(ker/loil), (Ca/Ebelt)]^{-0.2101} \\ &[(Axcp/Elsb), (Axmj/Elsb)]^{-0.0025} \\ &[(Lfrm/hfrm), (Wfrm/hfrm)]^{3.5409} \\ &[Temp/100]^{-0.4451} \\ &[(Humd)]^{-0.125} \\ &[((illmus \times Ags)^3)/Wt]^{-0.0609} \\ &[(Noise)]^{0.1131}\} \end{aligned} \tag{5.10}$$

5.1.2 Model Formulation for π_{D2}, Human Energy Consumed in Crankshaft Maintenance Activity by Identifying the Curve Fitting Constant and Various Indices of Pi Terms

The multiple regression analysis helps to identify the indices of different pi terms in the model aimed at, by considering eleven independent pi terms and one dependent pi term. Let the model be of the form for dependent pi term π_{D1} for human energy consumed in crankshaft maintenance activity.

$$Z_{2C} = \pi_{D2} = k_2 \times (\pi_1)^{a2} \times (\pi_2)^{b2} \times (\pi_3)^{c2} \times (\pi_4)^{d2} \times (\pi_5)^{e2} \times (\pi_6)^{f2} \times (\pi_7)^{g2} \times (\pi_8)^{h2} \times (\pi_9)^{i2} \times (\pi_{10})^{j2} \times (\pi_{11})^{k2} \quad (5.11)$$

To determine the values of K_2, a_2, b_2, c_2, d_2, e_2, f_2, g_2, h_2, i_2, j_2, and k_2 and to arrive at the regression hyperplane, the above equations are presented as follows:

$$\begin{aligned}
\sum z_{2C} &= nK_2 + a_2\sum A + b_2\sum B + c_2\sum C + d_2\sum D + e_2\sum E + f_2\sum F + g_2\sum G \\
&\quad + h_2\sum H + i_2\sum I + j_2\sum J + k_2\sum K \\
\sum z_{2C}A &= K_2\sum A + a_2\sum A^2 + b_2\sum AB + c_2\sum AC + d_2\sum AD + e_2\sum AE \\
&\quad + f_2\sum AF + g_2\sum AG + h_2\sum AH + i_2\sum AI + j_2\sum AJ + k_2\sum AK \\
\sum z_{2C}B &= K_2\sum B + a_2\sum AB + b_2\sum B^2 + c_2\sum BC + d_2\sum BD + e_2\sum BE \\
&\quad + f_2\sum BF + g_2\sum BG + h_2\sum BH + i_2\sum BI + j_2\sum BJ + k_2\sum BK \\
\sum z_{2C}C &= K_2\sum C + a_2\sum AC + b_2\sum CB + c_2\sum C^2 + d_2\sum CD + e_2\sum CE \\
&\quad + f_2\sum CF + g_2\sum CG + h_2\sum CH + i_2\sum CI + j_2\sum CJ + k_2\sum CK \\
\sum z_{2C}D &= K_2\sum D + a_2\sum AD + b_2\sum BD + c_2\sum CD + d_2\sum D^2 + e_2\sum DE \\
&\quad + f_2\sum DF + g_2\sum DG + h_2\sum DH + i_2\sum DI + j_2\sum DJ + k_2\sum DK \\
\sum z_{2C}E &= K_2\sum E + a_2\sum AE + b_2\sum BE + c_2\sum CE + d_2\sum DE + e_2\sum E^2 \\
&\quad + f_2\sum FE + g_2\sum GE + h_2\sum HE + i_2\sum IE + j_2\sum JE + k_2\sum EK \\
\sum z_{2C}F &= K_2\sum F + a_2\sum AF + b_2\sum FB + c_2\sum FC + d_2\sum FD + e_2\sum FE \\
&\quad + f_2\sum F^2 + g_2\sum FG + h_2\sum FH + i_2\sum FI + j_2\sum FJ + k_2\sum FK \\
\sum z_{2C}G &= K_2\sum G + a_2\sum AG + b_2\sum GB + c_2\sum GC + d_2\sum GD + e_2\sum GE \\
&\quad + f_2\sum GF + g_2\sum G^2 + h_2\sum GH + i_2\sum GI + j_2\sum GJ + k_2\sum GK \\
\sum z_{2C}H &= K_2\sum H + a_2\sum AH + b_2\sum HB + c_2\sum HC + d_2\sum HD + e_2\sum HE \\
&\quad + f_2\sum HF + g_2\sum HG + h_2\sum H^2 + i_2\sum HI + j_2\sum HJ + k_2\sum HK \\
\sum z_{2C}I &= K_2\sum I + a_2\sum AI + b_2\sum IB + c_2\sum IC + d_2\sum ID + e_2\sum IE + f_2\sum IF \\
&\quad + g_2\sum IG + h_2\sum IH + i_2\sum I^2 + j_2\sum IJ + k_2\sum IK \\
\sum z_{2C}J &= K_2\sum J + a_2\sum AJ + b_2\sum JB + c_2\sum JC + d_2\sum JD + e_2\sum JE + f_2\sum JF \\
&\quad + g_2\sum JG + h_2\sum JH + i_2\sum JI + j_2\sum J^2 + k_2\sum JK \\
\sum z_{2C}K &= K_2\sum K + a_2\sum AK + b_2\sum KB + c_2\sum KC + d_2\sum KD + e_2\sum KE \\
&\quad + f_2\sum KF + g_2\sum KG + h_2\sum KH + i_2\sum KI + j_2\sum KJ + k_2\sum K^2
\end{aligned} \quad (5.12)$$

Where n is number of sets of the values.

In the above set of equations the values of the multipliers K_2, a_2, b_2, c_2, d_2, e_2, f_2, g_2 h_2, i_2, j_2, and k_2 are substituted to compute the values of the unknowns (viz. K_2, a_2, b_2, c_2, d_2, e_2, f_2, g_2, h_2, i_2, j_2, and k_2). The values of the terms on LHS and the multipliers of a_2, b_2, c_2, d_2, e_2, f_2, g_2, h_2, i_2, j_2, and k_2 in the set of equation are simultaneously to get the values of K_2, a_2, b_2, c_2, d_2, e_2, f_2, g_2, h_2, i_2, j_2, and k_2. The above equations can be verified in the matrix form and further values of K_2, a_2, b_2, c_2, d_2, e_2, f_2, g_2 h_2, i_2, j_2, and k_2 can be obtained by using matrix analysis. The matrix obtained is given by $\mathbf{z_{2C}} = \mathbf{W_2} \times \mathbf{X_2}$

The matrix method of solving these equations using MATLAB is given below:

$$
Z_{2C} \times \begin{bmatrix} 1 \\ A \\ B \\ C \\ D \\ E \\ F \\ G \\ H \\ I \\ J \\ K \end{bmatrix} = \begin{bmatrix}
n & A & B & C & D & E & F & G & H & I & J & K \\
A & A^2 & BA & CA & DA & EA & FA & GA & HA & IA & JA & KA \\
B & AB & B^2 & CB & DB & EB & FB & GB & HB & IB & JB & KB \\
C & AC & BC & C^2 & DC & EC & FC & GC & HC & IC & JC & KC \\
D & AD & BD & CD & D^2 & ED & FD & GD & HD & ID & JD & KD \\
E & AE & BE & CE & DE & E^2 & FE & GE & HE & IE & JE & KE \\
F & AF & BF & CF & DF & EF & F^2 & GF & HF & IF & JF & KF \\
G & AG & BG & CG & DG & EG & FG & G^2 & HG & IG & JG & KG \\
H & AH & BH & CH & DH & EH & FH & GH & H^2 & IH & JH & KH \\
I & AI & BI & CI & DI & EI & FI & GI & HI & I^2 & JI & KI \\
J & AJ & BJ & CJ & DJ & EJ & FJ & GJ & HJ & IJ & J^2 & KJ \\
K & AK & BK & CK & DK & EK & FK & GK & HK & IK & JK & K^2
\end{bmatrix} \times \begin{bmatrix} K_2 \\ a_2 \\ b_2 \\ c_2 \\ d_2 \\ e_2 \\ f_2 \\ g_2 \\ h_2 \\ i_2 \\ j_2 \\ k_2 \end{bmatrix}
$$

W_2 = 12 × 12 matrix of the multipliers of K_2, a_2, b_2, c_2, d_2, e_2, f_2, g_2, h_2, i_2, j_2, and k_2
X_2 = 12 × 1 matrix of solutions of values of K_2, a_2, b_2, c_2, d_2, e_2, f_2, g_2, h_2, i_2, j_2, and k_2
Z_2 = 12 × 1 matrix of the terms on LHS

In the above matrix, n is the number of readings A, B, C, D, E, F, G, H, I, J, and K represent the independent pi terms π_1, π_2, π_3, π_4, π_5, π_6, π_7, π_8, π_9, π_{10}, and π_{11} while z_{2C} represents dependent pi term (human energy consumed in crankshaft maintenance activity). Next, calculate the values of independent pi terms for corresponding dependent pi terms which helps to form the equation in matrix form. It is recommended to use MATLAB software for this purpose for making simplifying the process.

After substituting the values from the 12 equations, which are to be solved simultaneously to get the values of K_2, a_2, b_2, c_2, d_2, e_2, f_2, g_2, h_2, i_2, j_2, and k_2, Equation 5.12 can be verified in the matrix form, and further values of K_2, a_2, b_2, c_2, d_2, e_2, f_2, g_2, h_2, i_2, j_2, and k_2 can be obtained using matrix analysis.

The matrix method of solving these equations using MATLAB is given below:

$$
\mathbf{z}_{2C} = \mathbf{W}_2 \times \mathbf{X}_2
$$

$$
\begin{bmatrix} 47.868 \\ 1.2286 \\ 23.502 \\ 156.94 \\ 83.209 \\ 100.59 \\ -39.39 \\ 30.261 \\ -23.43 \\ 78.796 \\ 258.33 \\ 60.324 \end{bmatrix} = \begin{bmatrix}
30 & 0.7 & 14.75 & 98.35 & 52.14 & 63.01 & -24.6 & 18.6 & -14.68 & 49.38 & 161.89 & 56.62 \\
0.7 & 0.38 & 0.331 & 2.29 & 1.22 & 1.47 & -0.52 & 0.44 & -0.34 & 1.13 & 3.71 & 1.29 \\
14.75 & 0.33 & 8.249 & 48.34 & 25.6 & 30.71 & -11.5 & 9.3 & -7.16 & 24.14 & 79.61 & 27.85 \\
98.36 & 2.3 & 48.35 & 322.4 & 171 & 206.6 & -80.9 & 62.1 & -48.14 & 161.9 & 530.77 & 185.6 \\
52.15 & 1.22 & 25.693 & 170.9 & 90.7 & 109.5 & -42.91 & 32.9 & -25.52 & 85.85 & 281.42 & 98.42 \\
63.02 & 1.47 & 30.72 & 206.6 & 110 & 132.87 & -52 & 39.3 & -30.8 & 103.7 & 340 & 118.9 \\
-24.6 & -.52 & -11.7 & -80.93 & -42.9 & -52 & 20.73 & -15.6 & 12.08 & -40.68 & -133.1 & -46.5 \\
18.97 & 0.44 & 9.3 & 62.1 & 33.97 & 39.83 & -15.60 & 11.9 & -9.28 & 31.2 & 102.3 & 35.79 \\
-14.7 & -0.35 & -7.17 & -48.1 & -25.5 & -30.8 & 12 & -9.28 & 7.30 & 24.29 & -79.2 & -27.7 \\
49.39 & 1.14 & 24.14 & 161.9 & 85.9 & 103.7 & 40.6 & 31.2 & -24.29 & 82.2 & 266.5 & 93.2 \\
161.9 & 3.71 & 79.62 & 530.7 & 281 & 340 & -133.2 & 102.3 & -79.2 & 266.5 & 873.8 & 3.5.5 \\
56.62 & 1.29 & 27.86 & 185.6 & 98.4 & 118.9 & -46.57 & 35.7 & -27.7 & 93.24 & 305.5 & 106.8
\end{bmatrix} \times \begin{bmatrix} K_2 \\ a_2 \\ b_2 \\ c_2 \\ d_2 \\ e_2 \\ f_2 \\ g_2 \\ h_2 \\ i_2 \\ j_2 \\ k_2 \end{bmatrix}
$$

After multiplying inverse of matrix W_2 with Z_2, we get X_2 matrix with K_2 and indices a_2, b_2, c_2, d_2, e_2, f_2, g_2, h_2, i_2, j_2, k_2.

$$\mathbf{X_2 = inv(W_2) \times z_{2C}}$$

But K_2 is log value so convert into normal value take antilog of K_2, thus, antilog (0.1991) = 1.2203

$$\begin{array}{llll} K_2 = 1.2203 & c_2 = -.2583 & f_2 = -.0314 & i_2 = 0.0103 \\ a_2 = 0.2769 & d_2 = 3.3045 & g_2 = -1.2341 & j_2 = 0.0002 \\ b_2 = 0.018 & e_2 = 0.0597 & h_2 = 0.0415 & k_2 = -0.3935 \end{array}$$

Hence the model for dependent term π_{D2}:

$$\begin{aligned} \pi_{D2} = k_1 &= (\pi_1)^{a2} \times (\pi_2)^{b2} \times (\pi_3)^{c2} \times (\pi_4)^{d2} \times (\pi_5)^{e2} \times (\pi_6)^{f2} \times (\pi_7)^{g2} \times (\pi_8)^{h2} \\ &\times (\pi_9)^{i2} \times (\pi_{10})^{j2} \times (\pi_{11})^{k2} \\ z_{2C} = \pi_{D2} &= 1.2203(\pi_1)^{0.2769}(\pi_2)^{0.018}(\pi_3)^{-0.2583}(\pi_4)^{3.3045}(\pi_5)^{0.0597}(\pi_6)^{-0.0314}(\pi_7)^{-1.2341} \\ &(\pi_8)^{0.0415}(\pi_9)^{0.0103}(\pi_{10})^{0.0002}(\pi_{11})^{-0.3935} \end{aligned} \tag{5.13}$$

Human energy consumed in crankshaft maintenance activity,
Dependent pi term,

$$\begin{aligned} z_{2C} = 1.2203 \times \{ &[(a \times c \times e \times g)/(b \times d \times f \times h)]^{0.2769}, \\ &[(Ags/Exs),(sks/Ens),(hls/Hbs)]^{0.018}, \\ &[(Dsps/Dsns) \times (Lsps/Dsns) \times (Dsmab/Dsns) \times (Lsns/Dsns) \times (Dsbs/Dsns) \\ &\times (Lsbs/Dsns) \times (Lsmab/Dsns) \times (Dsiab/Dsns) \times (Lsiab/Dsns) \\ &\times (Dsc/Dsns) \times (Lsc/Dsns) \times (Dmj/Dsns) \times (Lmj/Dsns) \times (Dcpj/Dsns) \\ &\times (Lcpj/Dsns) \times (tmjb/Dsns) \times (Dmjb/Dsns) \times (Dcpjb/Dsns) \\ &\times (tcpjb/Dsns)]^{-0.2583}, \\ &[(Dsmsp/Dbr),(Lbr/Dbr),(Lsmsp/Dbr),(Dsisp/Dbr),(Lsisp/Dbr), \\ &(Dbrr/Dbr),(Lbrr/Dbr)]^{3.3045}, \\ &[(ker/loil),(Ca/Ebelt)]^{0.0597} \\ &[(Axcp/Elsb),(Axmj/Elsb)]^{-0.0314} \\ &[(Lfrm/hfrm),(Wfrm/hfrm)]^{-1.2341} \\ &[Temp/100]^{0.0415} \\ &[(Humd)]^{0.0103} \\ &[((illmus \times Ags)^3)/Wt]^{0.0002} \\ &[(Noise)]^{-0.3935} \} \end{aligned} \tag{5.14}$$

5.1.3 Model Formulation for π_{D3}, Productivity of Crankshaft Maintenance Activity by Identifying the Curve Fitting Constant and Various Indices of Pi Terms

The multiple regression analysis helps to identify the indices of independent pi terms in the model aimed at, by considering eleven independent pi terms and one dependent pi term. Let the model be of the form for dependent pi term π_{D3} for productivity of crankshaft maintenance activity.

Productivity:

$$z_{3C} = p_{D3} = k_3 \times (\pi_1)^{a3} \times (\pi_2)^{b3} \times (\pi_3)^{c3} \times (\pi_4)^{d3} \times (\pi_5)^{e3} \times (\pi_6)^{f3} \times (\pi_7)^{g3} \times (\pi_8)^{h3} \times (\pi_9)^{i3} \times (\pi_{10})^{j3} \times (\pi_{11})^{k3} \quad (5.15)$$

To determine the values of K_3, a_3, b_3, c_3, d_3, e_3, f_3, g_3 h_3, i_3, j_3, and k_3 and to arrive at the regression hyperplane, the above equation are presented as follows:

$$\begin{aligned}
\Sigma z_{3C} &= nK_3 + a_3\Sigma A + b_3\Sigma B + c_3\Sigma C + d_3\Sigma D + e_3\Sigma E + f_3\Sigma F + g_3\Sigma G + h_3\Sigma H \\
&\quad + i_3\Sigma I + j_3\Sigma J + k_3\Sigma K \\
\Sigma z_{3C}A &= K_3\Sigma A + a_3\Sigma A^2 + b_3\Sigma AB + c_3\Sigma AC + d_3\Sigma AD + e_3\Sigma AE + f_3\Sigma AF \\
&\quad + g_3\Sigma AG + h_3\Sigma AH + i_3\Sigma AI + j_3\Sigma AJ + k_3\Sigma AK \\
\Sigma z_{3C}B &= K_3\Sigma B + a_3\Sigma AB + b_3\Sigma B^2 + c_3\Sigma BC + d_3\Sigma BD + e_3\Sigma BE + f_3\Sigma BF \\
&\quad + g_3\Sigma BG + h_3\Sigma BH + i_3\Sigma BI + j_3\Sigma BJ + k_3\Sigma BK \\
\Sigma z_{3C}C &= K_3\Sigma C + a_3\Sigma AC + b_3\Sigma CB + c_3\Sigma C^2 + d_3\Sigma CD + e_3\Sigma CE + f_3\Sigma CF \\
&\quad + g_3\Sigma CG + h_3\Sigma CH + i_3\Sigma CI + j_3\Sigma CJ + k_3\Sigma CK \\
\Sigma z_{3C}D &= K_3\Sigma D + a_3\Sigma AD + b_3\Sigma BD + c_3\Sigma CD + d_3\Sigma D^2 + e_3\Sigma DE + f_3\Sigma DF \\
&\quad + g_3\Sigma DG + h_3\Sigma DH + i_3\Sigma DI + j_3\Sigma DJ + k_3\Sigma DK \\
\Sigma z_{3C}E &= K_3\Sigma E + a_3\Sigma AE + b_3\Sigma BE + c_3\Sigma CE + d_3\Sigma DE + e_3\Sigma E^2 + f_3\Sigma FE \\
&\quad + g_3\Sigma GE + h_3\Sigma HE + i_3\Sigma IE + j_3\Sigma JE + k_3\Sigma EK \\
\Sigma z_{3C}F &= K_3\Sigma F + a_3\Sigma AF + b_3\Sigma FB + c_3\Sigma FC + d_3\Sigma FD + e_3\Sigma FE + f_3\Sigma F^2 \\
&\quad + g_3\Sigma FG + h_3\Sigma FH + i_3\Sigma FI + j_3\Sigma FJ + k_3\Sigma FK \\
\Sigma z_{3C}G &= K_3\Sigma G + a_3\Sigma AG + b_3\Sigma GB + c_3\Sigma GC + d_3\Sigma GD + e_3\Sigma GE + f_3\Sigma GF \\
&\quad + g_3\Sigma G^2 + h_3\Sigma GH + i_3\Sigma GI + j_3\Sigma GJ + k_3\Sigma GK \\
\Sigma z_{3C}H &= K_3\Sigma H + a_3\Sigma AH + b_3\Sigma HB + c_3\Sigma HC + d_3\Sigma HD + e_3\Sigma HE + f_3\Sigma HF \\
&\quad + g_3\Sigma HG + h_3\Sigma H^2 + i_3\Sigma HI + j_3\Sigma HJ + k_3\Sigma HK \\
\Sigma z_{3C}I &= K_3\Sigma I + a_3\Sigma AI + b_3\Sigma IB + c_3\Sigma IC + d_3\Sigma ID + e_3\Sigma IE + f_3\Sigma IF \\
&\quad + g_3\Sigma IG + h_3\Sigma IH + i_3\Sigma I^2 + j_3\Sigma IJ + k_3\Sigma IK \\
\Sigma z_{3C}J &= K_3\Sigma J + a_3\Sigma AJ + b_3\Sigma JB + c_3\Sigma JC + d_3\Sigma JD + e_3\Sigma JE + f_3\Sigma JF \\
&\quad + g_3\Sigma JG + h_3\Sigma JH + i_3\Sigma JI + j_3\Sigma J^2 + k_3\Sigma JK \\
\Sigma z_{3C}K &= K_3\Sigma K + a_3\Sigma AK + b_3\Sigma KB + c_3\Sigma KC + d_3\Sigma KD + e_3\Sigma KE \\
&\quad + f_3\Sigma KF + g_3\Sigma KG + h_3\Sigma KH + i_3\Sigma KI + j_3\Sigma KJ + k_3\Sigma K^2
\end{aligned} \quad (5.16)$$

In the above set of equations, the values of the multipliers K_3, a_3, b_3, c_3, d_3, e_3, f_3, g_3, h_3, i_3, j_3, and k_3 are substituted to compute the values of the unknowns (viz. K_3, a_3, b_3, c_3, d_3, e_3, f_3, g_3, h_3, i_3, j_3, and k_3). The values of the terms on LHS and the multipliers of a_3, b_3, c_3, d_3, e_3, f_3, g_3, h_3, i_3, j_3, and k_3 in the set of equation are calculated we will get a set of 12 equations, which are to be solved simultaneously to get the values of K_3, a_3, b_3, c_3, d_3, e_3, f_3, g_3, h_3, i_3, j_3, and k_3. The above equations can be verified in the matrix form and further values of K_3, a_3, b_3, c_3, d_3, e_3, f_3, g_3, h_3, i_3, j_3, and k_3 can be obtained by using matrix analysis.

The matrix obtained is given by,

$$\mathbf{z_{3C}} = \mathbf{W_3} \times \mathbf{X_3}$$

$$Z_{3C} \times \begin{bmatrix} 1 \\ A \\ B \\ C \\ D \\ E \\ F \\ G \\ H \\ I \\ J \\ K \end{bmatrix} = \begin{bmatrix} n & A & B & C & D & E & F & G & H & I & J & K \\ A & A^2 & BA & CA & DA & EA & FA & GA & HA & IA & JA & KA \\ B & AB & B^2 & CB & DB & EB & FB & GB & HB & IB & JB & KB \\ C & AC & BC & C^2 & DC & EC & FC & GC & HC & IC & JC & KC \\ D & AD & BD & CD & D^2 & ED & FD & GD & HD & ID & JD & KD \\ E & AE & BE & CE & DE & E^2 & FE & GE & HE & IE & JE & KE \\ F & AF & BF & CF & DF & EF & F^2 & GF & HF & IF & JF & KF \\ G & AG & BG & CG & DG & EG & FG & G^2 & HG & IG & JG & KG \\ H & AH & BH & CH & DH & EH & FH & GH & H^2 & IH & JH & KH \\ I & AI & BI & CI & DI & EI & FI & GI & HI & I^2 & JI & KI \\ J & AJ & BJ & CJ & DJ & EJ & FJ & GJ & HJ & IJ & J^2 & KJ \\ K & AK & BK & CK & DK & EK & FK & GK & HK & IK & JK & K^2 \end{bmatrix} \times \begin{bmatrix} K_3 \\ a_3 \\ b_3 \\ c_3 \\ d_3 \\ e_3 \\ f_3 \\ g_3 \\ h_3 \\ i_3 \\ j_3 \\ k_3 \end{bmatrix}$$

W_3 = 12 × 12 matrix of the multipliers of K_3, a_3, b_3, c_3, d_3, e_3, f_3, g_3, h_3, i_3, j_3, and k_3
X_3 = 12 × 1 matrix of solutions of values of K_3, a_3, b_3, c_3, d_3, e_3, f_3, g_3, h_3, i_3, j_3, and k_3
Z_3 = 12 × 1 matrix of the terms on LHS

In the above matrix, n is the number of sets of values. A, B, C, D, E, F, G, H, I, J, and K represent the independent pi terms π_1, π_2, π_3, π_4, π_5, π_6, π_7, π_8, π_9, π_{10}, and π_{11} while z_{3C} represents dependent pi term. Next, calculate the values of independent pi terms for corresponding dependent pi terms which helps to form the equation in matrix form. It is recommended to use MATLAB software for this purpose for making this process of model formulation quickest and least cumbersome.

After substituting values, we will get a set of 12 equations, which are to be solved simultaneously to get the values of K_3, a_3, b_3, c_3, d_3, e_3, f_3, g_3, h_3, i_3, j_3, and k_3. The equation (5.16) can be verified in the matrix form and further values of K_3, a_3, b_3, c_3, d_3, e_3, f_3, g_3, h_3, i_3, j_3, and k_3 can be obtained by using matrix analysis.

The matrix method of solving these equations using MATLAB is given below:

$$\mathbf{z_{3C} = W_3 \times X_3}$$

$$\begin{bmatrix} -36.184 \\ -.8089 \\ -17.636 \\ -118.63 \\ -62.898 \\ -76.127 \\ 29.87 \\ -22.875 \\ 17.689 \\ -59.652 \\ -195.29 \\ -68.287 \end{bmatrix} = \begin{bmatrix} 30 & 0.7 & 14.75 & 98.35 & 52.14 & 63.01 & -24.6 & 18.96 & -14.68 & 49.38 & 161.89 & 56.62 \\ 0.7 & 0.38 & 0.331 & 2.29 & 1.22 & 1.47 & -0.52 & 0.44 & -0.34 & 1.13 & 3.71 & 1.29 \\ 14.75 & 0.33 & 8.249 & 48.34 & 25.6 & 30.71 & -11.5 & 9.3 & -7.16 & 24.14 & 79.61 & 27.85 \\ 98.36 & 2.3 & 48.35 & 322.4 & 171 & 206.6 & -80.9 & 62.1 & -48.14 & 161.9 & 530.77 & 185.6 \\ 52.15 & 1.22 & 25.63 & 170.9 & 90.7 & 109.5 & -42.91 & 32.9 & -25.52 & 85.85 & 281.42 & 98.42 \\ 63.02 & 1.47 & 30.72 & 206.6 & 110 & 132.87 & -52 & 39.3 & -30.8 & 103.7 & 340 & 118.9 \\ -24.6 & -.52 & -11.7 & -80.93 & -42.9 & -52 & 20.73 & -15.6 & 12.08 & -40.68 & -133.1 & -46.5 \\ 18.97 & 0.44 & 9.3 & 62.1 & 33.97 & 39.83 & -15.60 & 11.9 & -9.28 & 31.2 & 102.3 & 35.79 \\ -14.7 & -0.35 & -7.17 & -48.1 & -25.5 & -30.8 & 12 & -9.28 & 7.30 & 24.29 & -79.2 & -27.7 \\ 49.39 & 1.14 & 24.14 & 161.9 & 85.9 & 103.7 & -40.6 & 31.2 & -24.29 & 82.2 & 266.5 & 93.2 \\ 161.9 & 3.71 & 79.62 & 530.7 & 281 & 340 & -133.2 & 102.3 & -79.2 & 266.5 & 873.8 & 305.5 \\ 56.62 & 1.29 & 27.86 & 185.6 & 98.4 & 118.9 & -46.57 & 35.7 & -27.7 & 93.24 & 305.5 & 106.8 \end{bmatrix} \times \begin{bmatrix} K \\ a_3 \\ b_3 \\ c_3 \\ d_3 \\ e_3 \\ f_3 \\ g_3 \\ h_3 \\ i_3 \\ j_3 \\ k_3 \end{bmatrix}$$

After multiplying inverse of matrix W_3 with Z_3 we get X_3 matrix with result values of K_3 and indices a_3, b_3, c_3, d_3, e_3, f_3, g_3, h_3, i_3, j_3, k_3.

$$\mathbf{X_3 = inv(W3) \times z_{3C}}$$

But K_3 is log value so convert into normal value take antilog of K_3, thus, antilog (0.521) = 1.6837

$K_3 = 1.6837$	$c_3 = -0.1837$	$f_3 = 0.0025$	$i_3 = 0.125$
$a_3 = -0.0873$	$d_3 = -1.2275$	$g_3 = 0.9916$	$j_3 = 0.0609$
$b_3 = -0.1026$	$e_3 = 0.2101$	$h_3 = 0.4451$	$k_3 = -0.1131$

Hence the model for dependent term π_{D3}:

$$\pi_{D3} = k_3 \times (\pi_1)^{a3} \times (\pi_2)^{b3} \times (\pi_3)^{c3} \times (\pi_4)^{d3} \times (\pi_5)^{e3} \times (\pi_6)^{f3} \times (\pi_7)^{g3} \times (\pi_8)^{h3} \times (\pi_9)^{i3} \times (\pi_{10})^{j3} \times (\pi_{11})^{k3}$$

$$Z_{3C} = \pi_{D3} = 1.6837(\pi_1)^{-0.0873}(\pi_2)^{-0.1026}(\pi_3)^{-0.1837}(\pi_4)^{-1.2275}(\pi_5)^{0.2101}(\pi_6)^{0.0025}(\pi_7)^{0.9916} (\pi_8)^{0.4451}(\pi_9)^{0.125}(\pi_{10})^{0.0609}(\pi_{11})^{-0.1131} \quad (5.17)$$

Productivity of crankshaft maintenance activity, dependent pi term:

$$\begin{aligned} z_{3C} = 1.6837 \times \{ & [(a \times c \times e \times g)/(b \times d \times f \times h)]^{-0.0873}, \\ & [(Ags/Exs),(sks/Ens),(hls/Hbs)]^{-0.1026}, \\ & [(Dsps/Dsns) \times (Lsps/Dsns) \times (Dsmab/Dsns) \times (Lsns/Dsns) \times (Dsbs/Dsns) \\ & \times (Lsbs/Dsns) \times (Lsmab/Dsns) \times (Dsiab/Dsns) \times (Lsiab/Dsns) \\ & \times (Dsc/Dsns) \times (Lsc/Dsns) \times (Dmj/Dsns) \times (Lmj/Dsns) \times (Dcpj/Dsns) \\ & \times (Lcpj/Dsns) \times (tmjb/Dsns) \times (Dmjb/Dsns) \times (Dcpjb/Dsns) \\ & \times (tcpjb/Dsns)]^{-0.1837}, \\ & [(Dsmsp/Dbr),(Lbr/Dbr),(Lsmsp/Dbr),(Dsisp/Dbr),(Lsisp/Dbr), \\ & (Dbrr/Dbr),(Lbrr/Dbr)]^{-1.2275}, \\ & [(ker/loil),(Ca/Ebelt)]^{0.2101} \\ & [(Axcp/Elsb),(Axmj/Elsb)]^{0.0025} \\ & [(Lfrm/hfrm),(Wfrm/hfrm)]^{0.9916} \\ & [Temp/100]^{0.4451} \\ & [(Humd)]^{0.125} \\ & [((illmus \times Ags)^3)/Wt]^{0.0609} \\ & [(Noise)]^{-0.1131} \} \end{aligned} \quad (5.18)$$

5.1.4 Models Developed for the Dependent Variables – Crankshaft Maintenance Activity

The exact forms of models obtained for the dependent variables overhauling time of crankshaft maintenance activity (z_1), human energy consumed in crankshaft maintenance activity (z_2) and productivity of crankshaft maintenance activity (z_3) are as under:

$$z_{1C} = \pi_{D1} = 1.7198(\pi_1)^{0.0873}(\pi_2)^{0.1026}(\pi_3)^{2.8137}(\pi_4)^{0.2115}(\pi_5)^{-0.2101}(\pi_6)^{-0.0025}(\pi_7)^{3.5409} (\pi_8)^{-0.4451}(\pi_9)^{-0.125}(\pi_{10})^{-0.0609}(\pi_{11})^{0.1131} \quad (5.19)$$

$$z_{2C} = \pi_{D2} = 1.2203(\pi_1)^{0.2769}(\pi_2)^{0.018}(\pi_3)^{-0.2583}(\pi_4)^{3.3045}(\pi_5)^{0.0597}(\pi_6)^{-0.0314}(\pi_7)^{-1.2341} (\pi_8)^{0.0415}(\pi_9)^{0.0103}(\pi_{10})^{0.0002}(\pi_{11})^{-0.3935} \quad (5.20)$$

$$z_{3C} = \pi_{D3} = 1.6837(\pi_1)^{-0.0873}(\pi_2)^{-0.1026}(\pi_3)^{-0.1837}(\pi_4)^{-1.2275}(\pi_5)^{0.2101}(\pi_6)^{0.0025}(\pi_7)^{0.9916}(\pi_8)^{0.4451}(\pi_9)^{0.125}(\pi_{10})^{0.0609}(\pi_{11})^{-0.1131} \quad (5.21)$$

5.2 FORMULATION OF FIELD DATA-BASED MODEL FOR LINER PISTON MAINTENANCE ACTIVITY

For liner piston maintenance operation, eleven independent pi terms (π_1, π_2, π_3, π_4, π_5, π_6, π_7, π_8, π_9, π_{10}, π_{11}) and three dependent pi terms (π_{D1}, π_{D2}, π_{D3}) have been identified from field for the model formulation.

$$z_{1P} = f(\pi_{D1}) = f(\pi_1, \pi_2, \pi_3, \pi_4, \pi_5, \pi_6, \pi_7, \pi_8, \pi_9, \pi_{10}, \pi_{11})$$
$$z_{2P} = f(\pi_{D2}) = f(\pi_1, \pi_2, \pi_3, \pi_4, \pi_5, \pi_6, \pi_7, \pi_8, \pi_9, \pi_{10}, \pi_{11})$$
$$z_{3P} = f(\pi_{D3}) = f(\pi_1, \pi_2, \pi_3, \pi_4, \pi_5, \pi_6, \pi_7, \pi_8, \pi_9, \pi_{10}, \pi_{11})$$

f stands for "function of"

Where:

$z_{1P} = \pi_{D1}$, first dependent pi term, overhauling time of liner piston maintenance activity
$z_{2P} = \pi_{D2}$, second dependent pi term, human energy consumed in liner piston maintenance activity
$z_{3P} = \pi_{D3}$, third dependent pi term, productivity of liner piston maintenance activity

To evolve the probable mathematical form for the dimensional equations of the phenomenon, we can assume that relationship between dependent and independent pi terms to be of exponential form. We can write the general equation as follows:

$$z_P = k \times (\pi_1)^a \times (\pi_2)^b \times (\pi_3)^c \times (\pi_4)^d \times (\pi_5)^e \times (\pi_6)^f \times (\pi_7)^g \times (\pi_8)^h \times (\pi_9)^i \times (\pi_{10})^j \times (\pi_{11})^k \quad (5.22)$$

Liner piston maintenance activity, dependent pi term:

$$\begin{aligned}
z_P = K \times \{ & [(a \times c \times e \times g)/(b \times d \times f \times h)]^a, \\
& [(Ags/Exs), (sks/Ens), (hls/Hbs)]^b, && \text{--- anthropometric data} \\
& [(Dlins/Lcr) \times (l.lin/L.cr) \times (D.pis/L.cr) \times (L.pis/L.cr) \times (D.prng/Lc.r) \\
& \times (t.prng/L.crs) \times (l.pstrk/L.cr) \times (D.ppin/L.cr) \times (L.ppin/L.cr) \times (Dbg.br/L.cr) \\
& \times (t.bg.br/L.cr) \times (Dlt.brj/L.cr) \times (t.lt.brj/L.cr) \times (D.b.nj/L.cr) \times (L.b.nt/L.cr) \\
& \times (d.b.bl/L.cr) \times (l.b.bl/L.cr)]^c, && \text{--- workers data} \\
& [(Dbr/Lbr), (Dsmsp/Lbr), (Lsmsp/Lbr), (Dsisp/Lbr), (Lsisp/Lbr), \\
& (Dbrr/Lbr), (Lbrr/Lbr)]^d, && \text{--- specification of tools} \\
& [(ker/Ebelt), (orient11/Loil)]^e, && \text{--- specification of solvent} \\
& [(bcl/Bel) \times (Scl/Bel)]^f, && \text{--- axial clearance} \\
& [(h.crt/l.crt), (b.crt/l.crt)]^g, && \text{--- workstation data} \\
& [temp/100]^h, && \text{--- temperature at workplace} \\
& [(Humd)]^i, && \text{--- humidity at workplace} \\
& [((illmus \times Ags)^3)/Wt]^j, && \text{--- Illumination at workplace} \\
& [(Noise)]^k && \text{--- noise at workplace} \\
\} &
\end{aligned} \quad (5.23)$$

Equation 5.22 contains 12 unknowns, Curve fitting constant K and indices a, b, c, d, e, f, g, h, i, j, and k. To get the values of these unknowns we need minimum a 12 set of values of (π_1, π_2, π_3, π_4, π_5, π_6, π_7, π_8, π_9, π_{10}, π_{11}, π_D).

As per the plan in the design of experimentation, we have 30 sets of such values. If any arbitrary 12 sets are selected then the values of unknown K_1 and indices a, b, c, d, e, f, g, h, i, j, and k can be computed, we may not have unique solution which represents a best curve fit for the remaining sets of values. To be specific, one can find out ${}^{n}C_{r}$ number combinations of r number of sets chosen out of n number of sets of the values or ${}^{30}C_{12}$ in our case. Solving these many sets and finding their solutions is a Herculean task. Hence, we employ the curve fitting technique (Gopalkrishnan and Banerji 2015). To implement this method to our experiment, it is imperative to have the equation in the form:

$$z_P = a + bX + cY + dZ + \ldots\ldots \tag{5.24}$$

Equation (5.22) can be rendered in the form of Equation (6.2.3) by taking log on both sides of Equation (5.22).

$$\begin{aligned}\log\pi_D = \log k + a\times\log\pi_1 + b\times\log\pi_2 + c\times\log\pi_3 + d\times\log\pi_4 + e\times\log\pi_5 + f\times\log\pi_6 \\ + g\times\log\pi_7 + h\times\log\pi_8 + i\times\log\pi_9 + j\times\log\pi_{10} + k\times\log\pi_{11}\end{aligned} \tag{5.25}$$

Let us denote:

$$\begin{array}{llll} z_P = \log\pi_D & B = \log\pi_2 & E = \log\pi_5 & H = \log\pi_8 \\ K_1 = \log k & C = \log\pi_3 & F = \log\pi_6 & I = \log\pi_9 \\ A = \log\pi_1 & D = \log\pi_4 & G = \log\pi_7 & J = \log\pi_{10} \\ & & & K = \log\pi_{11} \end{array}$$

Putting these values in equations 5.6.3.4, the same can be written as:

$$z_P = K_1 + a\times A + b\times B + c\times C + d\times D + e\times E + f\times F + g\times G + h\times H + i\times I + j\times J + k\times K \tag{5.26}$$

Equation 5.26 is a regression equation of Z on A, B, C, D, E, F, G, H, I, J, and K in an n dimensional co–ordinate system. This represents a regression hyperplane. Substituting the data collected from field, we will get one such equation for 30 such equations. The curve fitting method, in order to include every set of readings, instead of any 12 randomly chosen sets of readings, instructs us to add all these equations. This results into the following equation:

$$\sum z_P = nK + a\sum A + b\sum B + c\sum C + d\sum D + e\sum E + f\sum F + g\sum G + h\sum H + i\sum I + j\sum J + k\sum K \tag{5.27}$$

Where number of readings, n = 30.

As mentioned earlier, there are 12 unknowns in this equation, K and indices a, b, c, d, e, f, g, h, i, j. Hence, we will require 12 different equations to solve. In order to achieve these equations, we multiply equation 6.2.6 with A, B, C, D, E, F, G, H, I, J, and K to get a set of 12 different equations as shown below:

$$\begin{aligned}\sum z_P A = K\sum A + a\sum A^2 + b\sum AB + c\times\sum AC + d\sum AD + e\sum AE + f\sum AF + g\sum AG \\ + h\sum AH + i\sum AI + j\sum AJ + k\sum AK\end{aligned}$$

$$\begin{aligned}\sum z_P B = K\sum B + a\sum AB + b\sum B^2 + c\times\sum BC + d\sum BD + e\sum BE + f\sum BF + g\sum BG \\ + h\sum BH + i\sum BI + j\sum BJ + k\sum BK\end{aligned}$$

$$\Sigma z_pC = K\Sigma C + a\Sigma AC + b\Sigma CB + c\times\Sigma C^2 + d\Sigma CD + e\Sigma CE + f\Sigma CF + g\Sigma CG + h\Sigma CH + i\Sigma CI + j\Sigma CJ + k\Sigma CK$$

$$\Sigma z_pD = K\Sigma D + a\Sigma AD + b\Sigma BD + c\times\Sigma CD + d\Sigma D^2 + e\Sigma DE + f\Sigma DF + g\Sigma DG + h\Sigma DH + i\Sigma DI + j\Sigma DJ + k\Sigma DK$$

$$\Sigma z_pE = K\Sigma E + a\Sigma AE + b\Sigma BE + c\times\Sigma CE + d\Sigma DE + e\Sigma E^2 + f\Sigma FE + g\Sigma GE + h\Sigma HE + i\Sigma IE + j\Sigma JE + k\Sigma EK$$

$$\Sigma z_pF = K\Sigma F + a\Sigma AF + b\Sigma FB + c\times\Sigma FC + d\Sigma FD + e\Sigma FE + f\Sigma F^2 + g\Sigma FG + h\Sigma FH + i\Sigma FI + j\Sigma FJ + k\Sigma FK$$

$$\Sigma z_pG = K\Sigma G + a\Sigma AG + b\Sigma GB + c\times\Sigma GC + d\Sigma GD + e\Sigma GE + f\Sigma GF + g\Sigma G^2 + h\Sigma GH + i\Sigma GI + j\Sigma GJ + k\Sigma GK$$

$$\Sigma z_pH = K\Sigma H + a\Sigma AH + b\Sigma HB + c\times\Sigma HC + d\Sigma HD + e\Sigma HE + f\Sigma HF + g\Sigma HG + h\Sigma H^2 + i\Sigma HI + j\Sigma HJ + k\Sigma HK$$

$$\Sigma z_pI = K\Sigma I + a\Sigma AI + b\Sigma IB + c\times\Sigma IC + d\Sigma ID + e\Sigma IE + f\Sigma IF + g\Sigma IG + h\Sigma IH + i\Sigma I^2 + j\Sigma IJ + k\Sigma IK$$

$$\Sigma z_pJ = K\Sigma J + a\Sigma AJ + b\Sigma JB + c\times\Sigma JC + d\Sigma JD + e\Sigma JE + f\Sigma JF + g\Sigma JG + h\Sigma JH + i\Sigma JI + j\Sigma J^2 + k\Sigma JK$$

$$\Sigma z_pK = K\Sigma K + a\Sigma AK + b\Sigma KB + c\times\Sigma KC + d\Sigma KD + e\Sigma KE + f\Sigma KF + g\Sigma KG + h\Sigma KH + i\Sigma KI + j\Sigma KJ + k\Sigma K^2$$

Including equation 5.27, we have a set of 12 equations for the 12 unknowns. To solve these equations, we shall use matrix analysis. The matrix format of an equation is of the form:

$$\mathbf{z_P = W \times X}$$

In the above equation X is the unknown matrix. We solve this equation to determine the value of X by taking an inverse of matrix W. The final equation solved by using MATLAB is as follows.

$$\mathbf{X = inv(W) \times Z_P}$$

Solving this equation for every dependent pi term, the model is formed.

5.2.1 Model Formulation for π_{D1}, Overhauling Time of Liner Piston Maintenance Activity by Identifying the Curve Fitting Constant and Various Indices of Pi Terms

The multiple regression analysis helps to identify the indices of independent pi terms in the model aimed at, by considering eleven independent pi terms and one dependent pi term. Let the model be of the form for dependent pi term π_{D1}, overhauling time of liner piston maintenance activity.

$$z_{1P} = \pi_{D1} = k_1 \times (\pi_1)^{a1} \times (\pi_2)^{b1} \times (\pi_3)^{c1} \times (\pi_4)^{d1} \times (\pi_5)^{e1} \times (\pi_6)^{f1} \times (\pi_7)^{g1} \times (\pi_8)^{h1} \times (\pi_9)^{i1} \times (\pi_{10})^{j1} \times (\pi_{11})^{k1} \quad (5.28)$$

To determine the values of K_1, a_1, b_1, c_1, d_1, e_1, f_1, g_1 h_1, i_1, j_1, and k1 and to arrive at the regression hyperplane, the above equations are presented as follows:

$$\begin{aligned}
\Sigma z_{1P} &= nK_1 + a_1 \Sigma A + b_1 \Sigma B + c_1 \Sigma C + d_1 \Sigma D + e_1 \Sigma E + f_1 \Sigma F + g_1 \Sigma G + h_1 \Sigma H \\
&\quad + i_1 \Sigma I + j_1 \Sigma J + k_1 \Sigma K \\
\Sigma z_{1P}A &= K_1 \Sigma A + a_1 \Sigma A^2 + b_1 \Sigma AB + c_1 \Sigma AC + d_1 \Sigma AD + e_1 \Sigma AE + f_1 \Sigma AF \\
&\quad + g_1 \Sigma AG + h_1 \Sigma AH + i_1 \Sigma AI + j_1 \Sigma AJ + k_1 \Sigma AK \\
\Sigma z_{1P}B &= K_1 \Sigma B + a_1 \Sigma AB + b_1 \Sigma B^2 + c_1 \Sigma BC + d_1 \Sigma BD + e_1 \Sigma BE + f_1 \Sigma BF \\
&\quad + g_1 \Sigma BG + h_1 \Sigma BH + i_1 \Sigma BI + j_1 \Sigma BJ + k_1 \Sigma BK \\
\Sigma z_{1P}C &= K_1 \Sigma C + a_1 \Sigma AC + b_1 \Sigma CB + c_1 \Sigma C^2 + d_1 \Sigma CD + e_1 \Sigma CE + f_1 \Sigma CF \\
&\quad + g_1 \Sigma CG + h_1 \Sigma CH + i_1 \Sigma CI + j_1 \Sigma CJ + k_1 \Sigma CK \\
\Sigma z_{1P}D &= K_1 \Sigma D + a_1 \Sigma AD + b_1 \Sigma BD + c_1 \Sigma CD + d_1 \Sigma D^2 + e_1 \Sigma DE + f_1 \Sigma DF \\
&\quad + g_1 \Sigma DG + h_1 \Sigma DH + i_1 \Sigma DI + j_1 \Sigma DJ + k_1 \Sigma DK \\
\Sigma z_{1P}E &= K_1 \Sigma E + a_1 \Sigma AE + b_1 \Sigma BE + c_1 \Sigma CE + d_1 \Sigma DE + e_1 \Sigma E^2 + f_1 \Sigma FE \\
&\quad + g_1 \Sigma GE + h_1 \Sigma HE + i_1 \Sigma IE + j_1 \Sigma JE + k_1 \Sigma EK \\
\Sigma z_{1P}F &= K_1 \Sigma F + a_1 \Sigma AF + b_1 \Sigma FB + c_1 \Sigma FC + d_1 \Sigma FD + e_1 \Sigma FE + f_1 \Sigma F^2 \\
&\quad + g_1 \Sigma FG + h_1 \Sigma FH + i_1 \Sigma FI + j_1 \Sigma FJ + k_1 \Sigma FK \\
\Sigma z_{1P}G &= K_1 \Sigma G + a_1 \Sigma AG + b_1 \Sigma GB + c_1 \Sigma GC + d_1 \Sigma GD + e_1 \Sigma GE + f_1 \Sigma GF \\
&\quad + g_1 \Sigma G^2 + h_1 \Sigma GH + i_1 \Sigma GI + j_1 \Sigma GJ + k_1 \Sigma GK \\
\Sigma z_{1P}H &= K_1 \Sigma H + a_1 \Sigma AH + b_1 \Sigma HB + c_1 \Sigma HC + d_1 \Sigma HD + e_1 \Sigma HE + f_1 \Sigma HF \\
&\quad + g_1 \Sigma HG + h_1 \Sigma H^2 + i_1 \Sigma HI + j_1 \Sigma HJ + k_1 \Sigma HK \\
\Sigma z_{1P}I &= K_1 \Sigma I + a_1 \Sigma AI + b_1 \Sigma IB + c_1 \Sigma IC + d_1 \Sigma ID + e_1 \Sigma IE + f_1 \Sigma IF + g_1 \Sigma IG \\
&\quad + h_1 \Sigma IH + i_1 \Sigma I^2 + j_1 \Sigma IJ + k_1 \Sigma IK \\
\Sigma z_{1P}J &= K_1 \Sigma J + a_1 \Sigma AJ + b_1 \Sigma JB + c_1 \Sigma JC + d_1 \Sigma JD + e_1 \Sigma JE + f_1 \Sigma JF + g_1 \Sigma JG \\
&\quad + h_1 \Sigma JH + i_1 \Sigma JI + j_1 \Sigma J^2 + k_1 \Sigma JK \\
\Sigma z_{1P}K &= K_1 \Sigma K + a_1 \Sigma AK + b_1 \Sigma KB + c_1 \Sigma KC + d_1 \Sigma KD + e_1 \Sigma KE + f_1 \Sigma KF \\
&\quad + g_1 \Sigma KG + h_1 \Sigma KH + i_1 \Sigma KI + j_1 \Sigma KJ + k_1 \Sigma K^2
\end{aligned} \tag{5.29}$$

Where n is the number of sets of the values.

In the above set of equations, the values of the multipliers K_1, a_1, b_1, c_1, d_1, e_1, f_1, g_1, h_1, i_1, j_1, and k_1 are substituted to compute the values of the unknowns (viz. K_1, a_1, b_1, c_1, d_1, e_1, f_1, g_1, h_1, i_1, j_1, and k_1). The values of the terms on LHS and the multipliers of a1, b_1, c_1, d_1, e_1, f_1, g_1, h_1, i_1, j_1, and k_1 in the set of equation are calculated we will get a set of 12 equations, which are to be solved simultaneously to get the values of K_1, a_1, b_1, c_1, d_1, e_1, f_1, g_1, h_1, i_1, j_1, and k_1. The above equations can be verified in the matrix form and further values of K_1, a_1, b_1, c_1, d_1, e_1, f_1, g_1, h_1, i_1, j_1, and k_1 can be obtained by using matrix analysis. The matrix obtained is given by:

$$\mathbf{z_{1P}} = \mathbf{W_1} \times \mathbf{X_1}$$

$$Z_{1P} \times \begin{bmatrix} 1 \\ A \\ B \\ C \\ D \\ E \\ F \\ G \\ H \\ I \\ J \\ K \end{bmatrix} = \begin{bmatrix} n & A & B & C & D & E & F & G & H & I & J & K \\ A & A^2 & BA & CA & DA & EA & FA & GA & HA & IA & JA & KA \\ B & AB & B^2 & CB & DB & EB & FB & GB & HB & IB & JB & KB \\ C & AC & BC & C^2 & DC & EC & FC & GC & HC & IC & JC & KC \\ D & AD & BD & CD & D^2 & ED & FD & GD & HD & ID & JD & KD \\ E & AE & BE & CE & DE & E^2 & FE & GE & HE & IE & JE & KE \\ F & AF & BF & CF & DF & EF & F^2 & GF & HF & IF & JF & KF \\ G & AG & BG & CG & DG & EG & FG & G^2 & HG & IG & JG & KG \\ H & AH & BH & CH & DH & EH & FH & GH & H^2 & IH & JH & KH \\ I & AI & BI & CI & DI & EI & FI & GI & HI & I^2 & JI & KI \\ J & AJ & BJ & CJ & DJ & EJ & FJ & GJ & HJ & IJ & J^2 & KJ \\ K & AK & BK & CK & DK & EK & FK & GK & HK & IK & JK & K^2 \end{bmatrix} \times \begin{bmatrix} K_1 \\ a_1 \\ b_1 \\ c_1 \\ d_1 \\ e_1 \\ f_1 \\ g_1 \\ h_1 \\ i_1 \\ j_1 \\ k_1 \end{bmatrix}$$

W_1 = 12 × 12 matrix of the multipliers of K_1, b_1, c_1, d_1, e_1, f_1, g_1, h_1, i_1, j_1, and k_1
X_1 = 12 × 1 matrix of solutions of values of K_1, a_1, b_1, c_1, d_1, e_1, f_1, g_1, h_1, i_1, j_1, and k_1
Z_1 = 12 × 1 matrix of the terms on LHS

In the above matrix, n is the number of readings A, B, C, D, E, F, G, H, I, J, and K represent the independent pi terms π_1, π_2, π_3, π_4, π_5, π_6, π_7, π_8, π_9, π_{10}, and π_{11} while Z_1 represents dependent pi term, overhauling time of liner piston maintenance activity. Next, calculate the values of independent pi terms for corresponding dependent pi terms which helps to form the equation in matrix form. It is recommended to use MATLAB software for this purpose for making this process of model formulation quickest and least cumbersome.

$$\mathbf{z_{1P} = W_1 \times X_1}$$

$$\begin{bmatrix} z_1 \\ \\ 329.36 \\ 7.6784 \\ 142.86 \\ 3412 \\ -111.1 \\ 612.69 \\ -281.2 \\ -220.6 \\ -159.6 \\ 551.9 \\ 1767 \\ 624.9 \end{bmatrix} = \begin{bmatrix} & & & & & & w_1 & & & & & \\ & & & & & & & & & & & \\ 30 & 0.7 & 13.03 & 310.85 & -10.121 & 55.80 & -25.62 & -20.10 & -14.54 & 50.28 & 160.97 & 56.62 \\ 0.7 & 0.38 & 0.27 & 7.25 & -0.2362 & 1.39 & -0.59 & -0.47 & -0.36 & 1.14 & 3.70 & 1.33 \\ 13.03 & 0.267 & 6.13 & 134.93 & -4.3941 & 24.34 & -11.12 & -8.73 & -6.31 & 21.86 & 70.06 & 24.7 \\ 310.79 & 7.253 & 134.93 & 3219.63 & -104.85 & 578.03 & -265.3 & -208.2 & -150.58 & 520.86 & 1667.5 & 589.6 \\ -10.12 & -0.253 & -4.39 & -104.85 & 3.141 & -18.82 & 8.64 & 6.78 & 4.9 & -16.96 & -54.3 & -19.20 \\ 55.80 & 1.390 & 24.34 & 578.029 & -18.823 & 104.39 & -47.62 & -37.38 & -27.1 & 93.51 & 299.3 & 105.8 \\ -25.62 & -.588 & -11.12 & -265.37 & 8.641 & -47.62 & 21.93 & 17.16 & 12.39 & -42.9 & -137.5 & -48.6 \\ -20.10 & 0.469 & -8.73 & -208.2 & 6.779 & -37.38 & 17.16 & 13.46 & 9.74 & -33.68 & -107.83 & -38.13 \\ -14.54 & -0.357 & -6.31 & -150.52 & 4.903 & -27.10 & 12.39 & 9.74 & 7.17 & -24.58 & -77.98 & -27.5 \\ 50.28 & 1.137 & 21.86 & 520.82 & -16.961 & 93.51 & -42.9 & -33.68 & -24.58 & 85.06 & 269.98 & 95.3 \\ 160.97 & 3.704 & 70.06 & 1667.55 & -54.303 & 299.34 & -137.54 & -107.8 & -77.98 & 269.98 & 864.6 & 305.3 \\ 56.92 & 1.326 & 24.70 & 589.65 & -19.202 & 105.86 & -48.6 & -38.13 & -27.58 & 95.39 & 305.3 & 108 \end{bmatrix} \times \begin{bmatrix} x_1 \\ \\ K_1 \\ a_1 \\ b_1 \\ c_1 \\ d_1 \\ e_1 \\ f_1 \\ g_1 \\ h_1 \\ i_1 \\ j_1 \\ k_1 \end{bmatrix}$$

After multiplying inverse of matrix W_1 with Z_1 we get X_1 matrix, with result values of K_1 and indices a_1, b_1, c_1, d_1, e_1, f_1, g_1, h_1, i_1, j_1, k_1.

$$\mathbf{X_1 = inv(W1) \times z_{1P}}$$

But K_1 is log value so convert into normal value take antilog of K_1, thus, antilog (0.0122) = 1.0122

$$K_1 = 1.0122 \quad a_1 = -0.0973 \quad b_1 = -0.1917 \quad c_1 = 0.8772$$
$$d_1 = -6.0098 \quad f_1 = 0.378 \quad h_1 = -0.5615 \quad j_1 = 0.0887$$
$$e_1 = 0.0896 \quad g_1 = 0.6869, \quad i_1 = -0.2282 \quad k_1 = 0.0838$$

Hence the model for dependent term π_{D1}, overhauling time of liner piston maintenance operation is given as:

$$\pi_{D1} = k_1 \times (\pi_1)^{a2} \times (\pi_2)^{b2} \times (\pi_3)^{c2} \times (\pi_4)^{d2} \times (\pi_5)^{e2} \times (\pi_6)^{f2} \times (\pi_7)^{g2} \times (\pi_8)^{h2} \times (\pi_9)^{i2} \times (\pi_{10})^{j2} \times (\pi_{11})^{k2}$$

$$z_{1P} = \pi_{D1} = 1.0122(\pi_1)^{-0.0973}(\pi_2)^{-0.1917}(\pi_3)^{0.8772}(\pi_4)^{-6.0098}(\pi_5)^{0.0896}(\pi_6)^{0.378}(\pi_7)^{0.6869} (\pi_8)^{-0.5615}(\pi_9)^{-0.2282}(\pi_{10})^{0.0887}(\pi_{11})^{0.0838} \quad (5.30)$$

Overhauling time of liner piston maintenance activity, dependent pi term:

$$\begin{aligned} z_{1P} = 1.0122 \times \{ & [(a \times c \times e \times g)/(b \times d \times f \times h)]^{-0.0973}, \\ & [(Ags/Exs), (sks/Ens), (hls/Hbs)]^{-0.1917} \\ & [(Dlins/Lcr) \times (l.lin/L.cr) \times (D.pis/L.cr) \times (L.pis/L.cr) \times (D.prng/Lc.r) \\ & \times (t.prng/L.crs) \times (l.pstrk/L.cr) \times (D.ppin/L.cr) \times (L.ppin/L.cr) \\ & \times (Dbg.br/L.cr) \times (t.bg.br/L.cr) \times (Dlt.brj/L.cr) \times (t.lt.brj/L.cr) \\ & \times (D.b.nj/L.cr) \times (L.b.nt/L.cr) \times (d.b.bl/L.cr) \times (l.b.bl/L.cr)]^{0.8772}, \\ & [(Dbr/Lbr), (Dsmsp/Lbr), (Lsmsp/Lbr), (Dsisp/Lbr), (Lsisp/Lbr), \\ & (Dbrr/Lbr), (Lbrr/Lbr)]^{-6.0098}, \\ & [(ker/E\,belt), (orient\,11/L\,oil)]^{0.0896}, \\ & [(bcl/Bel) \times (Scl/Bel)]^{0.378}, \\ & [(h.crt/l.crt), (b.crt/l.crt)]^{0.6869,} \\ & [Temp/100]^{-0.5615,} \\ & [(Humd)]^{0.2282}, \\ & [(illmus \times Ags)^3/Wt]^{0.0887,} \\ & [(Noise)]^{0.0838} \} \end{aligned} \quad (5.31)$$

5.2.2 Model Formulation for π_{D2}, Human Energy Consumed in Liner Piston Maintenance Activity by Identifying the Curve Fitting Constant and Various Indices of Pi Terms

The multiple regression analysis helps to identify the indices of independent pi terms in the model aimed at, by considering eleven independent pi terms and one dependent pi term. Let the model be of the form for dependent pi term π_{D2}, human energy consumed in liner piston maintenance activity.

$$z_{2P} = \pi_{D2} = k_2 \times (\pi_1)^{a2} \times (\pi_2)^{b2} \times (\pi_3)^{c2} \times (\pi_4)^{d2} \times (\pi_5)^{e2} \times (\pi_6)^{f2} \times (\pi_7)^{g2} \times (\pi_8)^{h2} \times (\pi_9)^{i2} \times (\pi_{10})^{j2} \times (\pi_{11})^{k2} \quad (5.32)$$

To determine the values of K_2, a_2, b_2, c_2, d_2, e_2, f_2, g_2 h_2, i_2, j_2, and k_2 and to arrive at the regression hyperplane, the above equations are presented as follows:

$$\begin{aligned}
\Sigma z_{2P} &= nK_2+a_2\Sigma A+b_2\Sigma B+c_2\Sigma C+d_2\Sigma D+e_2\Sigma E+f_2\Sigma F+g_2\Sigma G+h_2\Sigma H \\
&\quad +i_2\Sigma I+j_2\Sigma J+k_2\Sigma K \\
\Sigma z_{2P}A &= K_2\Sigma A+a_2\Sigma A^2+b_2\Sigma AB+c_2\Sigma AC+d_2\Sigma AD+e_2\Sigma AE+f_2\Sigma AF \\
&\quad +g_2\Sigma AG+h_2\Sigma AH+i_2\Sigma AI+j_2\Sigma AJ+k_2\Sigma AK \\
\Sigma z_{2P}B &= K_2\Sigma B+a_2\Sigma AB+b_2\Sigma B^2+c_2\Sigma BC+d_2\Sigma BD+e_2\Sigma BE+f_2\Sigma BF \\
&\quad +g_2\Sigma BG+h_2\Sigma BH+i_2\Sigma BI+j_2\Sigma BJ+k_2\Sigma BK \\
\Sigma z_{2P}C &= K_2\Sigma C+a_2\Sigma AC+b_2\Sigma CB+c_2\Sigma C^2+d_2\Sigma CD+e_2\Sigma CE+f_2\Sigma CF \\
&\quad +g_2\Sigma CG+h_2\Sigma CH+i_2\Sigma CI+j_2\Sigma CJ+k_2\Sigma CK \\
\Sigma z_{2P}D &= K_2\Sigma D+a_2\Sigma AD+b_2\Sigma BD+c_2\Sigma CD+d_2\Sigma D^2+e_2\Sigma DE+f_2\Sigma DF \\
&\quad +g_2\Sigma DG+h_2\Sigma DH+i_2\Sigma DI+j_2\Sigma DJ+k_2\Sigma DK \\
\Sigma z_{2P}E &= K_2\Sigma E+a_2\Sigma AE+b_2\Sigma BE+c_2\Sigma CE+d_2\Sigma DE+e_2\Sigma E^2+f_2\Sigma FE \\
&\quad +g_2\Sigma GE+h_2\Sigma HE+i_2\Sigma IE+j_2\Sigma JE+k_2\Sigma EK \\
\Sigma z_{2P}F &= K_2\Sigma F+a_2\Sigma AF+b_2\Sigma FB+c_2\Sigma FC+d_2\Sigma FD+e_2\Sigma FE+f_2\Sigma F^2 \\
&\quad +g_2\Sigma FG+h_2\Sigma FH+i_2\Sigma FI+j_2\Sigma FJ+k_2\Sigma FK \\
\Sigma z_{2P}G &= K_2\Sigma G+a_2\Sigma AG+b_2\Sigma GB+c_2\Sigma GC+d_2\Sigma GD+e_2\Sigma GE+f_2\Sigma GF \\
&\quad +g_2\Sigma G^2+h_2\Sigma GH+i_2\Sigma GI+j_2\Sigma GJ+k_2\Sigma GK \\
\Sigma z_{2P}H &= K_2\Sigma H+a_2\Sigma AH+b_2\Sigma HB+c_2\Sigma HC+d_2\Sigma HD+e_2\Sigma HE+f_2\Sigma HF \\
&\quad +g_2\Sigma HG+h_2\Sigma H^2+i_2\Sigma HI+j_2\Sigma HJ+k_2\Sigma HK \\
\Sigma z_{2P}I &= K_2\Sigma I+a_2\Sigma AI+b_2\Sigma IB+c_2\Sigma IC+d_2\Sigma ID+e_2\Sigma IE+f_2\Sigma IF+g_2\Sigma IG \\
&\quad +h_2\Sigma IH+i_2\Sigma I^2+j_2\Sigma IJ+k_2\Sigma IK \\
\Sigma z_{2P}J &= K_2\Sigma J+a_2\Sigma AJ+b_2\Sigma JB+c_2\Sigma JC+d_2\Sigma JD+e_2\Sigma JE+f_2\Sigma JF+g_2\Sigma JG \\
&\quad +h_2\Sigma JH+i_2\Sigma JI+j_2\Sigma J^2+k_2\Sigma JK \\
\Sigma z_{2P}K &= K_2\Sigma K+a_2\Sigma AK+b_2\Sigma KB+c_2\Sigma KC+d_2\Sigma KD+e_2\Sigma KE+f_2\Sigma KF \\
&\quad +g_2\Sigma KG+h_2\Sigma KH+i_2\Sigma KI+j_2\Sigma KJ+k_2\Sigma K^2
\end{aligned} \quad (5.33)$$

Where, n is number of sets of the values.

In the above set of equations the values of the multipliers K_2, a_2, b_2, c_2, d_2, e_2, f_2, g_2, h_2, i_2, j_2, and k_2 are substituted to compute the values of the unknowns (viz. K_2, a_2, b_2, c_2, d_2, e_2, f_2, g_2, h_2, i_2, j_2, and k_2). The values of the terms on LHS and the multipliers of a_2, b_2, c_2, d_2, e_2, f_2, g_2, h_2, i_2, j_2, and k_2 in the set of equation are calculated, we will get a set of 12 equations, which are to be solved simultaneously to get the values of K_2, a_2, b_2, c_2, d_2, e_2, f_2, g_2, h_2, i_2, j_2, and k_2. The above equations can be verified in the matrix form and further values of K_2, a_2, b_2, c_2, d_2, e_2, f_2, g_2, h_2, i_2, j_2, and k_2 can be obtained by using matrix analysis. The matrix obtained is given by, $\mathbf{z_{2P} = W_2 \times X_2}$

The matrix method of solving these equations using MATLAB is given below:

$$Z_{2C}\times\begin{bmatrix}1\\A\\B\\C\\D\\E\\F\\G\\H\\I\\J\\K\end{bmatrix}=\begin{bmatrix}n & A & B & C & D & E & F & G & H & I & J & K\\A & A^2 & BA & CA & DA & EA & FA & GA & HA & IA & JA & KA\\B & AB & B^2 & CB & DB & EB & FB & GB & HB & IB & JB & KB\\C & AC & BC & C^2 & DC & EC & FC & GC & HC & IC & JC & KC\\D & AD & BD & CD & D^2 & ED & FD & GD & HD & ID & JD & KD\\E & AE & BE & CE & DE & E^2 & FE & GE & HE & IE & JE & KE\\F & AF & BF & CF & DF & EF & F^2 & GF & HF & IF & JF & KF\\G & AG & BG & CG & DG & EG & FG & G^2 & HG & IG & JG & KG\\H & AH & BH & CH & DH & EH & FH & GH & H^2 & IH & JH & KH\\I & AI & BI & CI & DI & EI & FI & GI & HI & I^2 & JI & KI\\J & AJ & BJ & CJ & DJ & EJ & FJ & GJ & HJ & IJ & J^2 & KJ\\K & AK & BK & CK & DK & EK & FK & GK & HK & IK & JK & K^2\end{bmatrix}\times\begin{bmatrix}K_2\\a_2\\b_2\\c_2\\d_2\\e_2\\f_2\\g_2\\h_2\\i_2\\j_2\\k_2\end{bmatrix}$$

$W_2 = 12 \times 12$ matrix of the multipliers of K_2, a_2, b_2, c_2, d_2, e_2, f_2, g_2, h_2, i_2, j_2, and k_2
$X_2 = 12 \times 1$ matrix of solutions of values of K_2, a_2, b_2, c_2, d_2, e_2, f_2, g_2, h_2, i_2, j_2, and k_2
$Z_2 = 12 \times 1$ matrix of the terms on LHS

In the above matrix, n is the number of readings A, B, C, D, E, F, G, H, I, J, and K represent the independent pi terms π_1, π_2, π_3, π_4, π_5, π_6, π_7, π_8, π_9, π_{10}, and π_{11} while, z_{2P} represents, dependent pi term, human energy consumed in liner piston maintenance activity. Next, calculate the values of independent pi terms for corresponding dependent pi terms which helps to form the equation in matrix form. It is recommended to use MATLAB software for this purpose for making this process of model formulation quickest and least cumbersome.

After substituting values we will get a set of 12 equations, which are to be solved simultaneously to get the values of K_2, a_2, b_2, c_2, d_2, e_2, f_2, g_2, h_2, i_2, j_2, and k_2. The equation 5.6.2.12 can be verified in the matrix form and further values of K_2, a_2, b_2, c_2, d_2, e_2, f_2, g_2, h_2, i_2, j_2, and k_2 can be obtained by using matrix analysis. The matrix analysis of solving these equations using MATLAB is given below.

$$\mathbf{z_2 = W_2 \times X_2}$$

After multiplying inverse of matrix W_2 with Z_2 we get X_2 matrix with result values of K_2 and indices a_2, b_2, c_2, d_2, e_2, f_2, g_2, h_2, i_2, j_2, k_2

$$X_2 = \mathrm{inv}(W2) \times z_{2P}$$

But K_2 is log value so convert into normal value take antilog of K_2, thus, antilog (0.5808) = 1.7874

$K_2 = 1.7874$	$c_2 = 0.1681$	$f_2 = -0.131$	$i_2 = -0.1479$
$a_2 = 0.355$	$d_2 = 4.4021$	$g_2 = -0.3313$	$j_2 = 0.0643$
$b_2 = 0.3448$	$e_2 = -0.0249$	$h_2 = -01018$	$k_2 = 0.2221$

Hence the model for dependent term π_{D2}:

$$\pi_{D2} = k_1 \times (\pi_1)^{a2} \times (\pi_2)^{b2} \times (\pi_3)^{c2} \times (\pi_4)^{d2} \times (\pi_5)^{e2} \times (\pi_6)^{f2} \times (\pi_7)^{g2} \times (\pi_8)^{h2} \times (\pi_9)^{i2} \times (\pi_{10})^{j2} \times (\pi_{11})^{k2}$$

$$z_{2P} = \pi_{D2} = 1.7874(\pi_1)^{0.355}(\pi_2)^{0.3448}(\pi_3)^{0.1681}(\pi_4)^{4.4021}(\pi_5)^{-0.0249}(\pi_6)^{-0.131}(\pi_7)^{-0.3313} (\pi_8)^{-0.1018}(\pi_9)^{-0.1479}(\pi_{10})^{0.0643}(\pi_{11})^{0.2221} \quad (5.34)$$

Human energy consumed in liner piston maintenance activity, dependent pi term:

$$\begin{aligned} z2P = 1.7874 \times \{ & [(a \times c \times e \times g)/(b \times d \times f \times h)]^{0.355} \\ & [(Ags/Exs),(sks/Ens),(hls/Hbs)]^{0.3448}, \\ & [(Dlins/Lcr) \times (l.lin/L.cr) \times (D.pis/L.cr) \times (L.pis/L.cr) \times (D.prng/Lc.r) \\ & \times (t.prng/L.crs) \times (l.pstrk/L.cr) \times (D.ppin/L.cr) \times (L.ppin/L.cr) \\ & \times (Dbg.br/L.cr) \times (t.bg.br/L.cr) \times (Dlt.brj/L.cr) \times (t.lt.brj/L.cr) \\ & \times (D.b.nj/L.cr) \times (L.b.nt/L.cr) \times (d.b.bl/L.cr) \times (l.b.bl/L.cr)]^{0.1681}, \\ & [(Dbr/Lbr),(Dsmsp/Lbr),(Lsmsp/Lbr),(Dsisp/Lbr),(Lsisp/Lbr), \\ & (Dbrr/Lbr),(Lbrr/Lbr)]^{4.4021}, \\ & [(ker/E\,belt),(orient\,11/L\,oil)]^{-0.0249} \\ & [(bcl/Bel) \times (Scl/Bel)]^{-0.131}, \\ & [(h.crt/l.crt),(b.crt/l.crt)]^{-0.3313}, \\ & [Temp/100]^{-0.1018}, \\ & [(Humd)]^{-0.1479}, \\ & [(illmus \times Ags)^3/Wt]^{0.0643}, \\ & [(Noise)]^{0.2221} \} \end{aligned} \quad (5.35)$$

5.2.3 Model Formulation for π_{D3}, Productivity of Liner Piston Maintenance Activity by Identifying the Curve Fitting Constant and Various Indices of Pi Terms

The multiple regression analysis helps to identify the indices of independent pi terms in the model aimed at, by considering eleven independent pi terms and one dependent pi term. Let the model be of the form for dependent pi term π_{D3} for productivity of liner piston productivity:

$$z_{3P} = \pi_{D3} = k_3 \times (\pi_1)^{a3} \times (\pi_2)^{b3} \times (\pi_3)^{c3} \times (\pi_4)^{d3} \times (\pi_5)^{e3} \times (\pi_6)^{f3} \times (\pi_7)^{g3} \times (\pi_8)^{h3} \times (\pi_9)^{i3} \times (\pi_{10})^{j3} \times (\pi_{11})^{k3} \quad (5.36)$$

To determine the values of K_3, a_3, b_3, c_3, d_3, e_3, f_3, g_3, h_3, i_3, j_3, and k_3 and to arrive at the regression hyperplane, the above equation are presented as follows:

$$\begin{aligned} \Sigma z_{3P} &= nK_3 + a_3 \Sigma A + b_3 \Sigma B + c_3 \Sigma C + d_3 \Sigma D + e_3 \Sigma E + f_3 \Sigma F + g_3 \Sigma G + h_3 \Sigma H \\ &\quad + i_3 \Sigma I + j_3 \Sigma J + k_3 \Sigma K \\ \Sigma z_{3P}A &= K_3 \Sigma A + a_3 \Sigma A^2 + b_3 \Sigma AB + c_3 \Sigma AC + d_3 \Sigma AD + e_3 \Sigma AE + f_3 \Sigma AF \\ &\quad + g_3 \Sigma AG + h_3 \Sigma AH + i_3 \Sigma AI + j_3 \Sigma AJ + k_3 \Sigma AK \end{aligned}$$

$$\begin{aligned}
\sum z_{3P}B &= K_3\sum B + a_3\sum AB + b_3\sum B^2 + c_3\sum BC + d_3\sum BD + e_3\sum BE + f_3\sum BF \\
&\quad + g_3\sum BG + h_3\sum BH + i_3\sum BI + j_3\sum BJ + k_3\sum BK \\
\sum z_{3P}C &= K_3\sum C + a_3\sum AC + b_3\sum CB + c_3\sum C^2 + d_3\sum CD + e_3\sum CE + f_3\sum CF \\
&\quad + g_3\sum CG + h_3\sum CH + i_3\sum CI + j_3\sum CJ + k_3\sum CK \\
\sum z_{3P}D &= K_3\sum D + a_3\sum AD + b_3\sum BD + c_3\sum CD + d_3\sum D^2 + e_3\sum DE + f_3\sum DF \\
&\quad + g_3\sum DG + h_3\sum DH + i_3\sum DI + j_3\sum DJ + k_3\sum DK \\
\sum z_{3P}E &= K_3\sum E + a_3\sum AE + b_3\sum BE + c_3\sum CE + d_3\sum DE + e_3\sum E^2 + f_3\sum FE \\
&\quad + g_3\sum GE + h_3\sum HE + i_3\sum IE + j_3\sum JE + k_3\sum EK \\
\sum z_{3P}F &= K_3\sum F + a_3\sum AF + b_3\sum FB + c_3\sum FC + d_3\sum FD + e_3\sum FE + f_3\sum F^2 \\
&\quad + g_3\sum FG + h_3\sum FH + i_3\sum FI + j_3\sum FJ + k_3\sum FK \\
\sum z_{3P}G &= K_3\sum G + a_3\sum AG + b_3\sum GB + c_3\sum GC + d_3\sum GD + e_3\sum GE + f_3\sum GF \\
&\quad + g_3\sum G^2 + h_3\sum GH + i_3\sum GI + j_3\sum GJ + k_3\sum GK \\
\sum z_{3P}H &= K_3\sum H + a_3\sum AH + b_3\sum HB + c_3\sum HC + d_3\sum HD + e_3\sum HE + f_3\sum HF \\
&\quad + g_3\sum HG + h_3\sum H^2 + i_3\sum HI + j_3\sum HJ + k_3\sum HK \\
\sum z_{3P}I &= K_3\sum I + a_3\sum AI + b_3\sum IB + c_3\sum IC + d_3\sum ID + e_3\sum IE + f_3\sum IF + g_3\sum IG \\
&\quad + h_3\sum IH + i_3\sum I^2 + j_3\sum IJ + k_3\sum IK \\
\sum z_{3P}J &= K_3\sum J + a_3\sum AJ + b_3\sum JB + c_3\sum JC + d_3\sum JD + e_3\sum JE + f_3\sum JF + g_3\sum JG \\
&\quad + h_3\sum JH + i_3\sum JI + j_3\sum J^2 + k_3\sum JK \\
\sum z_{3P}K &= K_3\sum K + a_3\sum AK + b_3\sum KB + c_3\sum KC + d_3\sum KD + e_3\sum KE + f_3\sum KF \\
&\quad + g_3\sum KG + h_3\sum KH + i_3\sum KI + j_3\sum KJ + k_3\sum K^2
\end{aligned} \tag{5.37}$$

In the above set of equations the values of the multipliers K_3, a_3, b_3, c_3, d_3, e_3, f_3, g_3 h_3, i_3, j_3, and k_3 are substituted to compute the values of the unknowns (viz. K_3, a_3, b_3, c_3, d_3, e_3, f_3, g_3 h_3, i_3, j_3, and k_3). The values of the terms on LHS and the multipliers of a_3, b_3, c_3, d_3, e_3, f_3, g_3 h_3, i_3, j_3, and k_3 in the set of equation are calculated, we will get a set of 12 equations, which are to be solved simultaneously to get the values of K_3, a_3, b_3, c_3, d_3, e_3, f_3, g_3 h_3, i_3, j_3, and k_3. The above equations can be verified in the matrix form and further values of K_3, a_3, b_3, c_3, d_3, e_3, f_3, g_3 h_3, i_3, j_3, and k_3 can be obtained by using matrix analysis.

The matrix obtained is given by:

$$\mathbf{z_{3P}} = \mathbf{W_3} \times \mathbf{X_3}$$

$$Z_{3P} \times \begin{bmatrix} 1 \\ A \\ B \\ C \\ D \\ E \\ F \\ G \\ H \\ I \\ J \\ K \end{bmatrix} = \begin{bmatrix}
n & A & B & C & D & E & F & G & H & I & J & K \\
A & A^2 & BA & CA & DA & EA & FA & GA & HA & IA & JA & KA \\
B & AB & B^2 & CB & DB & EB & FB & GB & HB & IB & JB & KB \\
C & AC & BC & C^2 & DC & EC & FC & GC & HC & IC & JC & KC \\
D & AD & BD & CD & D^2 & ED & FD & GD & HD & ID & JD & KD \\
E & AE & BE & CE & DE & E^2 & FE & GE & HE & IE & JE & KE \\
F & AF & BF & CF & DF & EF & F^2 & GF & HF & IF & JF & KF \\
G & AG & BG & CG & DG & EG & FG & G^2 & HG & IG & JG & KG \\
H & AH & BH & CH & DH & EH & FH & GH & H^2 & IH & JH & KH \\
I & AI & BI & CI & DI & EI & FI & GI & HI & I^2 & JI & KI \\
J & AJ & BJ & CJ & DJ & EJ & FJ & GJ & HJ & IJ & J^2 & KJ \\
K & AK & BK & CK & DK & EK & FK & GK & HK & IK & JK & K^2
\end{bmatrix} \times \begin{bmatrix} K_3 \\ a_3 \\ b_3 \\ c_3 \\ d_3 \\ e_3 \\ f_3 \\ g_3 \\ h_3 \\ i_3 \\ j_3 \\ k_3 \end{bmatrix}$$

W_3 = 12 × 12 matrix of the multipliers of K_3, a_3, b_3, c_3, d_3, e_3, f_3, g_3 h_3, i_3, j_3, and k_3
X_3 = 12 × 1 matrix of solutions of values of K_3, a_3, b_3, c_3, d_3, e_3, f_3, g_3 h_3, i_3, j_3, and k_3
Z_3 = 12 × 1 matrix of the terms on LHS

In the above matrix, n is the number of sets of values. A, B, C, D, E, F, G, H, I, J, and K represent the independent pi terms π_1, π_2, π_3, π_4, π_5, π_6, π_7, π_8, π_9, π_{10}, and π_{11} while z_{3P} represents dependent pi term. Next, calculate the values of independent pi terms for corresponding dependent pi terms which helps to form the equation in matrix form.

After substituting values, we will get a set of 12 equations, which are to be solved simultaneously to get the values of K_3, a_3, b_3, c_3, d_3, e_3, f_3, g_3 h_3, i_3, j_3, and k_3.

The matrix analysis of solving these equations using MATLAB is given below:

$$\mathbf{z_{3P} = W_3 \times X_3}$$

$$\begin{bmatrix} z_3 \\ \\ -31.33 \\ -0.735 \\ -13.67 \\ -324.6 \\ 10.571 \\ -58.21 \\ 26.774 \\ 20.99 \\ 15.15 \\ -52.54 \\ -168.2 \\ -59.45 \end{bmatrix} = \begin{bmatrix} & & & & & W_3 & & & & & & \\ \\ 30 & 0.7 & 13.03 & 310.85 & -10.121 & 55.80 & -25.62 & -20.10 & -14.54 & 50.28 & 160.97 & 56.62 \\ 0.7 & 0.38 & 0.27 & 7.25 & -0.2362 & 1.39 & -0.59 & -0.47 & -0.36 & 1.14 & 3.70 & 1.33 \\ 13.03 & 0.267 & 6.13 & 134.93 & -4.3941 & 24.34 & -11.12 & -8.73 & -6.31 & 21.86 & 70.06 & 24.7 \\ 310.79 & 7.253 & 134.93 & 3219.63 & -104.85 & 578.03 & -265.3 & -208.2 & -150.58 & 520.86 & 1667.5 & 598.6 \\ -10.12 & -0.253 & -4.39 & -104.85 & 3.414 & -18.82 & 8.64 & 6.78 & 4.9 & -16.96 & -54.3 & -19.20 \\ 55.80 & 1.390 & 24.34 & 578.029 & -18.823 & 104.39 & -47.62 & -37.38 & -27.1 & 93.51 & 299.3 & 105.8 \\ -25.62 & -.588 & -11.12 & -265.37 & 8.641 & -47.62 & 21.93 & 17.16 & 12.39 & -42.9 & -137.5 & -48.6 \\ -20.10 & 0.469 & -8.73 & -208.2 & 6.779 & -37.38 & 17.16 & 13.46 & 9.74 & -33.68 & -107.83 & -38.13 \\ -14.54 & -0.357 & -6.31 & -150.58 & 4.903 & -27.10 & 12.39 & 9.74 & 7.17 & -24.58 & -77.98 & -27.5 \\ 50.28 & 1.137 & 21.86 & 520.58 & -16.961 & -93.51 & -42.9 & -33.68 & -24.58 & 85.06 & 269.98 & 95.3 \\ 160.97 & 3.704 & 70.06 & 1667.55 & -54.303 & 299.34 & -137.54 & -107.8 & -77.98 & 269.98 & 864.6 & 305.3 \\ 56.92 & 1.326 & 24.70 & 589.65 & -19.202 & 105.86 & -48.6 & -38.13 & -27.58 & 95.39 & 305.3 & 108 \end{bmatrix} \times \begin{bmatrix} X_3 \\ \\ K_3 \\ a_3 \\ b_3 \\ c_3 \\ d_3 \\ e_3 \\ f_3 \\ g_3 \\ h_3 \\ i_3 \\ j_3 \\ k_3 \end{bmatrix}$$

After multiplying inverse of matrix W_3 with Z_3, we get X_3 matrix with result values of K_3 and indices a_3, b_3, c_3, d_3, e_3, f_3, g_3 h_3, i_3, j_3, k_3

$$\mathbf{X_3 = inv(W3) * z_{3P}}$$

But K_3 is log value so convert into normal value take antilog of K_3
Thus, antilog (0.1085) = 1.1148

$$K_3 = 1.1148$$
$$a_3 = 0.0973$$
$$b_3 = 0.1917$$
$$c_3 = 0.0188$$
$$d_3 = -2.5291$$
$$e_3 = -0.0896$$
$$f_3 = -0.378$$
$$g_3 = 2.9719$$
$$h_3 = 0.5615$$
$$i_3 = 0.2282$$
$$j_3 = -0.0887$$
$$k_3 = -0.0838$$

Hence the model for dependent term π_{D3}:

$$\pi_{D3} = k_3 \times (\pi_1)^{a3} \times (\pi_2)^{b3} \times (\pi_3)^{c3} \times (\pi_4)^{d3} \times (\pi_5)^{e3} \times (\pi_6)^{f3} \times (\pi_7)^{g3} \times (\pi_8)^{h3} \times (\pi_9)^{i3} \times (\pi_{10})^{j3} \times (\pi_{11})^{k3}$$

$$Z_{3C} = \pi_{D3} = 1.1148(\pi_1)^{0.0973}(\pi_2)^{0.1917}(\pi_3)^{0.0188}(\pi_4)^{-2.5291}(\pi_5)^{-0.0896}(\pi_6)^{-0.378}(\pi_7)^{2.9719} (\pi_8)^{0.5615}(\pi_9)^{0.2282}(\pi_{10})^{-0.0887}(\pi_{11})^{-0.0838} \quad (5.38)$$

Productivity of liner piston maintenance activity, dependent pi term:

$$\begin{aligned} Z_3 = 1.1148 \times \{ & [(a \times c \times e \times g)/(b \times d \times f \times h)]^{0.0973}, \\ & [(Ags/Exs),(sks/Ens),(hls/Hbs)]^{0.1917}, \\ & [(Dlins/Lcr) \times (l.lin/L.cr) \times (D.pis/L.cr) \times (L.pis/L.cr) \times (D.prng/Lc.r) \\ & \times (t.prng/L.crs) \times (l.pstrk/L.cr) \times (D.ppin/L.cr) \times (L.ppin/L.cr) \times (Dbg.br/L.cr) \\ & \times (t.bg.br/L.cr) \times (Dlt.brj/L.cr) \times (t.lt.brj/L.cr) \times (D.b.nj/L.cr) \times (L.b.nt/L.cr) \\ & \times (d.b.bl/L.cr) \times (l.b.bl/L.cr)]^{0.0188}, \\ & [(Dbr/Lbr),(Dsmsp/Lbr),(Lsmsp/Lbr),(Dsisp/Lbr),(Lsisp/Lbr), \\ & (Dbrr/Lbr),(Lbrr/Lbr)]^{2.5291}, \\ & [(ker/E\ belt),(orient11/L\ oil)]^{-0.0896}, \\ & [(bcl/Bel) \times (Scl/Bel)]^{-0.378}, \\ & [(h.crt/l.crt),(b.crt/l.crt)]^{2.9719}, \\ & [Temp/100]^{0.5615}, \\ & [(Humd)]^{0.2282}, \\ & [((illmus \times Ags)^3)/Wt]^{-0.0887}, \\ & [(Noise)]^{-0.0838} \ \} \end{aligned} \quad (5.39)$$

5.2.4 Models Developed for the Dependent Variables – Liner Piston Maintenance Activity

The exact forms of models obtained for the dependent variables overhauling time of piston maintenance activity (z_1), human energy consumed in piston maintenance activity (z_2) and productivity of liner piston maintenance activity (z_3) are as under:

$$Z_{1P} = \pi_{D1} = 1.0122(\pi_1)^{-0.0973}(\pi_2)^{-0.1917}(\pi_3)^{0.8772}(\pi_4)^{-6.0098}(\pi_5)^{0.0896}(\pi_6)^{0.378}(\pi_7)^{0.6869} (\pi_8)^{-0.5615}(\pi_9)^{-0.2282}(\pi_{10})^{0.0887}(\pi_{11})^{0.0838} \quad (5.40)$$

$$Z_{2P} = \pi_{D2} = 1.7874(\pi_1)^{0.355}(\pi_2)^{0.3448}(\pi_3)^{0.1681}(\pi_4)^{4.4021}(\pi_5)^{-0.0249}(\pi_6)^{-0.131}(\pi_7)^{-0.3313} (\pi_8)^{-0.1018}(\pi_9)^{-0.1479}(\pi_{10})^{0.0643}(\pi_{11})^{0.2221} \quad (5.41)$$

$$Z_{3P} = \pi_{D3} = 1.1148\ (\pi_1)^{0.0973}(\pi_2)^{0.1917}(\pi_3)^{0.0188}(\pi_4)^{-2.5291}(\pi_5)^{-0.0896}(\pi_6)^{-0.378}(\pi_7)^{2.971} (\pi_8)^{0.5615}(\pi_9)^{0.2282}(\pi_{10})^{-0.0887}(\pi_{11})^{-0.0838} \quad (5.42)$$

5.3 FORMULATION OF FIELD DATA-BASED MODEL FOR DIESEL BLENDING

The diesel engine run was conducted with a single cylinder diesel engine. The result obtained was fuelled with blends of treated transformer oil and diesel fuel in varying proportion such as 10:90, 20:80, 25:75, 30:70, 40:60. The run were covered under varying loads of 10 kg, 15 kg, 20 kg. The performance of engine was evaluated on the basis brake thermal efficiency (BTE), Brake-Specific fuel consumption (BSFC).

5.3.1 Model Formulation by Identifying the Curve Fitting Constant and Various Indices of π Terms

By taking into account four independent variables and one dependent π term, multiple regression analysis assists in identifying the indices of the various π term in the model targeted. Let the model be of the form:

$$(Z_1) = K_1 \times [(\pi_1)^{a1} \times (\pi_2)^{b1} \times (\pi_3)^{c1} \times (\pi_4)^{d1}]$$
$$(Z_2) = K_2 \times [(\pi_1)^{a2} \times (\pi_2)^{b2} \times (\pi_3)^{c2} \times (\pi_4)^{d2}]$$

To find the values of a_1, b_1, c_1, and d_1 equation is presented as follows:

$$\Sigma Z_1 = nK_1 + a_1 \times \Sigma A + b_1 \times \Sigma B + c_1 \times \Sigma C + d_1 \times \Sigma D$$
$$\Sigma Z_1 \times A = K_1 \times \Sigma A + a_1 \times \Sigma A \times A + b_1 \times \Sigma B \times A + c_1 \times \Sigma C \times A + d_1 \times \Sigma D \times A$$
$$\Sigma Z_1 \times B = K_1 \times \Sigma B + a_1 \times \Sigma A \times B + b_1 \times \Sigma B \times B + c_1 \times \Sigma C \times B + d_1 \times \Sigma D \times B$$
$$\Sigma Z_1 \times C = K_1 \times \Sigma C + a_1 \times \Sigma A \times C + b_1 \times \Sigma B \times C + c_1 \times \Sigma C \times C + d_1 \times \Sigma D \times C$$
$$\Sigma Z_1 \times D = K_1 \times \Sigma D + a_1 \times \Sigma A \times D + b_1 \times \Sigma B \times D + c_1 \times \Sigma C \times D + d_1 \times \Sigma D \times D$$

In the above set of equations, the values of K_1, a_1, b_1, $c_1$and d_1 are substituted to compute the values of the unknowns. After substituting these values in the equations, one will get a set of five equations, which are to be solved simultaneously to get the values of K_1, a_1, b_1, c_1, and d_1. The above equations can be transfer used in the matrix form and subsequently values of K_1, a_1, b_1, c_1, and d_1 can be obtained by adopting matrix analysis.

$$\mathbf{X_1 = inv(W) \times P_1}$$

$W = 5 \times 5$ matrix of the multipliers of K_1, a_1, b_1, c_1, and d1
$P_1 = 5 \times 1$ matrix on LHS and
$X_1 = 5 \times 1$ matrix of solutions

Then, the matrix evaluated is given by:

Matrix

$$Z_1 \times \begin{bmatrix} 1 \\ A \\ B \\ C \\ D \end{bmatrix} = \begin{bmatrix} n & A & B & C & D \\ A & A^2 & BA & CA & DA \\ B & AB & B^2 & CB & DB \\ C & AC & BC & C^2 & DC \\ D & AD & BD & CD & D^2 \end{bmatrix} \times \begin{bmatrix} K_1 \\ a_1 \\ b_1 \\ c_1 \\ d_1 \end{bmatrix}$$

In the above equations, n is the number of sets of readings, A, B, C and D represent the independent π terms π_1, π_2, π_3 and π_4 while, Z represents dependent π term, values of K_1, a_1, b_1, c_1, and d_1 obtained.

5.3.2 Model Formulation for Brake Thermal Efficiency (Z_1)

By taking into account four independent variables and one dependent term, multiple regression analysis assists in identifying the indices of the various terms in the model targeted at. Let the model aspire to be of the form,

$$(Z_1) = k_1 \times [(\Pi_1)^{a1} \times (\Pi_2)^{b1} \times (\Pi_3)^{c1} \times (\Pi_4)^{d1}]$$

The above equations are shown below to find the values of a_1, b_1, c_1, and d_1 and to arrive at the regression hyperplane:

$$\begin{aligned}
&\Sigma Z_1 = nK_1 + a_1 \times \Sigma A + b_1 \times \Sigma B + c_1 \times \Sigma C + d_1 \times \Sigma D\\
&\Sigma Z_1 \times A = K_1 \times \Sigma A + a_1 \times \Sigma A \times A + b_1 \times \Sigma B \times A + c_1 \times \Sigma C \times A + d_1 \times \Sigma D \times A\\
&\Sigma Z_1 \times B = K_1 \times \Sigma B + a_1 \times \Sigma A \times B + b_1 \times \Sigma B \times B + c_1 \times \Sigma C \times B + d_1 \times \Sigma D \times B\\
&\Sigma Z_1 \times C = K_1 \times \Sigma C + a_1 \times \Sigma A \times C + b_1 \times \Sigma B \times C + c_1 \times \Sigma C \times C + d_1 \times \Sigma D \times C\\
&\Sigma Z_1 \times D = K_1 \times \Sigma D + a_1 \times \Sigma A \times D + b_1 \times \Sigma B \times D + c_1 \times \Sigma C \times D + d_1 \times \Sigma D \times D
\end{aligned}$$

By inserting these values into the equations, one is left with a set of five equations that must be solved concurrently to obtain the values of K_1, a_1, b_1, c_1, and d_1. The aforementioned equations may be validated in matrix notation, and additional values of K_1, a_1, b_1, c_1, and d_1 can be derived using matrix analysis.

$$\mathbf{Z_1 = W_1 \times X_1}$$

Here,

W_1 = 5 x 5 matrix of the multipliers of K_1, a_1, b_1, c_1, and d_1
P_1 = 5 x 1 matrix of the terms on LHS and
X_1 = 5 x 1 matrix of solutions of values of K_1, a_1, b_1, c_1, and d_1

Then, the matrix obtained is given by:

Matrix

$$Z_1 \times \begin{bmatrix} 1 \\ A \\ B \\ C \\ D \end{bmatrix} = \begin{bmatrix} n & A & B & C & D \\ A & A^2 & BA & CA & DA \\ B & AB & B^2 & CB & DB \\ C & AC & BC & C^2 & DC \\ D & AD & BD & CD & D^2 \end{bmatrix} \times \begin{bmatrix} K_1 \\ a_1 \\ b_1 \\ c_1 \\ d_1 \end{bmatrix}$$

In the above equations, n is the number of sets of readings, A, B, C and D represent the independent π terms π_1, π_2, π_3 and π_4 while Z_1 matrix represents dependent π term.

Substituting the values of A, A^2, BA, CA… up to D^2 in the above matrix the value of constant K_1 and indices a_1, b_1, c_1, and d_1 are evaluated by taking the inverse of matrix W_1 and multiplying with Z_1 matrix as shown:

$$\begin{bmatrix} 25.00063 \\ 23.17162 \\ -2.5585 \\ 48.02126 \\ -47.0445 \end{bmatrix} = \begin{bmatrix} 15 & 13.86193 & -1.29908 & 28.80479 & -28.226 \\ 13.86193 & 14.0445 & -1.055 & 26.68585 & -26.084 \\ -1.29908 & -1.055 & 0.697372 & -2.49815 & 2.444522 \\ 28.80479 & 26.68585 & -2.49815 & 55.31891 & -54.2029 \\ -28.226 & -26.0842 & 2.444522 & -54.2029 & 53.11378 \end{bmatrix} \begin{bmatrix} K_1 \\ a_1 \\ b_1 \\ c_1 \\ d_1 \end{bmatrix}.$$

$$K_1 = 1.93, a_1 = 0.1758, b_1 = -0.7380, c_1 = -0.6950 \text{ and } d_1 = -0.4490$$

Taking antilog of K
Antilog (1.93) = 85.1138
The accurate form of model evaluated is as under:

$$(Z_1) = 85.1138 \times (\Pi_1)^{0.1758} \times (\Pi_2)^{-0.7380} \times (\Pi_3)^{-0.6950} \times (\Pi_4)^{-0.4490}$$

Brake thermal efficiency (Z_1)

$$(Bte) = 85.1138 \times (\Pi_1)^{0.1758} \times (\Pi_2)^{-0.7380} \times (\Pi_3)^{-0.6950} \times (\Pi_4)^{-0.4490}$$

5.3.3 Model Formulation for Brake-Specific Fuel Consumption (Z_2)

Similarly for brake-specific fuel consumption, n is the number of sets of readings, A, B, C, and D represent the independent π terms π_1, π_2, π_3, and π_4 while Z_2 matrix represents dependent pi term.

Substituting the values of A, A^2, BA, CA… up to D^2 in the above matrix the value of constant K_1 and indices a_1, b_1, c_1, and d1 are evaluated by taking the inverse of matrix W_1 and multiplying with Z_2 matrix as shown:

$$\begin{bmatrix} 13.34391 \\ 12.26366 \\ -0.76234 \\ 25.61249 \\ -25.1097 \end{bmatrix} = \begin{bmatrix} 15 & 13.86193 & -1.29908 & 28.80479 & -28.226 \\ 13.86193 & 14.0445 & -1.055 & 26.68585 & -26.084 \\ -1.29908 & -1.055 & 0.697372 & -2.49815 & 2.444522 \\ 28.80479 & 26.68585 & -2.49815 & 55.31891 & -54.2029 \\ -28.226 & -26.0842 & 2.444522 & -54.2029 & 53.11378 \end{bmatrix} \begin{bmatrix} K_1 \\ a_1 \\ b_1 \\ c_1 \\ d_1 \end{bmatrix}.$$

$$K_1 = -0.1338, a_1 = -0.1016, b_1 = 0.7070, c_1 = -0.6336 \text{ and } d_1 = -1.2727$$

Taking antilog of K
Antilog (−0.1338) = 0.7348
The exact form of model obtained is as under:

$$(Z_2) = 0.7348 \times (\Pi_1)^{-0.1016} \times (\Pi_2)^{0.7070} \times (\Pi_3)^{-0.6336} \times (\Pi_4)^{-1.2727}$$

Brake thermal efficiency (Z_2)

$$(Bsfc) = 0.7348 \times (\Pi_1)^{-0.1016} \times (\Pi_2)^{0.7070} \times (\Pi_3)^{-0.6336} \times (\Pi_4)^{-1.2727}$$

5.4 FORMULATION OF FIELD DATA-BASED MODEL FOR SOLAR UPDRAFT TOWER

5.4.1 Model Formulation for Turbine Speed Developed by Identifying the Constant and Various Indices of π Terms

To determine the values of a_1, b_1, c_1, d_1, e1, and f_1, and to arrive at the regression hyperplane, the above equations are presented as follows:

$$\Sigma Z_1 = nK_1 + a_1 \times \Sigma A + b_1 \times \Sigma B + c_1 \times \Sigma C + d_1 \times \Sigma D + e_1 \times \Sigma E + f_1 \times \Sigma F$$

$$\Sigma Z_1 \times A = K_1 \times \Sigma A + a_1 \times \Sigma A \times A + b_1 \times \Sigma B \times A + c_1 \times \Sigma C \times A + d_1 \times \Sigma D \times A + e_1 \times \Sigma E \times A + f_1 \times \Sigma F \times A$$

$$\Sigma Z_1 \times B = K_1 \times \Sigma B + a_1 \times \Sigma A \times B + b_1 \times \Sigma B \times B + c_1 \times \Sigma C \times B + d_1 \times \Sigma D \times B + e_1 \times \Sigma E \times B + f_1 \times \Sigma F \times B$$

$$\Sigma Z_1 \times C = K_1 \times \Sigma C + a_1 \times \Sigma A \times C + b_1 \times \Sigma B \times C + c_1 \times \Sigma C \times C + d_1 \times \Sigma D \times C + e_1 \times \Sigma E \times C + f_1 \times \Sigma F \times C$$

$$\Sigma Z_1 \times D = K_1 \times \Sigma D + a_1 \times \Sigma A \times D + b_1 \times \Sigma B \times D + c_1 \times \Sigma C \times D + d_1 \times \Sigma D \times D + e_1 \times \Sigma E \times D + f_1 \times \Sigma F \times D$$

$$\Sigma Z_1 \times E = K_1 \times \Sigma E + a_1 \times \Sigma A \times E + b_1 \times \Sigma B \times E + c_1 \times \Sigma C \times E + d_1 \times \Sigma D \times E + e_1 \times \Sigma E \times E + f_1 \times \Sigma F \times E$$

$$\Sigma Z_1 \times F = K_1 \times \Sigma F + a_1 \times \Sigma A \times F + b_1 \times \Sigma B \times F + c_1 \times \Sigma C \times F + d_1 \times \Sigma D \times F + e_1 \times \Sigma E \times F + f_1 \times \Sigma F \times F$$

In the above set of equations, the values of the multipliers K_1, a_1, b_1, c_1, d_1, e_1, and f_1 are substituted to compute the values of the unknowns (viz. K_1, a_1, b_1, c_1, d_1, e_1, and f_1). The values of the terms on LHS and the multipliers of K_1, a_1, b_1, c_1, d_1, e_1, and f_1 in the set of equations are calculated. After substituting these values in the equations, one will get a set of seven equations, which are to be solved simultaneously to get the values of K_1, a_1, b_1, c_1, d_1, e_1, and f_1.The above equations can be verified in the matrix form and further values of K_1, a_1, b_1, c_1, d_1, e_1, and f_1 can be obtained by using matrix analysis.

$$\mathbf{Z_1 = W_1 \times X_1}$$

Here:

$W = 7 \times 7$ matrix of the multipliers of K_1, a_1, b_1, c_1, d_1, e_1, and f_1 $P_1 = 7 \times 1$ matrix of the terms on LHS and

$X_1 = 7 \times 1$ matrix of solutions of values of K_1, a_1, b_1, c_1, d_1, e_1, and f_1 Then, the matrix obtained is given by:

Matrix

$$Z_1 \times \begin{bmatrix} 1 \\ A \\ B \\ C \\ D \\ E \\ F \end{bmatrix} = \begin{bmatrix} n & A & B & C & D & E & F \\ A & A^2 & BA & CA & DA & EA & FA \\ B & AB & B^2 & CB & DB & EB & FB \\ C & AC & BC & C^2 & BC & EC & FC \\ D & AD & BD & CD & D^2 & ED & FD \\ E & AE & BE & CE & DE & E^2 & FE \\ F & AF & BF & CF & DF & EF & F^2 \end{bmatrix} \times \begin{bmatrix} K_1 \\ a_1 \\ b_1 \\ c_1 \\ d_1 \\ e_1 \\ f_1 \end{bmatrix}$$

X_1 matrix with K_1 and indices a_1, b_1, c_1, d_1, e_1, and f_1 evaluated:

In the above equations, n is the number of sets of readings. A, B, C, D, E, and F represent the independent π terms such as π_1, π_2, π_3, π_4, π_5, and π_6 while Z represents dependent pi term.

Next, calculate the values of independent π term for corresponding dependent pi term, which helps to form the equation in matrix form. For making quickest and least cumbersome model formulation of this process, MATLAB software is recommended to use for this purpose.

$$\mathbf{P_1 = W_1 \times X_1}$$

$$\begin{matrix} 173.61 \\ -838.55 \\ -210.13 \\ -129.49 \\ 229.74 \\ 95.38 \\ 683.16 \end{matrix} = \begin{matrix} 80.00 & -386.55 & -96.84 & -59.69 & 105.77 & 43.95 & 315.08 \\ -386.55 & 1869.09 & 467.93 & 288.39 & -510.48 & -212.38 & -1525.15 \\ -96.84 & 467.93 & 117.54 & 72.25 & -128.00 & -53.21 & -381.42 \\ -59.69 & 288.39 & 72.25 & 44.62 & -78.80 & -32.79 & -235.07 \\ 105.77 & -510.48 & -128.00 & -78.80 & 140.37 & 58.11 & 415.43 \\ 43.95 & -212.38 & -53.21 & -32.79 & 58.11 & 24.15 & 173.11 \\ 315.08 & -1525.15 & -381.42 & -235.07 & 415.43 & 173.11 & 1246.44 \end{matrix} \times \begin{matrix} K \\ a_1 \\ b_1 \\ c_1 \\ d_1 \\ e_1 \\ f_1 \end{matrix}$$

$$[P_1]=[W_1][X_1]$$

Using Mat lab, $X_1= W_1\backslash P_1$, after solving X_1 matrix with K_1 and indices a_1, b_1, c_1, d_1, e_1, f_1 are as follows:

$$K_1=-0.2437, a_1=-0.9101, b_1=0.0424, c_1=0.0474, d_1=0.3036, e_1=-0.6156, f_1=-0.4977$$

But K_1 is log value so to convert it into normal value by taking antilog of K_1 Antilog (–0.2437) = 0.5705

Hence the model for dependent term π_{D1} i.e. turbine speed is:

$$\pi_{D1}=K_1\times(\pi_1)^{a1}\times(\pi_2)^{b1}\times(\pi_3)^{c1}\times(\pi_4)^{d1}\times(\pi_5)^{e1}\times(\pi_6)^{f1}$$

5.4.2 Model Formulation for Turbine Power Developed by Identifying the Constant and Various Indices of π Terms

Similar process is adopted for dependent pi terms, i.e. π_{D2} (power developed) and corresponding constant are determined.

$$P_2 = W_2 \times X_2$$

$$\begin{bmatrix}46.13\\-224.60\\-55.76\\-34.31\\60.56\\25.35\\185.02\end{bmatrix}=\begin{bmatrix}80.00 & -386.55 & -96.84 & -59.69 & 105.77 & 43.95 & 315.08\\-386.55 & 1869.09 & 467.93 & 288.39 & -510.48 & -212.38 & -1525.15\\-96.84 & 467.93 & 117.54 & 72.25 & -128.00 & -53.21 & -381.42\\-59.69 & 288.39 & 72.25 & 44.62 & -78.80 & -32.79 & -235.07\\105.77 & -510.48 & -128.00 & -78.80 & 140.37 & 58.11 & 415.43\\43.95 & -212.38 & -53.21 & -32.79 & 58.11 & 24.15 & 173.11\\315.08 & -1525.15 & -381.42 & -235.07 & 415.43 & 173.11 & 1246.44\end{bmatrix}\times\begin{bmatrix}K\\a_1\\b_1\\c_1\\d_1\\e_1\\f_1\end{bmatrix}$$

$$[P_2]=[W_2][X_2]$$

Using Mat lab, $X_2 = W_2\backslash P_2$, after solving X_2 matrix with K_2 and indices a_2, b_2, c_2, d_2, e_2, f_2 are as follows

$$K_2=-18.1439, a_2=-3.9024, b_2=0.1755, c_2=0.2861, d_2=0.8756, e_2=6.6344, f_2=-1.1456$$

But K_2 is log value so to convert into normal value taking antilog of K_2 antilog (–18.1439) = 7.1796×10^{-19}

Hence the model for dependent term π_{D2} i.e. power developed is:

$$\pi_{D2}=K_2\times(\pi_1)^{a2}\times(\pi_2)^{b2}\times(\pi_3)^{c2}\times(\pi_4)^{d2}\times(\pi_5)^{e2}\times(\pi_6)^{f2}$$

$$\pi_{D2}=\left(\frac{P_C}{K\times D_C}\right)$$

$$=7.1796\times10^{-19}\left[\begin{matrix}\left(\frac{H_{gc}\times H_{gc}\times\theta_c}{D_c^2}\right)^{-3.9024}\left(\frac{H_{ch}\times D_{ch}\times N_b}{D_c^2}\right)^{0.1755}(H_u)^{0.2861}\\ \left(\frac{(T_c\times T_c)(V_c\times V_c)\times D_c^2}{g^2}\right)^{0.8756}\left(\frac{\left(\sqrt{g}\times T_c\times A_{ci}\right)}{D_c^{5/2}}\right)^{6.6344}\left(\frac{D_c\times Q}{K}\right)^{-1.1456}\end{matrix}\right] \quad (5.43)$$

$$(P_d) = (K \times D_C)$$

$$= 7.1796 \times 10^{-19} \left[\frac{\left(\dfrac{H_{gc} \times T_{cc} \times \theta_c}{D_c^2}\right)^{-3.9024} \left(\dfrac{H_{ch} \times D_{ch} \times N_b}{D_c^2}\right)^{0.1755} (H_u)^{0.2861}}{\left(\dfrac{(T_a \times T_c)(V_i \times V_c) \times D_c^2}{g^2}\right)^{0.8756} \left(\dfrac{\left(\sqrt{g} \times T_h \times A_{ci}\right)}{D_c^{5/2}}\right)^{6.6344} \left(\dfrac{D_c \times Q}{K}\right)^{-1.1456}} \right]$$

6 Artificial Neural Network Simulation

6.1 INTRODUCTION

The field data-based modelling has been achieved based on field data for the three dependent pi terms for crankshaft maintenance activity and liner piston maintenance activity. In such complex phenomenon involving non-linear system where in the validation of field data-based models is not in close proximity, it becomes necessary to formulate ANN simulation of the observed data. An ANN is an information processing paradigm that is inspired by the way biological nervous systems, such as the brain, process information. The output of this network can be evaluated by comparing it with observed data and the calculated data of mathematical model. For development of ANN, the designer must recognize the inherent patterns. Once, this is accomplished training the network is mostly a fine-tuning process. An ANN simulation consists of three layers of nodes. The first layer is known as input layer. The number of neurons in the input layer is equal to the number of independent variables. Second layer is known as the hidden layer. It consists of 11 neurons. The third layer is the output layer. It contains one neuron as one of dependent variables at a time. Multilayer feed forward topology is decided for the network.

6.2 PROCEDURE FOR FORMULATION OF ANN SIMULATION

MATLAB software is selected for developing ANN simulation. The various steps followed in developing the algorithm to form ANN are as follows.

The collected data from the field is separated into two parts viz. input data or the data of independent pi terms and the output data or the data of dependent pi terms. The input data and output data are imported to the program respectively.

1. The collected data from the field is separated into two parts: input data (independent pi terms) and output data (dependent pi terms). The input and output data are imported to the program.
2. The input and output data are read by preset functions and appropriately sized.
3. In the pre-processing step, the input and output data are normalized using mean and standard deviation.
4. The input and output data is then categorized in three categories: testing, validation, and training. From 30 observations, an initial 75% of them are selected for training, and 75% data for validation and middle overlapping 50% data for testing.
5. The data is then stored in structures for training, testing, and validation.
6. Looking at the pattern of the data, a feed-forward back propagation neural network is chosen.
7. This network is then trained using the training data. The computation errors in the actual and target data are computed and then the network is simulated.
8. The regression analysis and the representation are done through the standard functions. The values of regression coefficient and the equation of regression lines are represented on the 11 different graphs plotted for the 11 dependent pi terms, viz. π_{D1}, π_{D2}, π_{D3}, for both the operations maintenance activity for crankshaft and liner piston.

DOI: 10.1201/9781003318699-6

6.3 ANN PROGRAM FOR CRANKSHAFT MAINTENANCE ACTIVITY

MATLAB software is selected for developing ANN for dependent pi terms z_1, z_2, and z_3 of crankshaft maintenance activity. The programs executed for dependent pi terms z_1, z_2, and z_3 of crankshaft maintenance activity are as follows.

6.3.1 ANN Program for Overhauling Time of Crankshaft Maintenance Activity (z_{1C})

```
inputs3 = [
1.1432 2.8641 1898.91 54.739 133.333 0.1296 4.2872 0.3120 31 211692 62.75
1.1312 2.6224 1898.91 54.739 235.294 0.1346 4.2872 0.2600 31 205207 72.00
1.3626 3.2008 1898.91 54.739 126.316 0.1358 4.2872 0.3280 27 206213 70.00
0.5313 3.5412 1898.91 54.739 173.913 0.1235 4.2872 0.3400 27 284469 85.00
1.2537 3.3497 1898.91 54.739  72.917 0.5617 4.2872 0.3600 30 265978 80.00
1.2433 2.9728 1898.91 54.739  62.500 0.1346 4.2872 0.3810 30 217171 84.00
1.3533 2.5071 1898.91 54.739 158.730 0.1512 4.2872 0.4300 24 218019 79.00
1.0600 2.8143 1898.91 54.739 168.421 0.1296 4.2872 0.3530 24 326646 73.00
1.1228 2.6171 1898.91 54.739  97.222 0.1235 4.2872 0.3930 27 307085 76.00
1.1709 3.2954 1898.91 54.739 111.111 0.1674 4.2872 0.3980 27 242484 77.00
0.5586 3.5854 1898.91 54.739  81.522 0.1296 4.2872 0.4120 55 232822 74.00
1.0607 2.5363 1898.91 54.739 187.500 0.1296 4.2872 0.4550 55 316768 72.40
1.2704 3.0764 1898.91 54.739  95.238 0.1516 4.2872 0.3000 77 256797 75.00
0.9949 2.5547 1898.91 54.739 134.615 0.1296 4.2872 0.2970 77 195158 78.00
0.9451 3.1174 1898.91 54.739 152.381 0.1368 4.2872 0.3120 80 232616 83.00
1.0917 2.9659 1898.91 54.739 139.130 0.1607 4.2872 0.2750 80 249653 82.00
1.1757 3.8440 1898.91 54.739 133.929 0.1296 4.2872 0.3360 74 257049 83.00
0.9396 2.8082 1898.91 54.739 121.739 0.2016 4.2872 0.3080 74 249714 81.00
0.8789 3.2340 1898.91 54.739 133.929 0.1666 4.2872 0.3060 61 353501 80.00
0.8891 3.5544 1898.91 54.739 114.286 0.1584 4.2872 0.2940 61 284282 82.00
1.3606 2.0908 1898.91 54.739  92.593 0.1296 4.2872 0.3100 55 285949 81.00
1.6588 3.2228 1898.91 54.739 142.857 0.1420 4.2872 0.2970 55 211200 70.00
1.2315 2.3053 1898.91 54.739 187.500 0.1358 4.2872 0.2920 56 226664 78.00
0.8703 2.5929 1898.91 54.739  83.333 0.1296 4.2872 0.2870 56 202145 80.00
0.8537 2.9993 1898.91 54.739 145.833 0.1825 4.2872 0.2860 53 313577 70.00
1.6588 2.5675 1898.91 54.739 138.889 0.1512 4.2872 0.2840 53 215631 75.00
1.2315 2.2235 1898.91 54.739 133.333 0.1674 4.2872 0.3320 42 243432 72.00
0.8782 2.6099 1898.91 54.739 133.333 0.1296 4.2872 0.3120 42 236606 80.00
0.8584 3.0012 1898.91 54.739  83.333 0.1666 4.2872 0.2750 29 226571 82.00
0.8585 2.9701 1898.91 54.739 166.667 0.1420 4.2872 0.2940 29 297496 83.00
]
a1 = inputs3
a2 = a1
input_data = a2;
output3 = [
```

```
  1.8185E+11
  1.8766E+11
  2.0639E+11
  1.4866E+11
  2.4824E+11
  2.4475E+11
  1.6054E+11
  1.6739E+11
  1.6532E+11
  1.5240E+11
  1.6002E+11
  1.1559E+11
  1.5499E+11
  1.6945E+11
  1.6106E+11
  1.5021E+11
  1.8172E+11
  1.7785E+11
  1.7113E+11
  1.8598E+11
  1.8624E+11
  1.5822E+11
  1.5563E+11
  1.8728E+11
  1.8986E+11
  1.7591E+11
  1.7927E+11
  1.8366E+11
  1.6855E+11
  1.8392E+11
]
y1 = output3
y2 = y1
size(a2);
size(y2);
p = a2';
sizep = size(p);
t = y2';
sizet = size(t);
[S Q] = size(t)
[pn,meanp,stdp,tn,meant,stdt] = prestd(p,t);
net=newff(minmax(pn),[30 1],{'logsig' 'purelin'},'trainlm');
net.performFcn = 'mse';
```

```
net.trainParam.goal = .99;
net.trainParam.show = 200;
net.trainParam.epochs = 50;
net.trainParam.mc = 0.05;
net = train(net,pn,tn);
an = sim(net,pn);
[a] = poststd(an,meant,stdt);
Error = t-a;
x1 = 1:30;
plot(x1,t,'rs-',x1,a,'b-')
legend('Experimental','Neural');
title('Output (Red) and Neural Network Prediction (Blue) Plot');
xlabel('Experiment No.');
ylabel('Output');
grid on;
figure
error_percentage = 100×error./t
plot(x1,error_percentage)
legend('percentage error');
axis([0 30 -100 100]);
title('Percentage Error Plot in Neural Network Prediction');
xlabel('Experiment No.');
ylabel('Error in %');
grid on;
for ii = 1:30
xx1 = input_data(ii,1);
yy2 = input_data(ii,2);
zz3 = input_data(ii,3);
xx4 = input_data(ii,4);
yy5 = input_data(ii,5);
zz6 = input_data(ii,6);
xx7 = input_data(ii,7);
yy8 = input_data(ii,8);
zz9 = input_data(ii,9);
xx10 = input_data(ii,10);
yy11 = input_data(ii,11);
pause
yyy(1,ii) = 1.7198×power(xx1,0.0873)×power(yy2,0.1026)
         ×power(zz3,2.8137)×power(xx4,0.2115)
         ×power(yy5,-0.2101)×power(zz6,-0.0025)
         ×power(xx7,3.5409)×power(yy8,-0.4451)
         ×power(zz9,-0.125)×power(xx10,-0.0609)
         ×power(yy11,0.1131);
yy_practical(ii) = (y2(ii,1));
yy_eqn(ii) = (yyy(1,ii))
yy_neur(ii) = (a(1,ii))
yy_practical_abs(ii) = (y2(ii,1));
yy_eqn_abs(ii) = (yyy(1,ii));
yy_neur_abs(ii) = (a(1,ii));
pause
end
```

```
figure;
plot(x1,yy_practical_abs,'r-',x1,yy_eqn_abs,'b-',x1,yy_neur_
abs,'k-');
legend('Practical','Equation','Neural');
title('Comparision between practical data, equation-based data and
neural based data');
xlabel('Experimental');
grid on;
figure;
plot(x1,yy_practical_abs,'r-',x1,yy_eqn_abs,'b-');
legend('Practical',' Equation');
title('Comparision between practical data, equation-based data and
neural based data');
xlabel('Experimental');
grid on;
figure;
plot(x1,yy_practical_abs,'r-',x1,yy_neur_abs,'k-');
legend('Practical','Neural');
title('Comparision between practical data, equation-based data and
neural based data');
xlabel('Experimental');
grid on;
error1 = yy_practical_abs-yy_eqn_abs
figure;
error_percentage1 = 100×error1./yy_practical_abs;
plot(x1,error_percentage,'k-',x1,error_percentage1,'b-');
legend('Neural','Equation');
axis([0 30 -100 100]);
title('Percentage Error Plot in Equation (blue), Neural Network
(black) Prediction');
xlabel('Experiment No.');
ylabel('Error in %');
grid on;
meanexp = mean(output3)
meanann = mean(a)
meanmath = mean(yy_eqn_abs)
mean_absolute_error_performance_function = mae(error)
mean_squared_error_performance_function = mse(error)
net = newff(minmax(pn),[30 1],{'logsig' 'purelin'},'trainlm','lear
ngdm','msereg');
an = sim(net,pn);
[a] = poststd(an,meant,stdt);
error = t(1,[1:30])-a(1,[1:30]);
net.performParam.ratio = 20/(20+1);
perf = msereg(error,net)
 rand('seed',1.818490882E9)
[ps] = minmax(p);
[ts] = minmax(t);
numInputs = size(p,1);
numHiddenNeurons = 30;
numOutputs = size(t,1);
```

```
net = newff(minmax(p), [numHiddenNeurons,numOutputs]);
[pn,meanp,stdp,tn,meant,stdt] = prestd(p,t);
[ptrans,transmit] = prepca(pn,0.001);
[R Q] = size(ptrans);
testSamples = 15:1:Q;
validateSamples = 20:1:Q;
trainSamples = 1:1:Q ;
validation.P = ptrans(:,validateSamples) ;
validation.T = tn(:,validateSamples) ;
testing.P = ptrans(:,testSamples) ;
testing.T = tn(:,testSamples)
ptr = ptrans(:,trainSamples) ;
ttr = tn(:,trainSamples);
net = newff(minmax(ptr),[30 1],{'logsig' 'purelin'},'trainlm');
[net,tr] = train(net,ptr,ttr,[] ,[],validation,testing);
plot(tr.epoch,tr.perf,'r',tr.epoch,tr.vperf,'g',tr.epoch,tr.tperf,
'h') ;
legend('Training', 'validation', 'Testing',-1) ;
ylabel('Error') ;
an = sim(net,ptrans);
a = poststd(an,meant,stdt);
pause;
figure;
[m,b,r] = postreg(a,t);
```

The various output and graphs obtained by ANN program for overhauling time of crankshaft maintenance activity are shown as below (Figures 6.1–6.7).

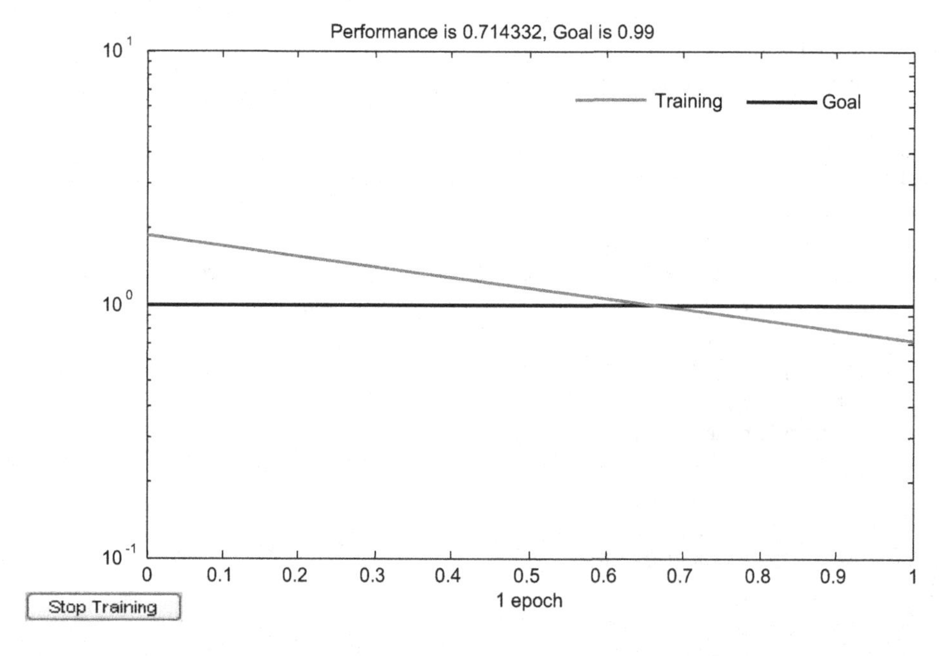

FIGURE 6.1 Training of the network for overhauling time of crankshaft maintenance activity (z_{1C}).

Training of the network for overhauling time of crankshaft maintenance activity.

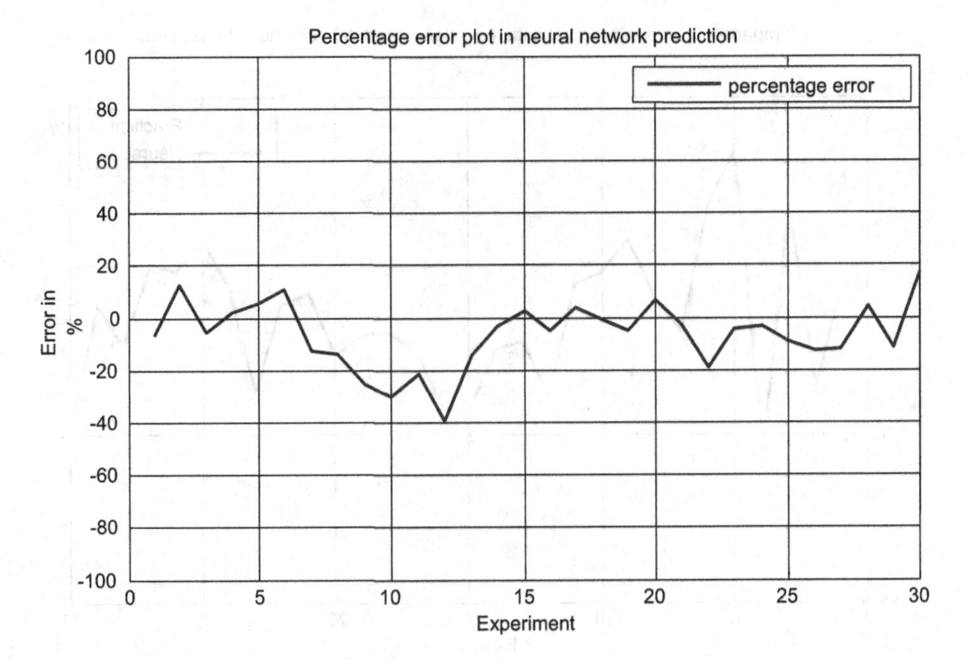

FIGURE 6.2 Percentage error plot prediction for the network for overhauling time of crankshaft maintenance activity (z_{1C}).

Percentage error plot prediction for the network for overhauling time of crankshaft maintenance activity.

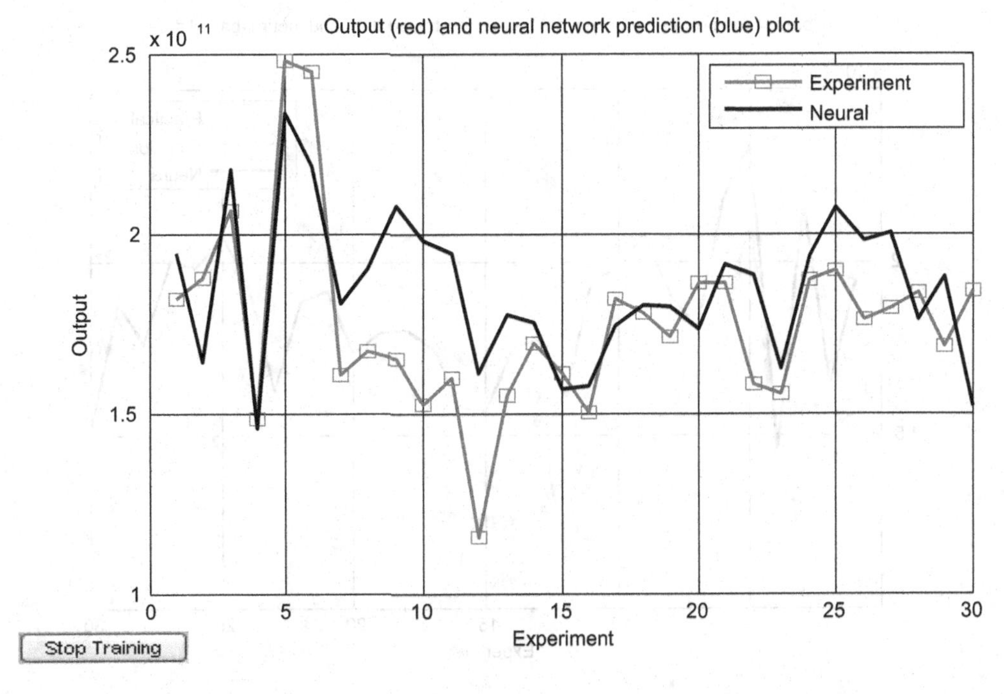

FIGURE 6.3 Graph of comparison with experimental/field database and neural prediction for the network for overhauling time of crankshaft maintenance activity (z_{1C}).

Comparison with experimental/field database and neural prediction for the network for overhauling time of crankshaft maintenance activity.

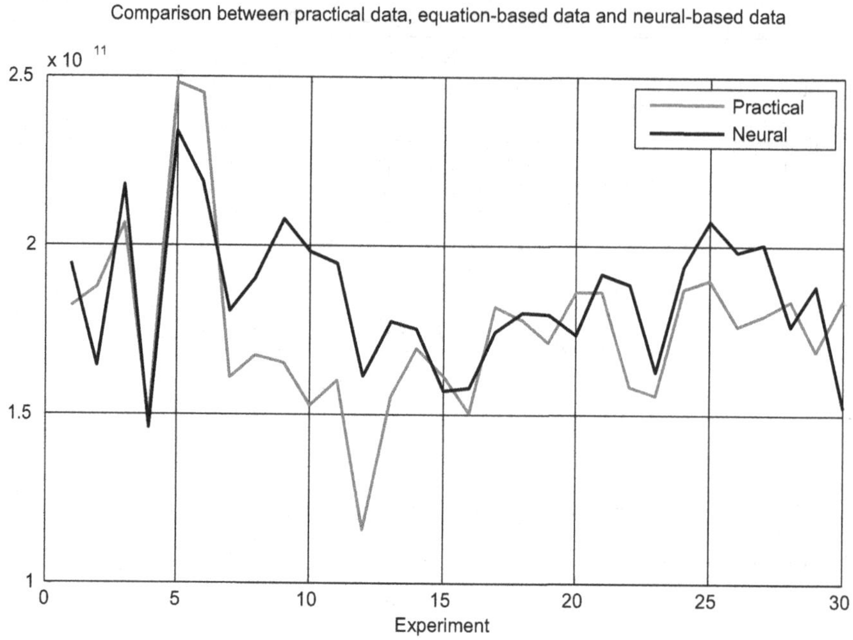

FIGURE 6.4 Graph of comparison with practical/field database and neural prediction for the network for overhauling time of crankshaft maintenance activity (z_{1C}).

Comparison of experimental data and neural data for the dependent variable Time.

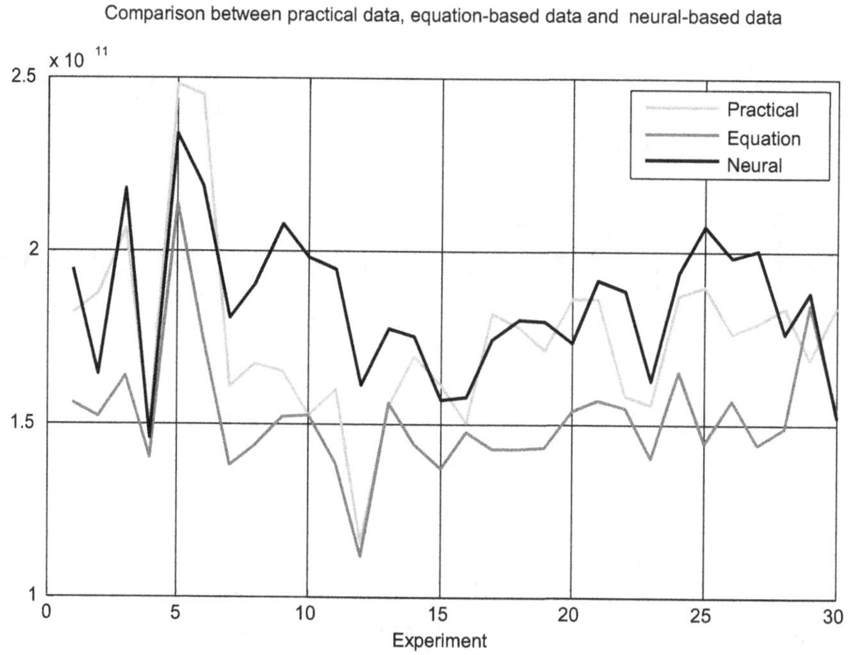

FIGURE 6.5 Graph of comparison with practical/field database, neural network prediction and equation-based prediction for the network for overhauling time of crankshaft maintenance activity (z_{1C}).

Comparison graph of output obtained from mathematical model, neural network and recorded experimental reading for the dependent variable Time.

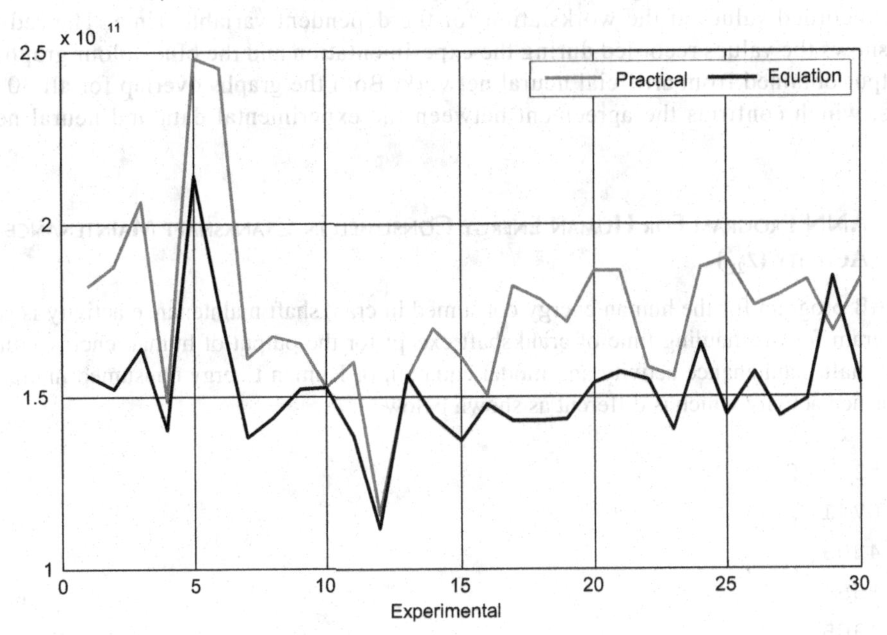

FIGURE 6.6 Graph of comparison with practical/field database and mathematical equation-based for the network for overhauling time of crankshaft maintenance activity (z_{1C}).

Comparison of experimental data and output obtained from mathematical model for the dependent variable Time.

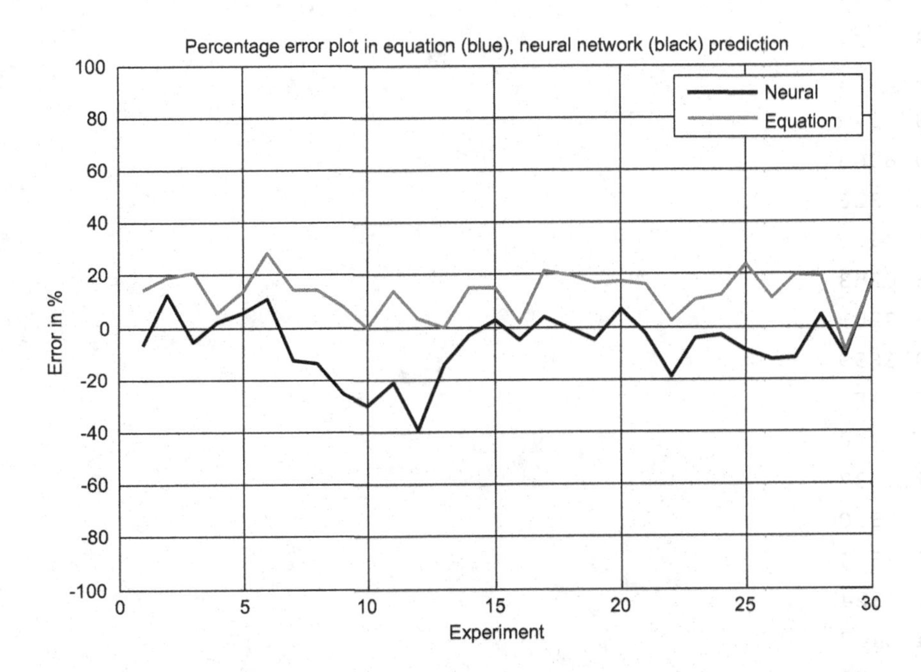

FIGURE 6.7 Graph of comparison of percentage error with mathematical equation-based and neural network prediction for the network for overhauling time of crankshaft maintenance activity (z_{1C}).

Comparison of percentage error plot prediction for the network for overhauling time of crankshaft maintenance activity.

A graph showing the comparison of the output obtained from neural network and experimental recorded values at the workstation for the dependent variable Time. The red colour graph shows the values recorded during the experimentation and the blue colour graph shows the output obtained from artificial neural network. Both the graphs overlap for all 30 observations, which confirms the agreement between the experimental data and neural network data.

6.3.2 ANN Program For Human Energy Consumed in Crankshaft Maintenance Activity (Z_{2C})

MATLAB program for the human energy consumed in crankshaft maintenance activity is same as the program for overhauling time of crankshaft except for the output of human energy consumed in crankshaft maintenance activity and model equation of Human Energy consumed in crankshaft maintenance activity which is different as shown below:

```
output3 = [
  55.0741
  45.4203
  42.9550
  43.4965
  39.1228
  39.3353
  43.1065
  35.3253
  38.3224
  33.5066
  28.7072
  50.6565
  45.7326
  31.0430
  26.0253
  46.7209
  40.3699
  45.4182
  53.2160
  34.8037
  53.0510
  58.3544
  38.1324
  34.2447
  29.6070
  33.5099
```

```
    39.1597
    34.5009
    36.3526
    31.2646
]
```

Equation of Human energy consumed in crankshaft maintenance activity:

```
yyy(1,ii) = 1.2203xpower(xx1,0.2769)xpower(yy2,0.018)
            xpower(zz3,-0.2583)xpower(xx4,3.3045)
            xpower(yy5,0.0597)xpower(zz6,-0.0314)
            xpower(xx7,-1.2341xpower(yy8,0.0415)
            xpower(zz9,0.0103)xpower(xx10,0.0002)
            xpower(yy11,-0.3935);
```

The above output and equation of human energy consumed in crankshaft maintenance activity is placed instead of output and equation in MATLAB program of overhauling time of crankshaft maintenance activity.

The various output and graphs obtained by ANN program for human energy consumed in crankshaft maintenance activity are shown as below (Figures 6.8–6.14).

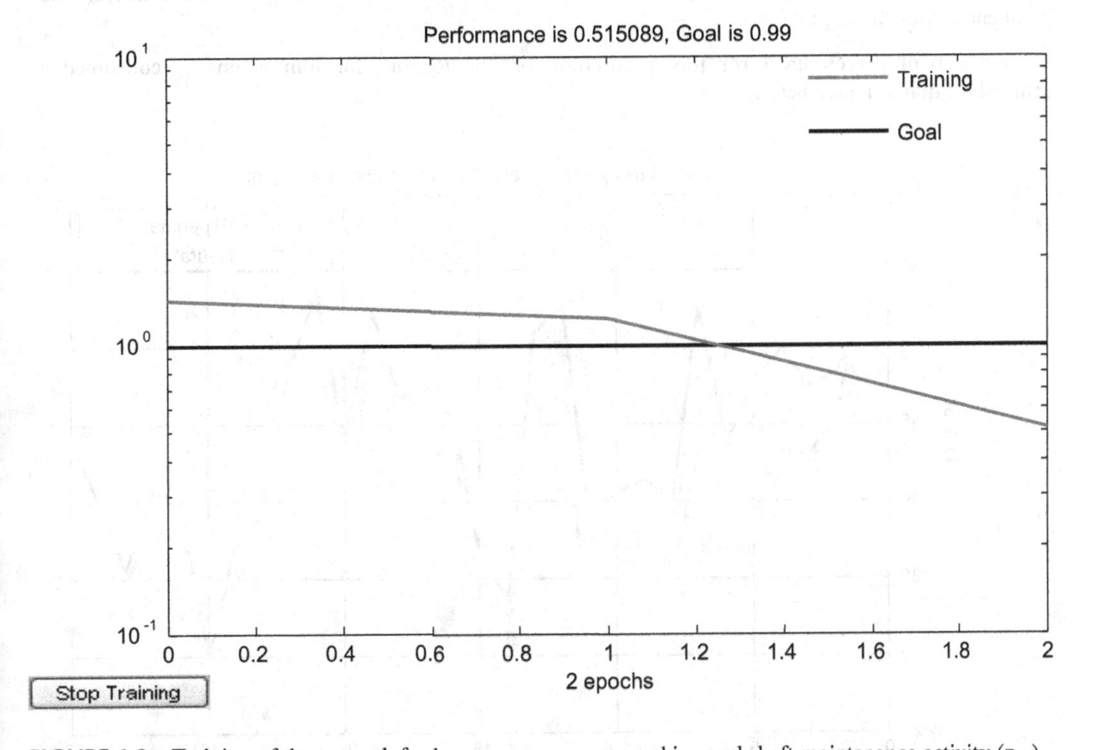

FIGURE 6.8 Training of the network for human energy consumed in crankshaft maintenance activity (z_{2C}).

Training of the network for human energy consumed in crankshaft maintenance of crankshaft maintenance activity.

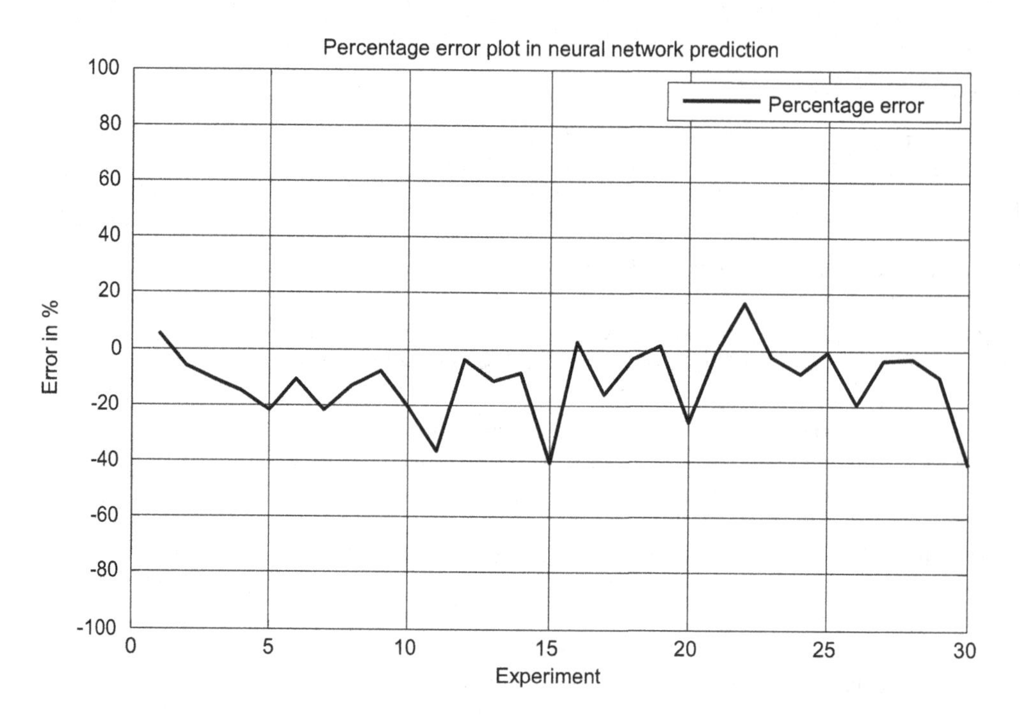

FIGURE 6.9 Percentage error plot prediction for the network for human energy consumed in crankshaft maintenance activity (z_{2C}).

Comparison of percentage error plot prediction for the network for human energy consumed by crankshaft maintenance activity.

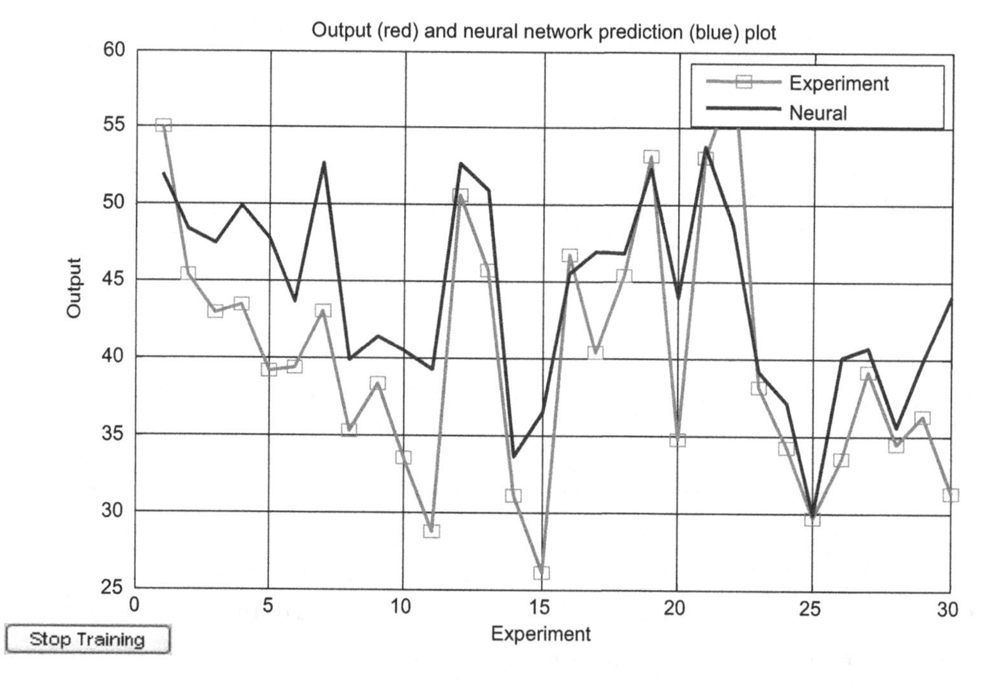

FIGURE 6.10 Graph of comparison with experimental/field database and neural prediction for the network for human energy consumed in crankshaft maintenance activity (z_{2C}).

Comparison of experimental data and neural data for the dependent variable for human energy consumed.

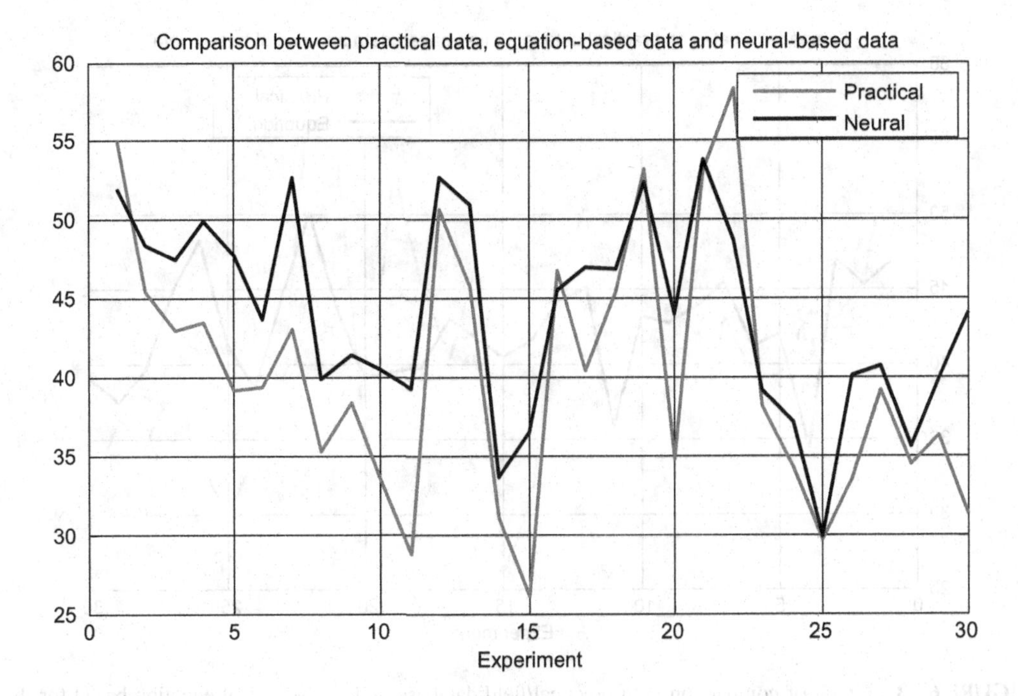

FIGURE 6.11 Graph of comparison with practical/field database and neural prediction for the network for human energy consumed in crankshaft maintenance activity (z_{2C}).

Comparison of experimental data and neural data for the dependent variable for human energy consumed.

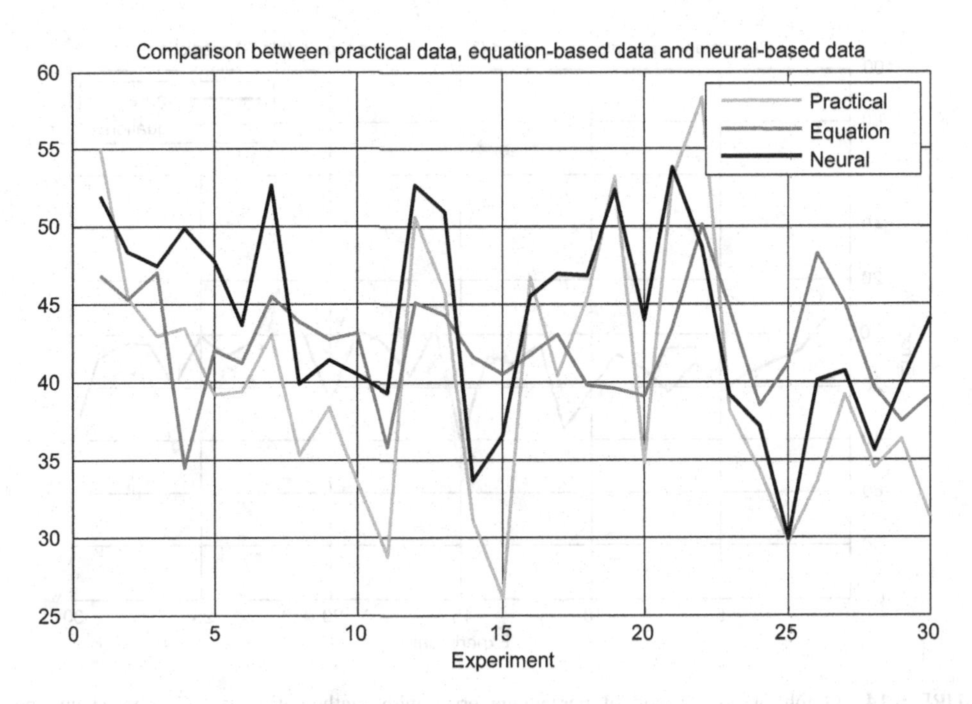

FIGURE 6.12 Graph of comparison with practical/field database, neural network prediction and equation-based prediction for the network for human energy consumed in crankshaft maintenance activity (z_{2C}).

Comparison graph of output obtained from a mathematical model, neural network and recorded experimental readings for the dependent variable for human energy consumed.

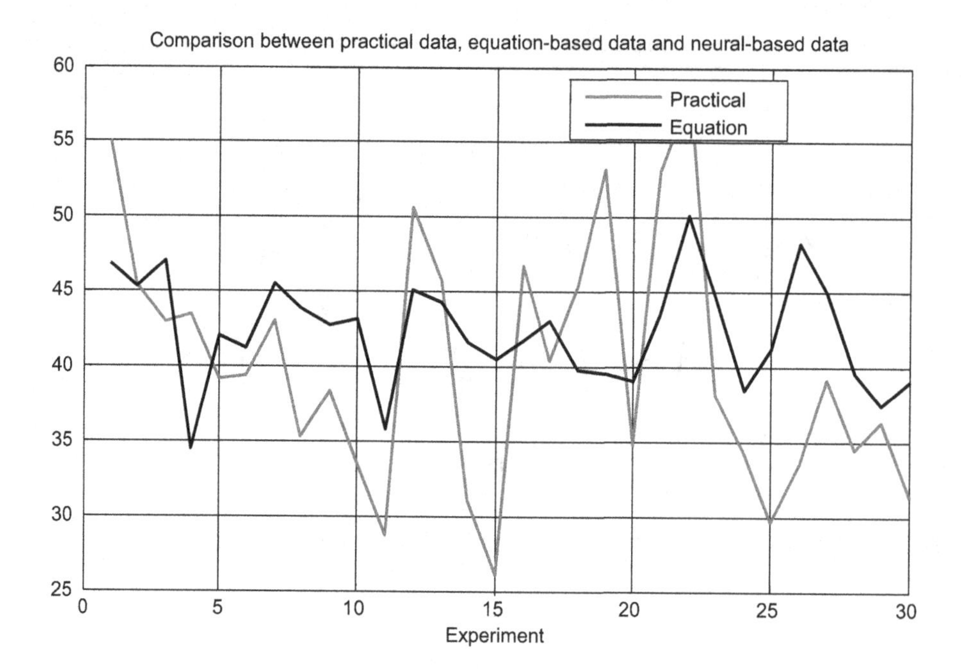

FIGURE 6.13 Graph of comparison with practical/field database and mathematical equation-based for the network for human energy consumed in crankshaft maintenance activity (z_{2C}).

Comparison of experimental data and output obtained from a mathematical model for the dependent variable Time.

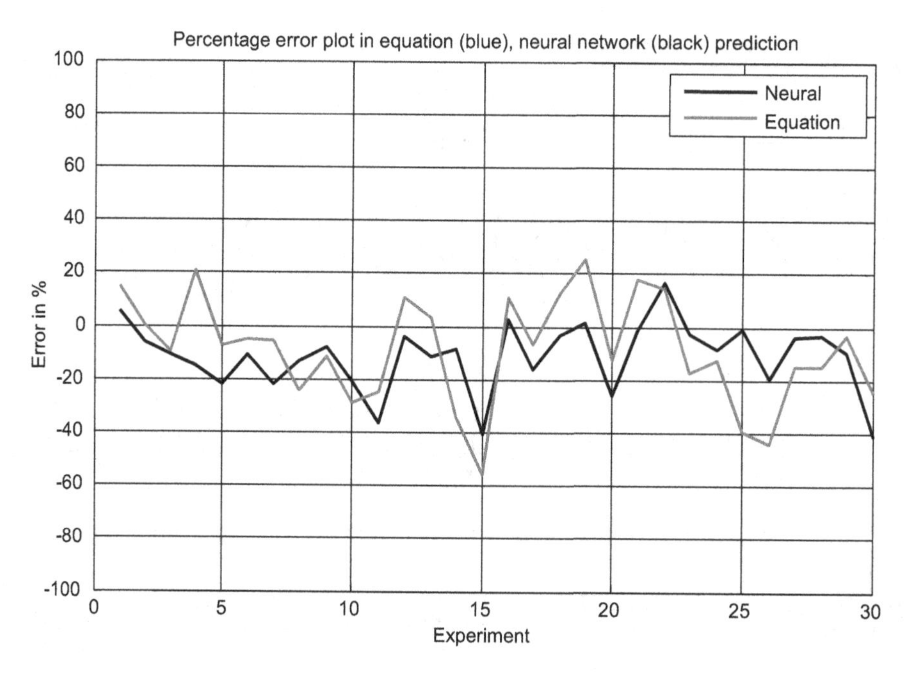

FIGURE 6.14 Graph of comparison of percentage error with mathematical equation-based and neural network prediction for the network for human energy consumed in crankshaft maintenance activity (z_{2C}).

Comparison of percentage error plot prediction for the network for human energy consumed by crankshaft maintenance activity.

6.3.3 ANN Program for Productivity of Crankshaft Maintenance Activity (Z_{3C})

MATLAB program for the productivity of crankshaft maintenance activity is the same as the program for overhauling time of crankshaft, except for the output of productivity of crankshaft maintenance activity, and the model equation of productivity of crankshaft maintenance activity, which is different, as shown below:

```
output3 = [
            0.065185
            0.067269
            0.073981
            0.053287
            0.088981
            0.087731
            0.057546
            0.060000
            0.059259
            0.054630
            0.057361
            0.041435
            0.055556
            0.060741
            0.057731
            0.053843
            0.065139
            0.063750
            0.061343
            0.066667
            0.066759
            0.056713
            0.055787
            0.067130
            0.068056
            0.063056
            0.064259
            0.065833
            0.060417
            0.065926
]
```

Equation for the productivity of crankshaft maintenance activity:

```
yyy(1, ii) = 1.6837xpower(xx1,-0.0873)xpower(yy2,-0.1026)
             xpower(zz3,-0.1837)xpower(xx4,-1.2275)
             xpower(yy5,0.2101)xpower(zz6,0.0025)
             xpower(xx7,0.9916)xpower(yy8,0.4451);
```

The above output and equation of productivity for crankshaft maintenance activity is used instead of that for the MATLAB program of overhauling time of crankshaft maintenance activity.

The various output and graphs obtained by ANN program for productivity of crankshaft maintenance activity are shown in Figures 6.15–6.21).

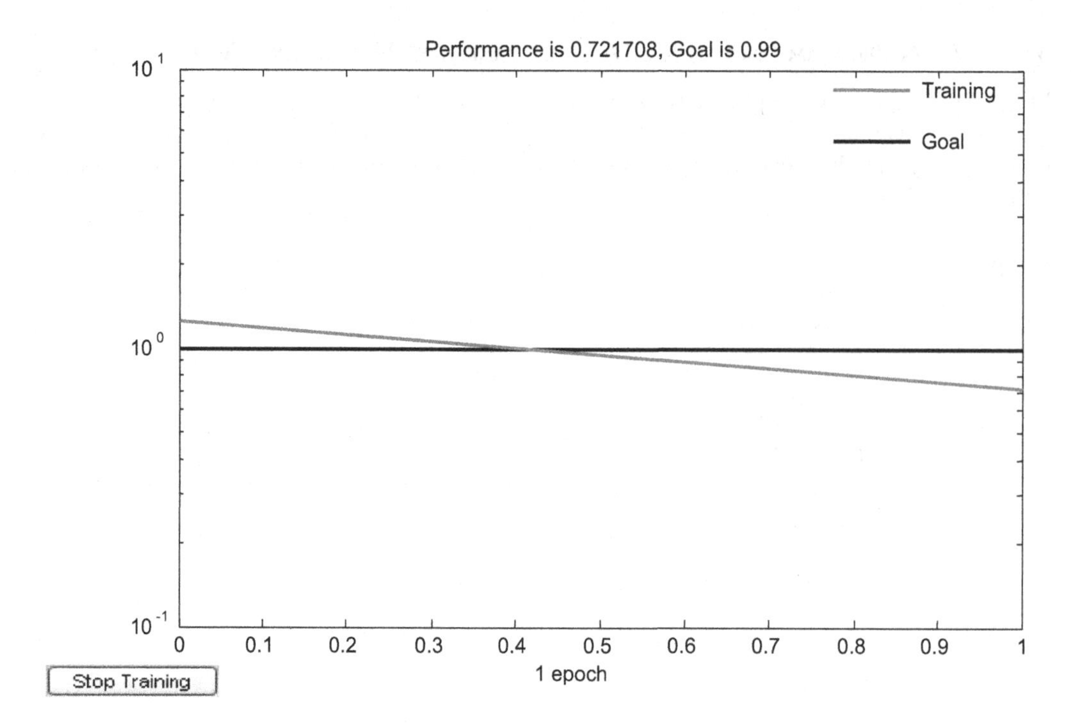

FIGURE 6.15 Training of the net for productivity of crankshaft maintenance activity (z_{3C}).

Training of the network for productivity of crankshaft maintenance activity.

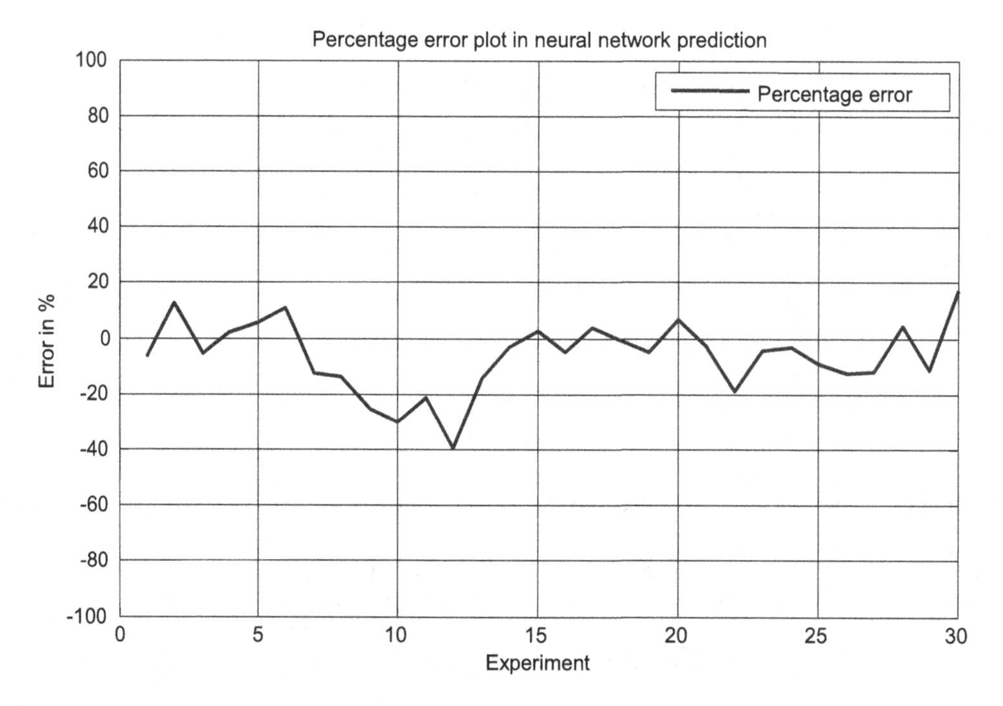

FIGURE 6.16 Percentage error plot prediction for the network for productivity of crankshaft maintenance activity (z_{3C}).

Comparison of percentage error plot prediction for the network for productivity of crankshaft maintenance activity.

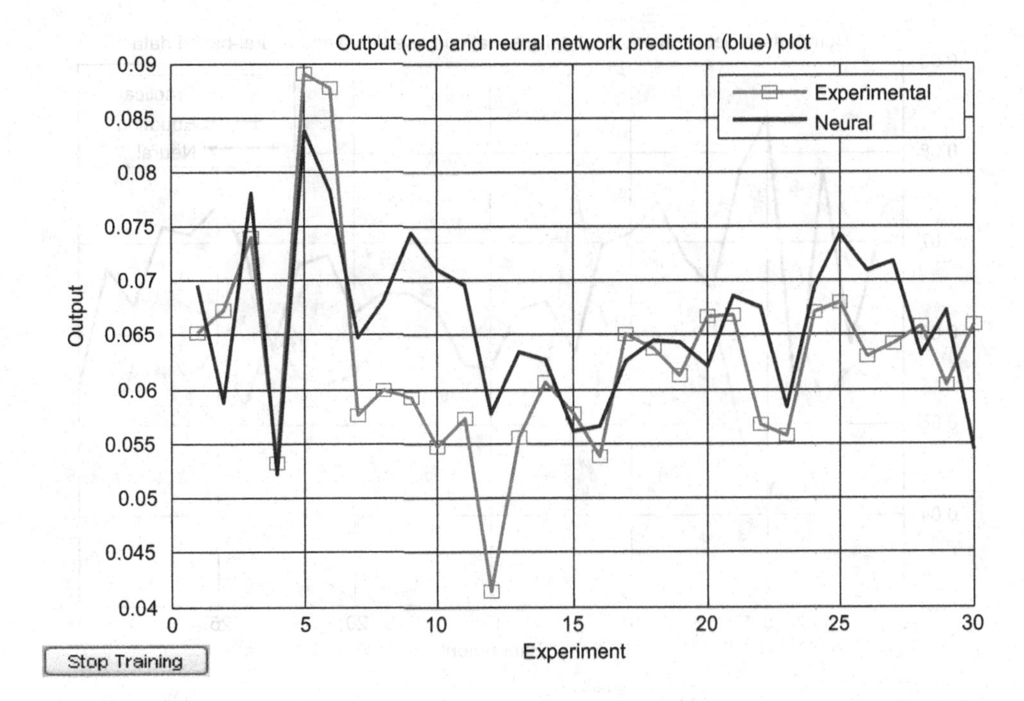

FIGURE 6.17 Graph of comparison with experimental/field database and neural prediction for the network for productivity of crankshaft maintenance activity (z_{3C}).

Comparison of experimental data and neural data for the dependent variable productivity.

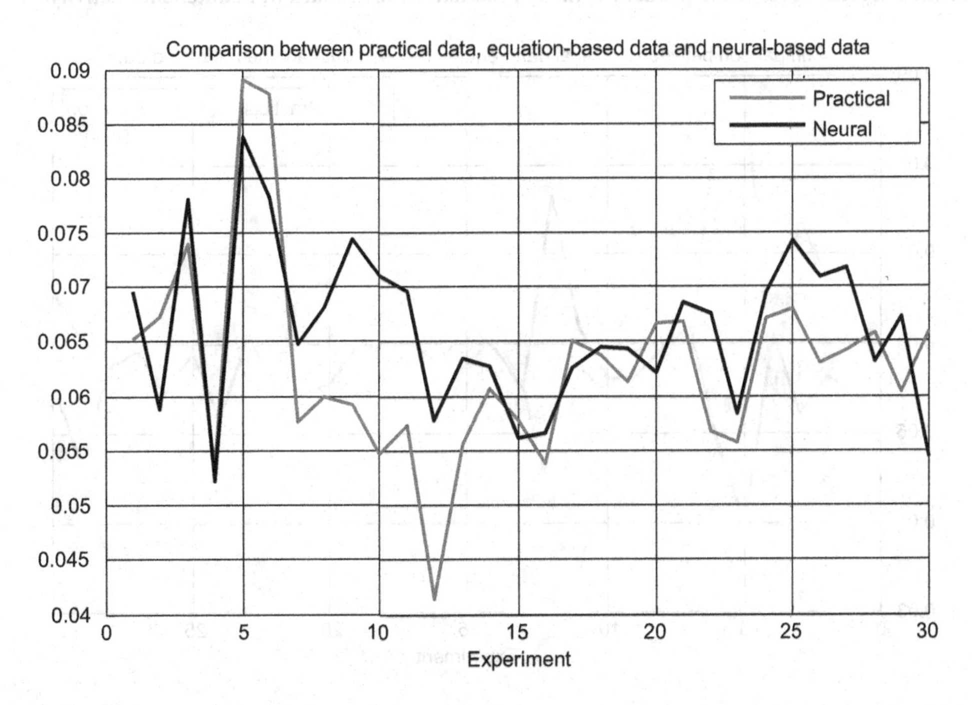

FIGURE 6.18 Graph of comparison with practical/field database and neural prediction for the network for productivity of crankshaft maintenance activity (z_{3C}).

Comparison of experimental data and neural data for the dependent variable productivity of crankshaft maintenance activity.

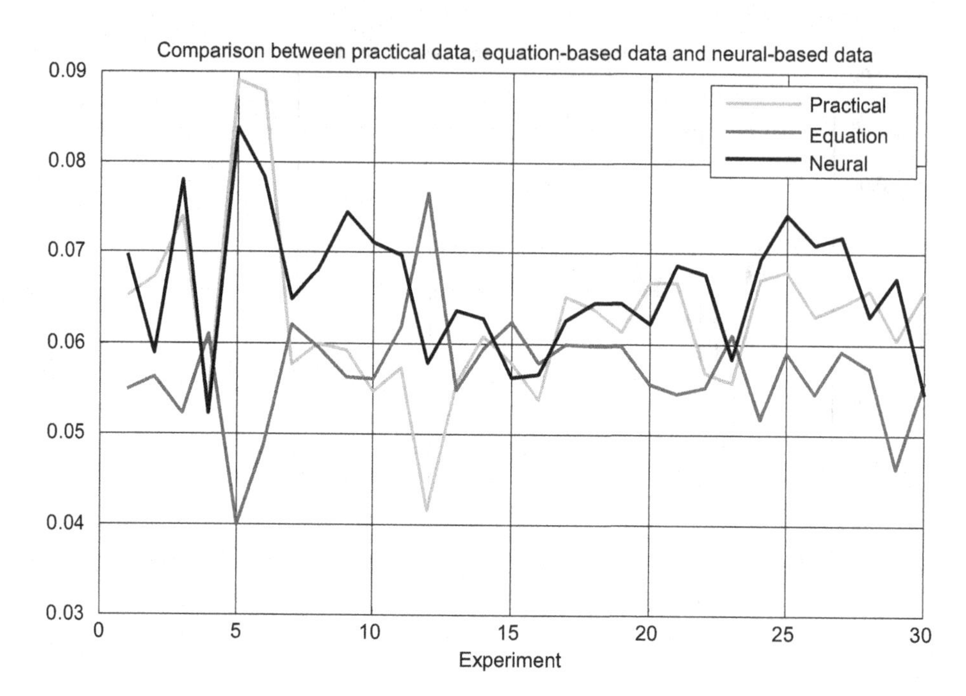

FIGURE 6.19 Graph of comparison with practical/field database, neural network prediction and equation-based prediction for the network for productivity of crankshaft maintenance activity (z_{3C}).

Comparison graph of output obtained from mathematical model, neural network and recorded experimental reading for the dependent variable productivity of crankshaft maintenance activity.

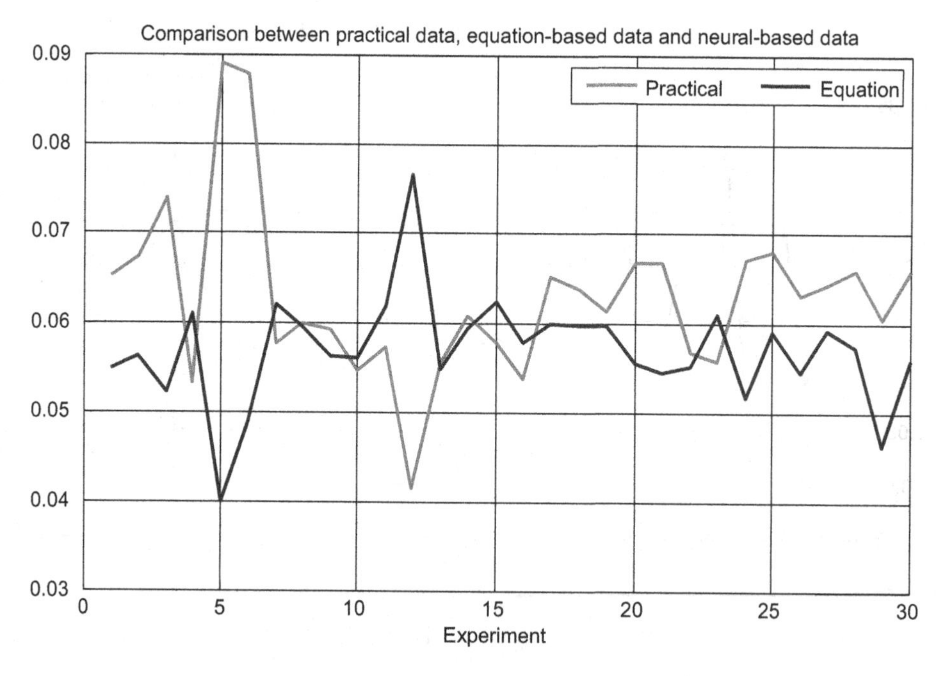

FIGURE 6.20 Graph of comparison with practical/field database and mathematical equation-based for the network for productivity of crankshaft maintenance activity(z_{3C}).

Comparison of experimental data and output obtained from mathematical model for the dependent variable productivity of crankshaft maintenance activity.

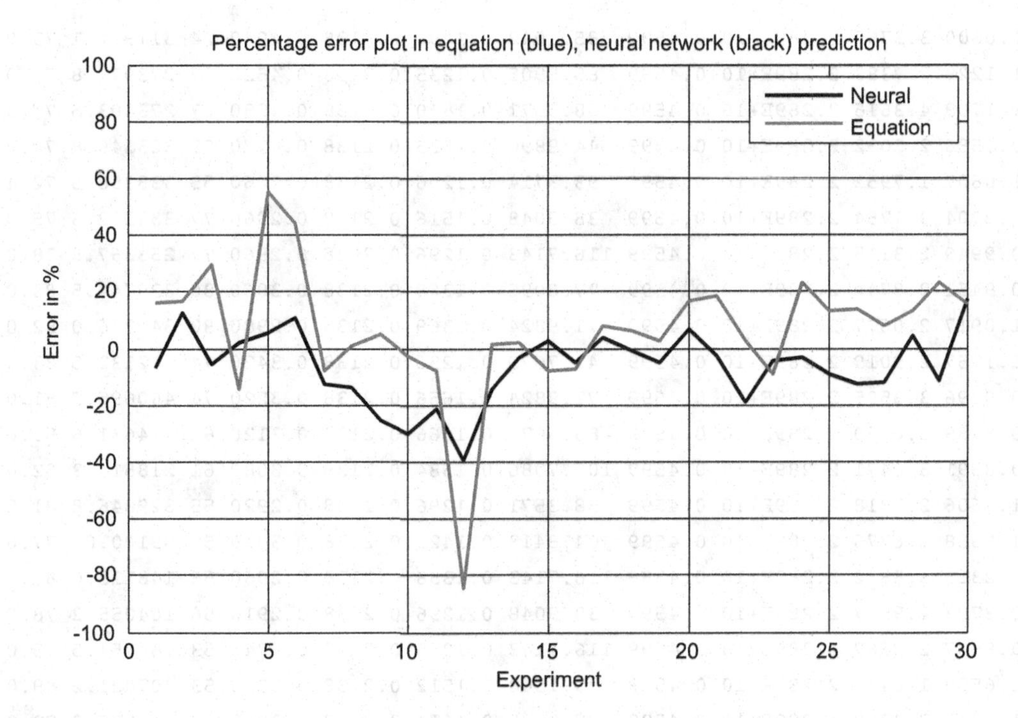

FIGURE 6.21 Graph of comparison of percentage error with mathematical equation-based and neural network prediction for the network for productivity of crankshaft maintenance activity (z_{3C}).

Comparison of percentage error plot prediction for the network for productivity of crankshaft maintenance activity.

From the above comparison of phenomenal response by a conventional approach and ANN simulation, it seems that the curve obtained by dependent pi terms for crankshaft maintenance activity for overhauling time (π_{D1}), human energy consumed (π_{D2}), and productivity (π_{D3}) of crankshaft, overlap due to fewer errors, which falls on the positive side, and gives an accurate relationship between ANN simulation and field data.

6.4 ANN PROGRAM FOR LINER PISTON MAINTENANCE ACTIVITY

The program executed for different dependent pi terms z_{1P}, z_{2P}, and z_{3P} of liner piston maintenance activity are as follows:

6.4.1 ANN Program for Overhauling Time of Liner Piston Maintenance Activity (z_{1P})

```
inputs3 = [
1.1431 2.6458 2.289E+10 0.4599  58.3571 0.1296 0.2138 0.3160 54 262984.7 73.0
1.1312 5.1701 2.289E+10 0.4599 105.0429 0.1346 0.2138 0.2600 54 299608.0 72.0
1.3626 2.6820 2.289E+10 0.4599 102.1250 0.1358 0.2138 0.3080 43 286392.9 70.0
0.5313 4.3452 2.289E+10 0.4599  57.5859 0.1282 0.2138 0.3220 43 357852.9 85.0
1.2537 3.8081 2.289E+10 0.4599  81.7000 0.1173 0.2138 0.3600 30 331656.8 80.0
1.2433 2.4495 2.289E+10 0.4599  93.3714 0.1346 0.2138 0.3980 30 245220.8 84.0
1.3533 1.6917 2.289E+10 0.4599  58.3571 0.1512 0.2138 0.4200 24 114463.6 79.0
```

```
1.0600 3.3765 2.289E+10 0.4599  85.1042 0.1296 0.2138 0.4040 24 311912.1 73.0
1.1228 2.7383 2.289E+10 0.4599  85.5905 0.1235 0.2138 0.4520 27 273976.8 76.0
1.1709 4.3616 2.289E+10 0.4599  58.3571 0.1540 0.2138 0.4580 27 273803.5 77.0
0.5586 2.2092 2.289E+10 0.4599  44.9890 0.1235 0.2138 0.4170 55 323346.8 74.0
1.0607 1.7952 2.289E+10 0.4599  93.3714 0.1296 0.2138 0.3060 55 338358.5 72.4
1.2704 3.1754 2.289E+10 0.4599  38.9048 0.1516 0.2138 0.2960 77 333392.6 75.0
0.9949 2.3115 2.289E+10 0.4599 116.7143 0.1296 0.2138 0.2960 77 253367.8 78.0
0.9451 2.8744 2.289E+10 0.4599  77.8095 0.1368 0.2138 0.3060 80 329744.5 83.0
1.0917 2.0677 2.289E+10 0.4599  71.9824 0.1369 0.2138 0.2980 80 340736.0 82.0
1.1757 2.9019 2.289E+10 0.4599  43.7679 0.1296 0.2138 0.3420 74 342132.5 83.0
0.9396 3.3505 2.289E+10 0.4599  71.9824 0.1656 0.2138 0.3220 74 400695.7 81.0
0.8789 3.0670 2.289E+10 0.4599  83.3674 0.1666 0.2138 0.3120 61 114631.6 80.0
0.8891 3.6471 2.289E+10 0.4599 102.7086 0.1584 0.2138 0.3080 61 218514.3 82.0
1.3606 2.7818 2.289E+10 0.4599  58.3571 0.1296 0.2138 0.2920 55 329846.8 81.0
1.6588 2.8776 2.289E+10 0.4599  64.8413 0.1420 0.2138 0.3000 55 89100.0  77.0
1.2315 3.5832 2.289E+10 0.4599 116.7143 0.1358 0.2138 0.2960 56 148113.6 88.0
0.8703 1.9527 2.289E+10 0.4599  38.9048 0.1296 0.2138 0.2910 56 164055.2 78.0
0.8537 2.8567 2.289E+10 0.4599 116.7143 0.1825 0.2138 0.2740 53 160551.5 75.0
1.6588 1.6415 2.289E+10 0.4599  89.1568 0.1512 0.2138 0.2900 53 307021.2 89.0
1.2315 2.4368 2.289E+10 0.4599  77.8095 0.1674 0.2138 0.3040 42 140875.0 78.0
0.8782 2.2223 2.289E+10 0.4599  84.0343 0.1296 0.2138 0.2770 42 158507.6 80.0
0.8584 1.6306 2.289E+10 0.4599  38.9048 0.1474 0.2138 0.3325 29 168150.4 84.0
0.8585 2.3931 2.289E+10 0.4599  70.0000 0.1420 0.2138 0.3925 29 111716.3 83.0
]
a1 = inputs3
a2 = a1
input_data = a2;
output3 = [
  89564135943
  94313208574
  87305714354
  91563911483
  87025439368
  83422869416
  90561263770
  89990800536
  97547410679
  95629701187
  92139331349
  92572079081
  76708886641
  82875386928
  84523241627
  80568694507
  80838380325
  82193568459
```

```
  84799461187
  76708886641
  77896619483
  83697286565
  75922134029
  96659948325
  2.20818E+11
  2.35324E+11
  90275806852
  2.18149E+11
  93005840670
  83834664115
]
y1 = output3
y2 = y1
size(a2);
size(y2);
p = a2';
sizep = size(p);
t = y2';
sizet = size(t);
[S Q] = size(t)
[pn,meanp,stdp,tn,meant,stdt] = prestd(p,t);
net = newff(minmax(pn),[30 1],{'logsig' 'purelin'},'trainlm');
net.performFcn = 'mse';
net.trainParam.goal = .99;
net.trainParam.show = 200;
net.trainParam.epochs = 50;
net.trainParam.mc = 0.05;
net = train(net,pn,tn);
an = sim(net,pn);
[a] = poststd(an,meant,stdt);
error = t-a;
x1 = 1:30;
plot(x1,t,'rs-',x1,a,'b-')
legend('Experimental','Neural');
title('Output (Red) and Neural Network Prediction (Blue) Plot');
xlabel('Experiment No.');
ylabel('Output');
grid on;
figure
error_percentage = 100×error./t
plot(x1,error_percentage)
legend('percentage error');
axis([0 30 -100 100]);
title('Percentage Error Plot in Neural Network Prediction');
xlabel('Experiment No.');
ylabel('Error in %');
grid on;
for ii=1:30
```

```
xx1 = input_data(ii,1);
yy2 = input_data(ii,2);
zz3 = input_data(ii,3);
xx4 = input_data(ii,4);
yy5 = input_data(ii,5);
zz6 = input_data(ii,6);
xx7 = input_data(ii,7);
yy8 = input_data(ii,8);
zz9 = input_data(ii,9);
xx10 = input_data(ii,10);
yy11 = input_data(ii,11);
pause
yyy(1,ii) = 1.0122×power(xx1,-0.0973)×power(yy2,-0.1917)
            ×power(zz3,0.8772)×power(xx4,-6.0098)
            ×power(yy5,0.0896)×power(zz6,0.378)
            ×power(xx7,0.6869)×power(yy8,-0.5615)
            ×power(zz9,-0.2282)×power(xx10,0.0887)
            ×power(yy11,0.0838);
yy_practical(ii) = (y2(ii,1));
yy_eqn(ii) = (yyy(1,ii))
yy_neur(ii) = (a(1,ii))
yy_practical_abs(ii) = (y2(ii,1));
yy_eqn_abs(ii) = (yyy(1,ii));
yy_neur_abs(ii) = (a(1,ii));
pause
end
figure;
plot(x1,yy_practical_abs,'r-',x1,yy_eqn_abs,'b-',x1,yy_neur_
abs,'k-');
legend('Practical','Equation','Neural');
title('Comparision between practical data, equation-based data and
neural based data');
xlabel('Experimental');
grid on;
figure;
plot(x1,yy_practical_abs,'r-',x1,yy_eqn_abs,'b-');
legend('Practical',' Equation');
title('Comparision between practical data, equation-based data and
neural based data');
xlabel('Experimental');
grid on;
figure;
plot(x1,yy_practical_abs,'r-',x1,yy_neur_abs,'k-');
legend('Practical','Neural');
title('Comparision between practical data, equation-based data and
neural based data');
xlabel('Experimental');
grid on;
error1 = yy_practical_abs-yy_eqn_abs
figure
error_percentage1 = 100×error1./yy_practical_abs;
plot(x1,error_percentage,'k-',x1,error_percentage1,'b-');
```

```
legend('Neural','Equation');
axis([0 30 -100 100]);
title('Percentage Error Plot in Equation (blue),  Neural Network
(black) Prediction');
xlabel('Experiment No.');
ylabel('Error in %');
grid on;
meanexp = mean(output3)
meanann = mean(a)
meanmath = mean(yy_eqn_abs)
mean_absolute_error_performance_function = mae(error)
mean_squared_error_performance_function = mse(error)
net = newff(minmax(pn),[30 1],{'logsig' 'purelin'},'trainlm','lear
ngdm','msereg');
an = sim(net,pn);
[a] = poststd(an,meant,stdt);
error = t(1,[1:30])-a(1,[1:30]);
net.performParam.ratio = 20/(20+1);
perf = msereg(error,net)
rand('seed',1.818490882E9)
[ps] = minmax(p);
[ts] = minmax(t);
numInputs = size(p,1);
numHiddenNeurons = 30;
numOutputs = size(t,1);
net = newff(minmax(p), [numHiddenNeurons,numOutputs]);
[pn,meanp,stdp,tn,meant,stdt] = prestd(p,t);
[ptrans,transmit] = prepca(pn,0.001);
[R Q] = size(ptrans);
testSamples = 15:1:Q;
validateSamples = 20:1:Q;
trainSamples = 1:1:Q ;
validation.P = ptrans(:,validateSamples) ;
validation.T = tn(:,validateSamples) ;
testing.P = ptrans(:,testSamples) ;
testing.T = tn(:,testSamples)
ptr = ptrans(:,trainSamples) ;
ttr = tn(:,trainSamples);
net = newff(minmax(ptr),[30 1],{'logsig' 'purelin'},'trainlm');
 [net,tr] = train(net,ptr,ttr,[] ,[],validation,testing);
plot(tr.epoch,tr.perf,  'r',tr.epoch,tr.vperf,  'g',tr.epoch,tr.
tperf, 'h') ;
legend('Training', 'validation', 'Testing',-1) ;
ylabel('Error') ;
an = sim(net,ptrans);
a = poststd(an,meant,stdt);
pause;
figure
 [m,b,r] = postreg(a,t);
```

The various output and graphs obtained by ANN program for overhauling time of liner piston maintenance activity are shown below in Figures 6.22–6.28.

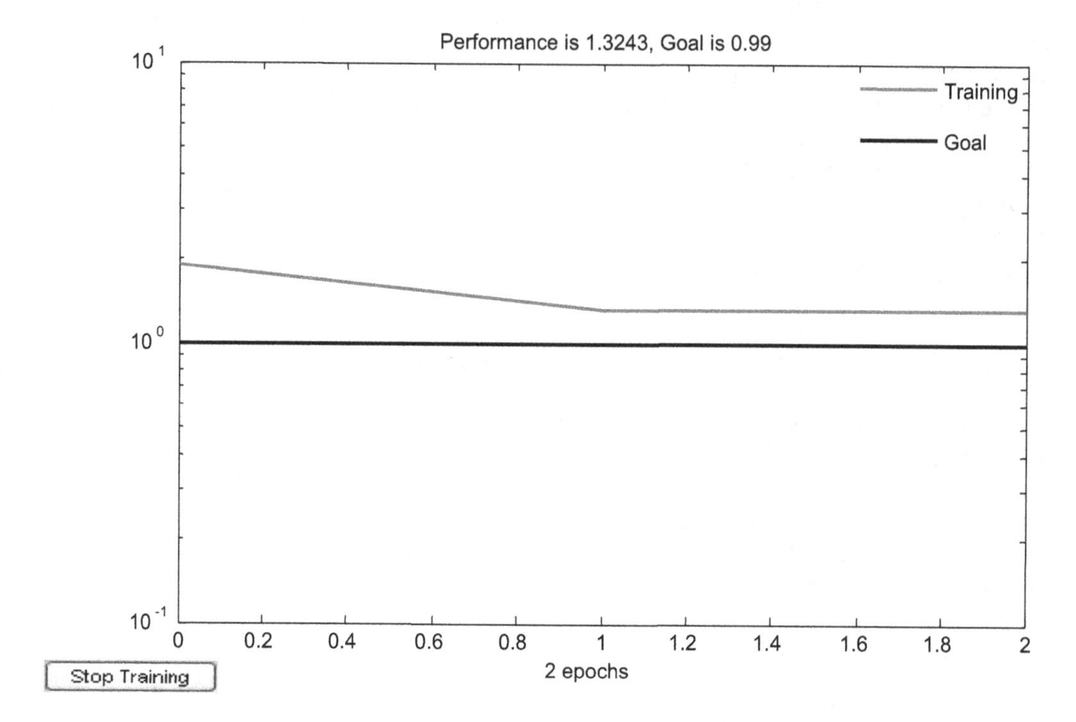

FIGURE 6.22 Training of the network for overhauling time of liner piston overhauling(z_{1P}).

Training of the network for productivity of overhauling time of liner piston overhauling.

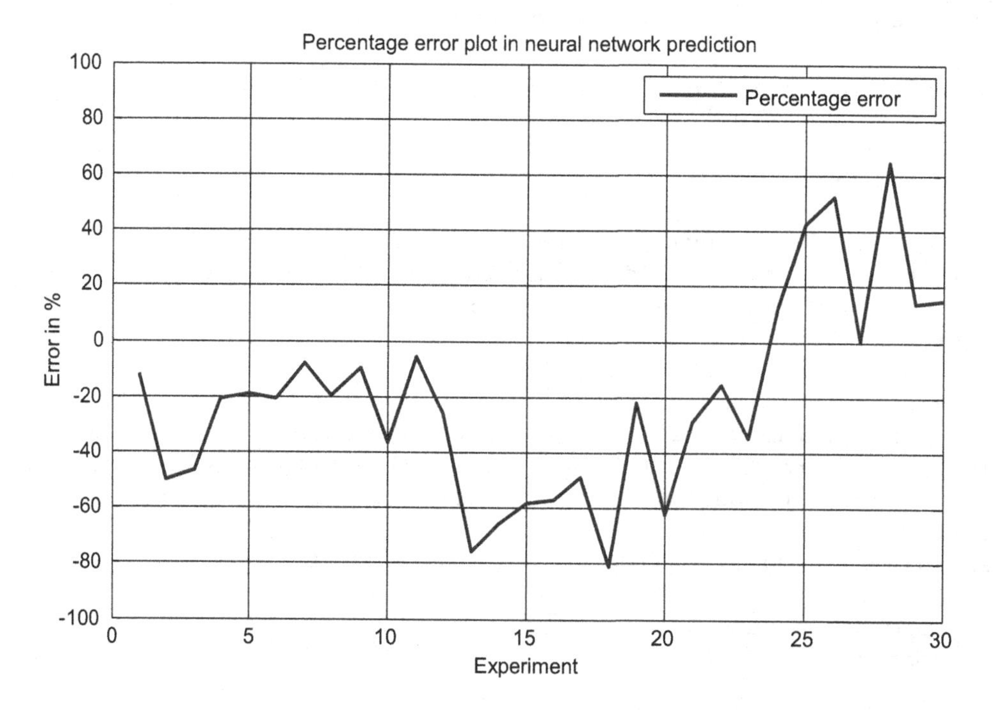

FIGURE 6.23 Percentage error plot prediction for the network for overhauling time-liner piston overhauling (z_{1P}).

Comparison of percentage error plot prediction for the network for overhauling time-liner piston overhauling.

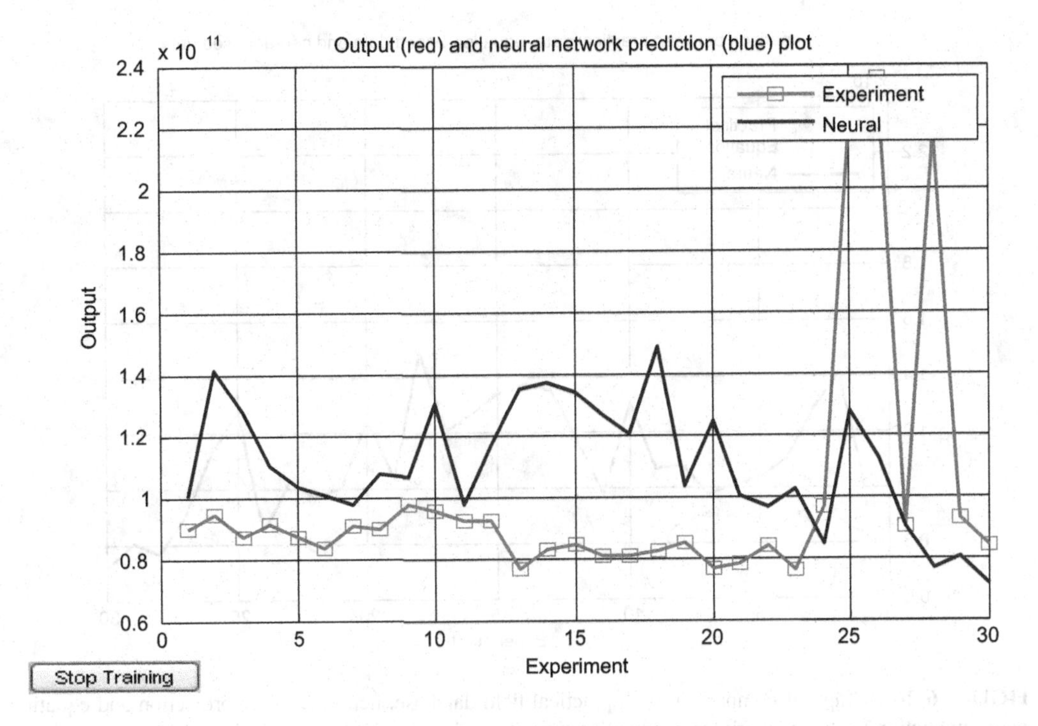

FIGURE 6.24 Graph of comparison with experimental/field database and neural prediction for the network for time-liner piston maintenance activity (z_{1P}).

Comparison of experimental data and neural data for the dependent variable Time.

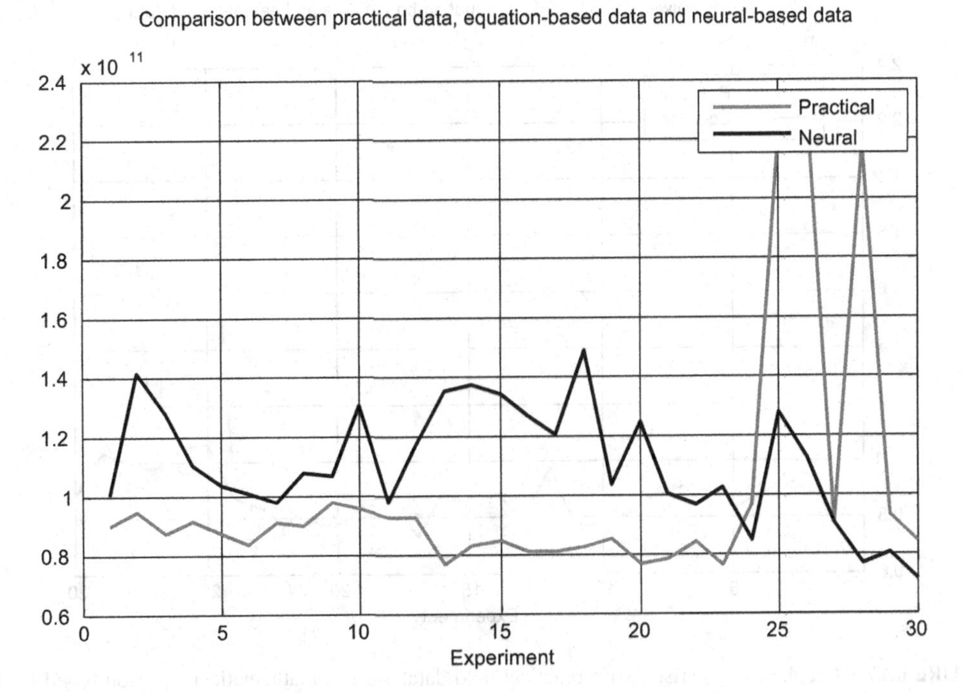

FIGURE 6.25 Graph of comparison with practical/field database and neural prediction for the network for overhauling time-liner piston maintenance activity (z_{1P}).

Comparison of experimental data and neural data for the dependent variable Time.

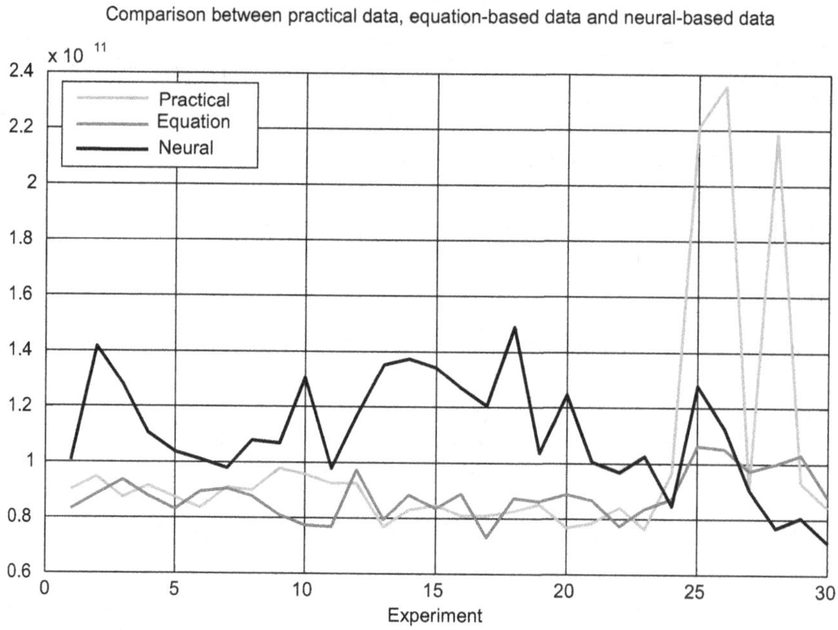

FIGURE 6.26 Graph of comparison with practical/field database, neural network prediction and equation-based prediction for the network for overhauling time-liner piston maintenance activity (z_{1P}).

Comparison graph of output obtained from mathematical model, neural network and recorded experimental reading for the dependent variable Time.

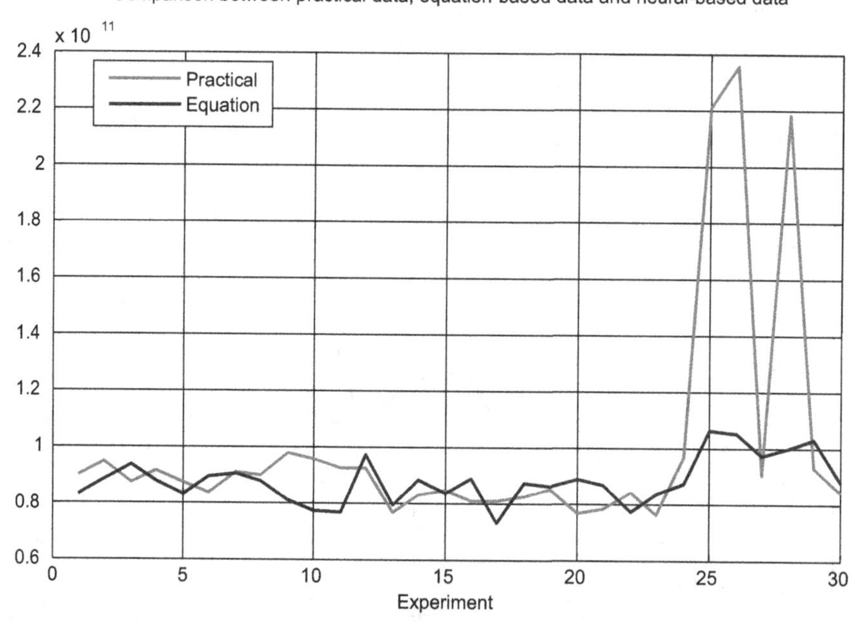

FIGURE 6.27 Graph of comparison with practical/field database and mathematical equation-based for the network of overhauling time-liner piston maintenance activity(z_{1P}).

Comparison of experimental data and output obtained from a mathematical model for the dependent variable Time.

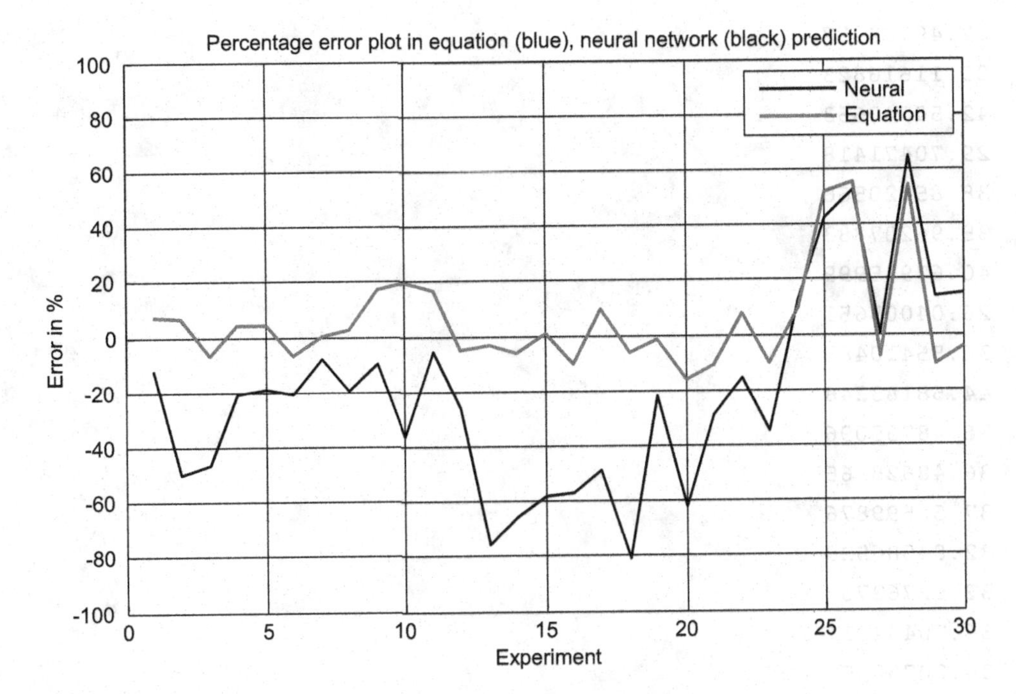

FIGURE 6.28 Graph of comparison of percentage error with mathematical equation-based and neural network prediction for the network of overhauling time-liner piston maintenance activity (z_{1P}).

Comparison of percentage error plot prediction for the network for time-liner piston overhauling.

6.4.2 ANN Program for Human Energy Consumed in Liner Piston Maintenance Activity (Z_{2P})

MATLAB program for the human energy consumed in liner piston maintenance activity is same as the program for overhauling time of liner piston except for the output of Human Energy consumed and model equation of human energy consumed in liner piston maintenance activity, which differs, as shown below in the output and equation of human energy consumed in liner piston maintenance activity.

```
output3 = [
  36.94814097
  46.70345859
  40.50902192
  35.85299040
  48.07656758
  45.23389851
  31.88721045
  47.73687669
  35.44014206
  35.79876092
```

```
    22.40702357
    29.11616823
    42.50740653
    29.70471418
    38.85420566
    35.96207883
    40.67975999
    26.04005651
    36.55410485
    44.58183248
    36.48755096
    36.48628365
    37.52599876
    32.84586528
    32.92762763
    36.70434811
    34.20705352
    33.05722722
    29.04490425
    36.38105280
]
```

Equation of Human Energy consumed in liner piston maintenance activity

```
yyy(1,ii)  = 1.7874xpower(xx1,0.355)xpower(yy2,0.3448)
             xpower(zz3,0.1681)xpower(xx4,4.4021)
             xpower(yy5,-0.0249)xpower(zz6,-0.131)
             xpower(xx7,-0.3313)xpower(yy8,-0.1018)
             xpower(zz9,-0.1479)xpower(xx10,0.0643)
             xpower(yy11,0.2221);
```

The above output and equation of human energy consumed in liner piston maintenance activity replaces output and equation in MATLAB of overhauling time of liner piston maintenance activity. The various graphs obtained for human energy consumed in liner piston maintenance activity are shown below by ANN in Figures 6.29–6.35.

6.4.3 ANN Program for Productivity of Liner Piston Maintenance Activity (z_{3P})

The MATLAB program for the productivity of liner piston maintenance activity is same as the program for overhauling time of liner piston, except for the output of productivity of liner piston Maintenance activity and model equation of productivity of liner piston Maintenance activity,

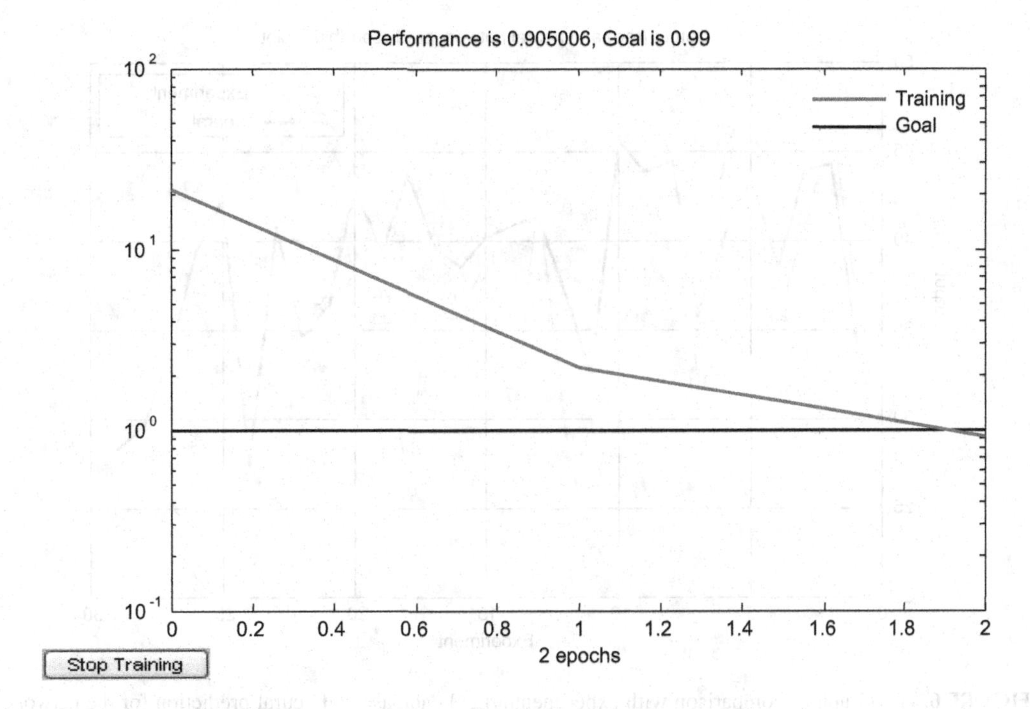

FIGURE 6.29 Training of the network of human energy consumed in liner piston maintenance activity (z_{2P}).

Training of the network of human energy consumed of liner piston overhauling.

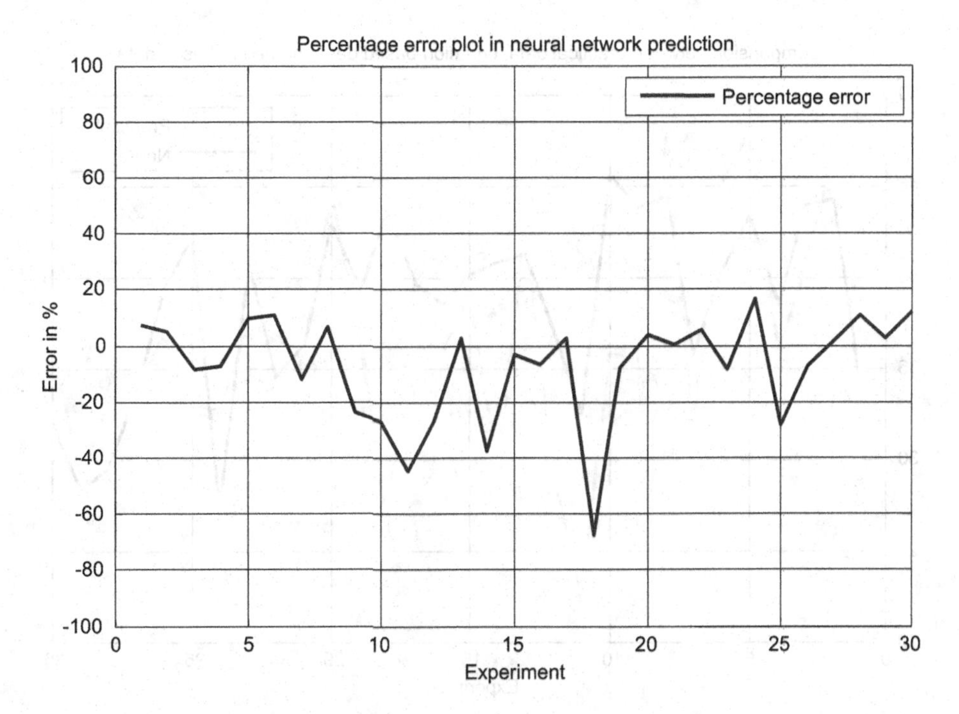

FIGURE 6.30 Percentage error plot prediction for the network for human energy consumed in liner piston maintenance activity (z_{2P}).

Comparison of percentage error plot prediction for the network for human energy consumed by liner piston overhauling.

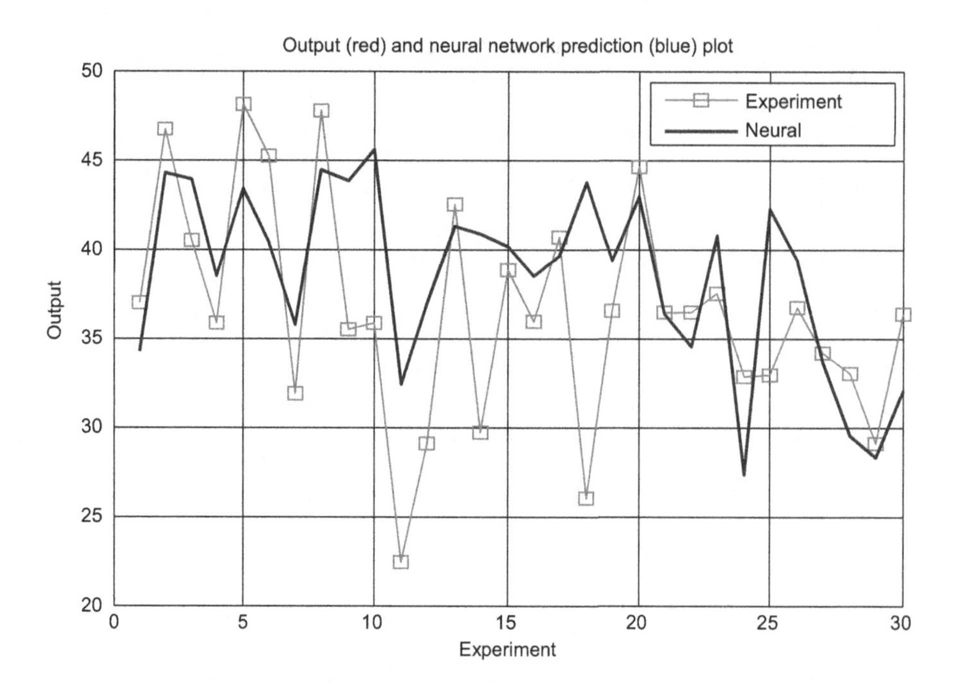

FIGURE 6.31 Graph of comparison with experimental/field database and neural prediction for the network for human energy consumed in liner piston maintenance activity (z_{2P}).

Comparison of experimental data and neural data for the dependent variable for human energy consumed.

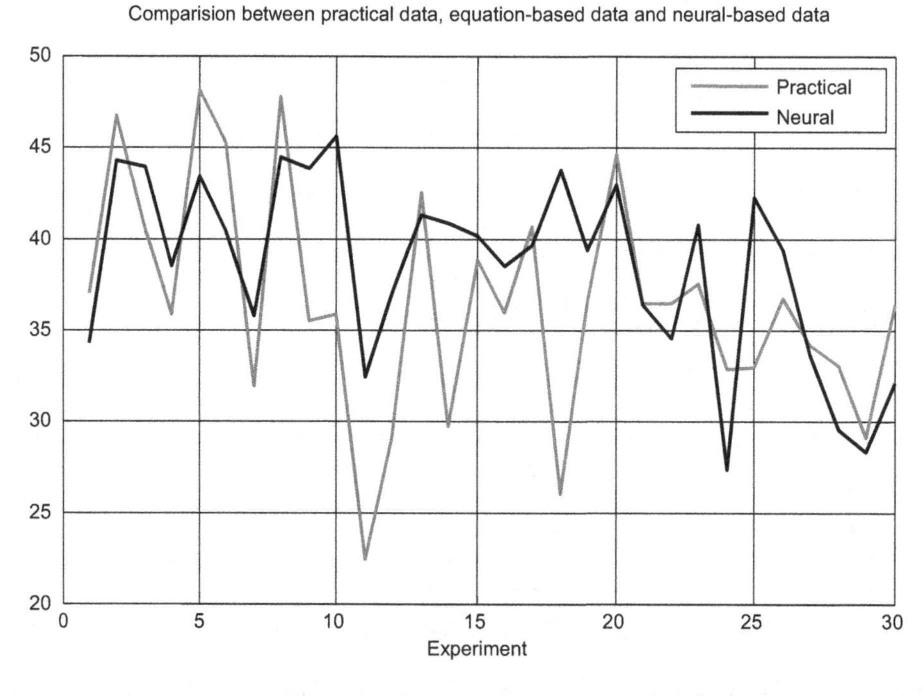

FIGURE 6.32 Graph of comparison with practical/field database and neural prediction for the network for human energy consumed in liner piston maintenance activity (z_{2P}).

Comparison of experimental data and neural data for the dependent variable for human energy consumed in liner piston maintenance activity.

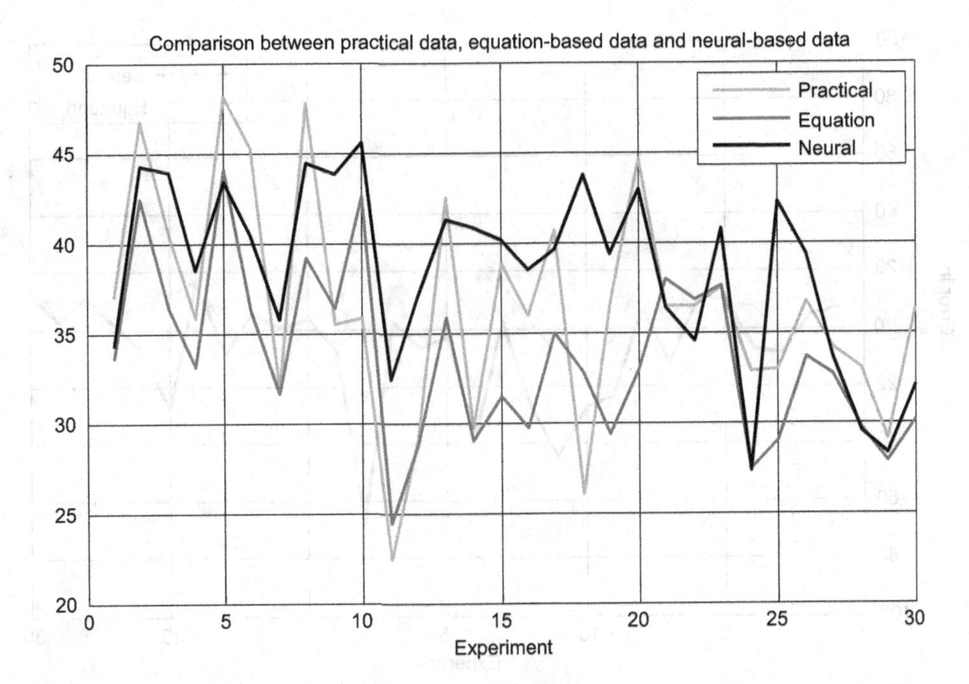

FIGURE 6.33 Graph of comparison with practical/field database, neural network prediction and equation-based prediction for the network for human energy consumed in liner piston maintenance activity (z_{2P}).

Comparison graph of output obtained from mathematical model, neural network and recorded experimental reading for the dependent variable for human energy consumed in liner piston maintenance activity.

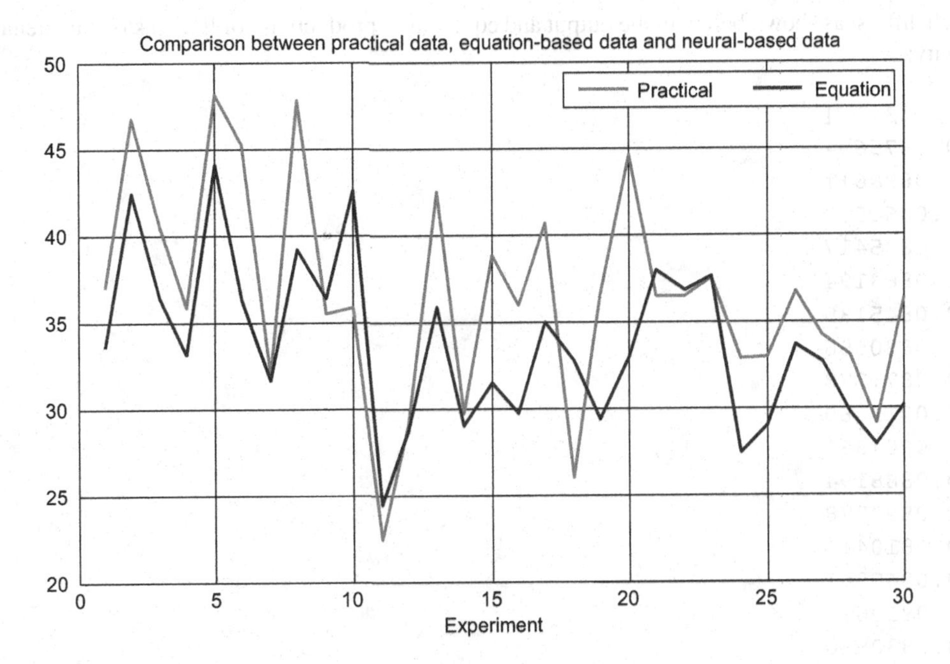

FIGURE 6.34 Graph of comparison with practical/field database and mathematical equation-based for the network for human energy consumed in liner piston maintenance activity (z_{2P}).

Comparison of experimental data and output obtained from mathematical model for the dependent variable for human energy consumed in liner piston maintenance activity.

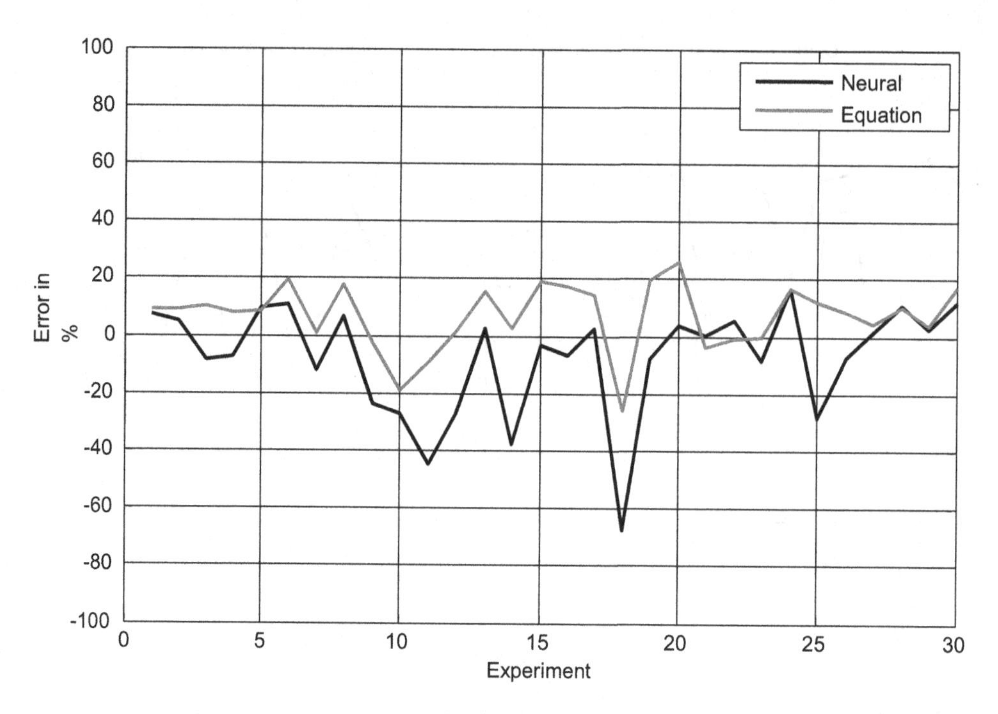

FIGURE 6.35 Graph of comparison of percentage error with mathematical equation-based and neural network prediction for the network for human energy consumed in liner piston maintenance activity (z_{2P}).

Comparison of percentage error plot prediction for the network for human energy consumed-liner piston overhauling.

which differs, as shown below in the output and equation of productivity of liner piston maintenance activity:

```
output3 = [
  0.0875694
  0.0898611
  0.0864583
  0.0885417
  0.0863194
  0.0845139
  0.0880556
  0.0877778
  0.0913889
  0.0904861
  0.0888194
  0.0890278
  0.0810417
  0.0842361
  0.0850694
  0.0830556
  0.0831944
  0.0838889
  0.0852083
  0.0810417
```

```
    0.0816667
    0.0846528
    0.0806250
    0.0909722
    0.1375000
    0.1419444
    0.0879167
    0.1366667
    0.0892361
    0.0847222
]
```

Equation for productivity of liner piston maintenance activity

```
yyy(1,ii) = 1.1148xpower(xx1,0.0973)xpower(yy2,0.1917)
            xpower(zz3,0.0188)xpower(xx4,-2.5291)
            xpower(yy5,-0.0896)xpower(zz6,-0.378)
            xpower(xx7,2.9719)xpower(yy8,0.5615)
            xpower(zz9,0.2282)xpower(xx10,-0.0887)
            xpower(yy11,-0.0838)
```

The above output and equation of productivity of liner piston maintenance activity replaces output and equation in MATLAB of overhauling time of liner piston maintenance activity. The various graphs obtained for productivity of liner piston maintenance activity are shown below by ANN (Figures 6.36–6.46).

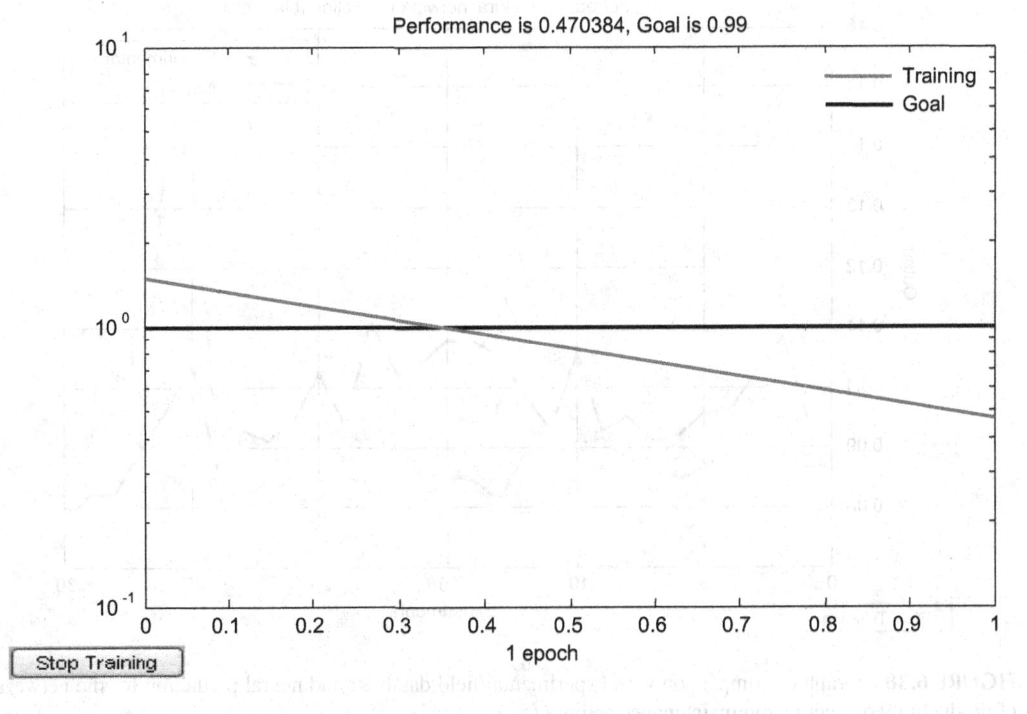

FIGURE 6.36 Training of the network for productivity of liner piston maintenance activity (z_{3P}).

Training of the network for productivity of liner piston overhauling.

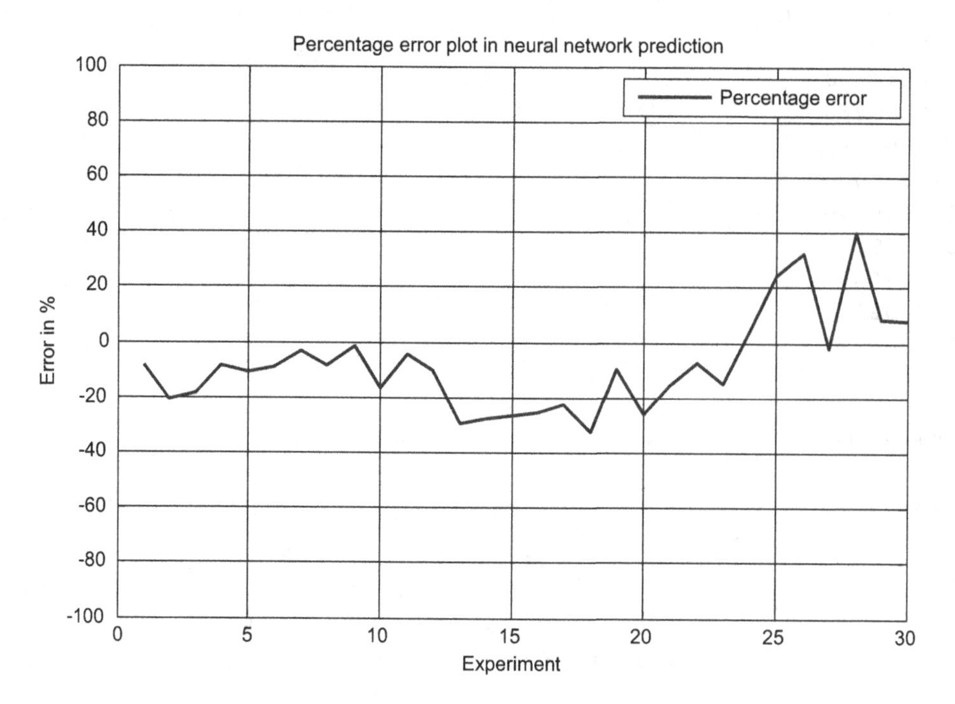

FIGURE 6.37 Percentage error plot prediction for the network for productivity of liner piston maintenance activity (z_{3P}).

Comparison of percentage error plot prediction for the network for productivity for liner piston overhauling.

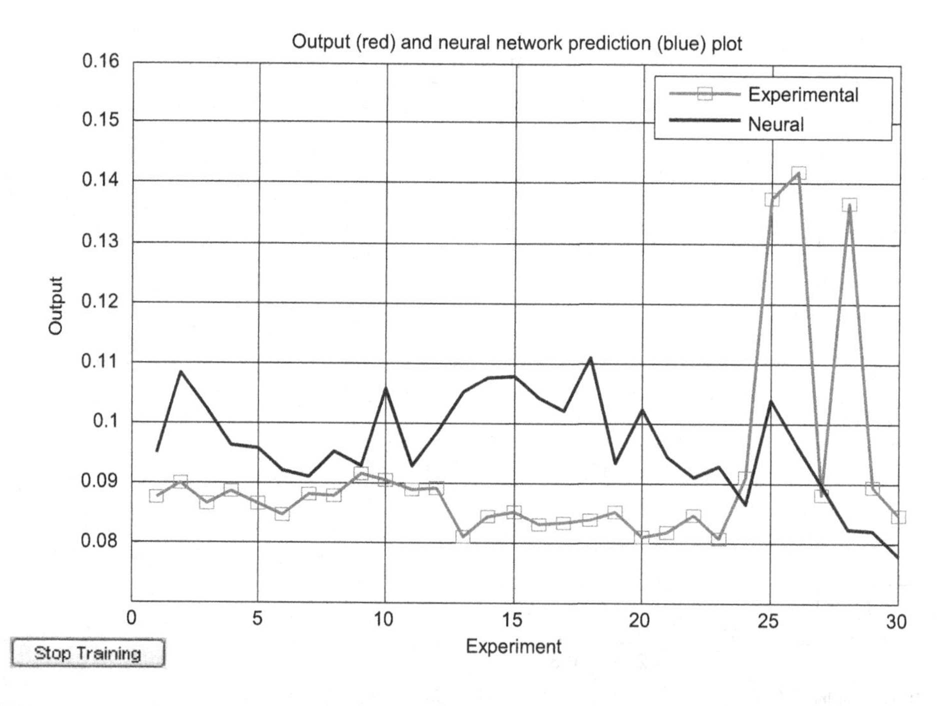

FIGURE 6.38 Graph of comparison with experimental/field database and neural prediction for the network of productivity of liner piston maintenance activity (z_{3P}).

Comparison of experimental data and neural data for the dependent variable productivity of liner piston maintenance activity.

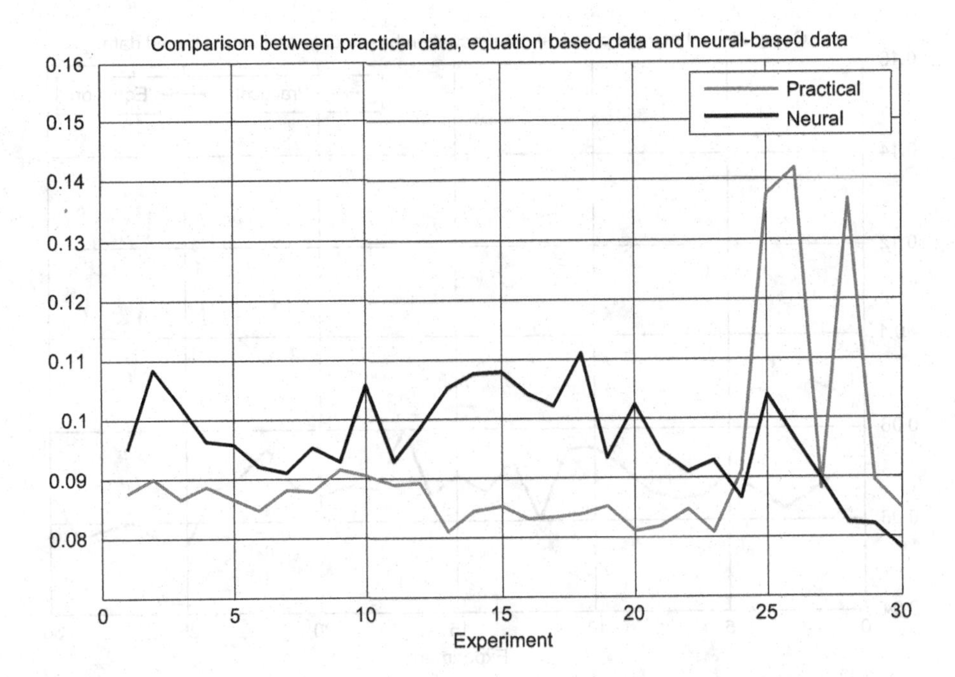

FIGURE 6.39 Graph of comparison with practical/field database and neural prediction for the network for productivity of liner piston maintenance activity (z_{3P}).

Comparison of experimental data and neural data for the dependent variable productivity of liner piston maintenance activity.

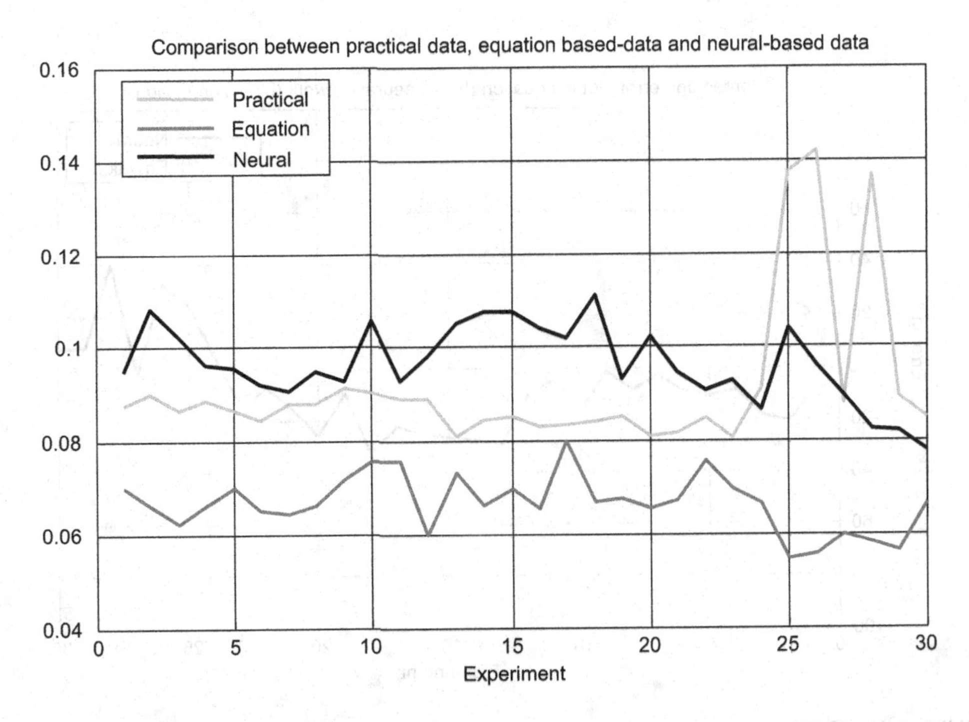

FIGURE 6.40 Graph of comparison with practical/field database, neural network prediction and equation-based prediction for the network for productivity of liner piston maintenance activity (z_{3P}).

Comparison graph of output obtained from mathematical model, neural network and recorded experimental reading for the dependent variable productivity of liner piston maintenance activity.

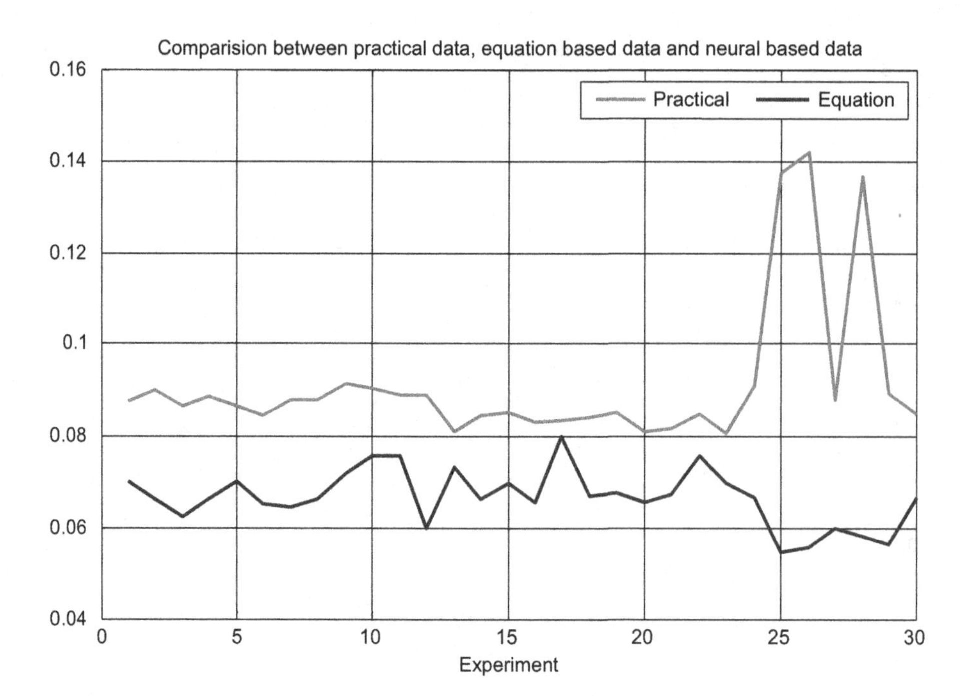

FIGURE 6.41 Graph of comparison with practical/field database and mathematical equation-based for the network for productivity of liner piston maintenance activity(z_{3P}).

Comparison of experimental data and output obtained from mathematical model for the dependent variable productivity of liner piston maintenance activity.

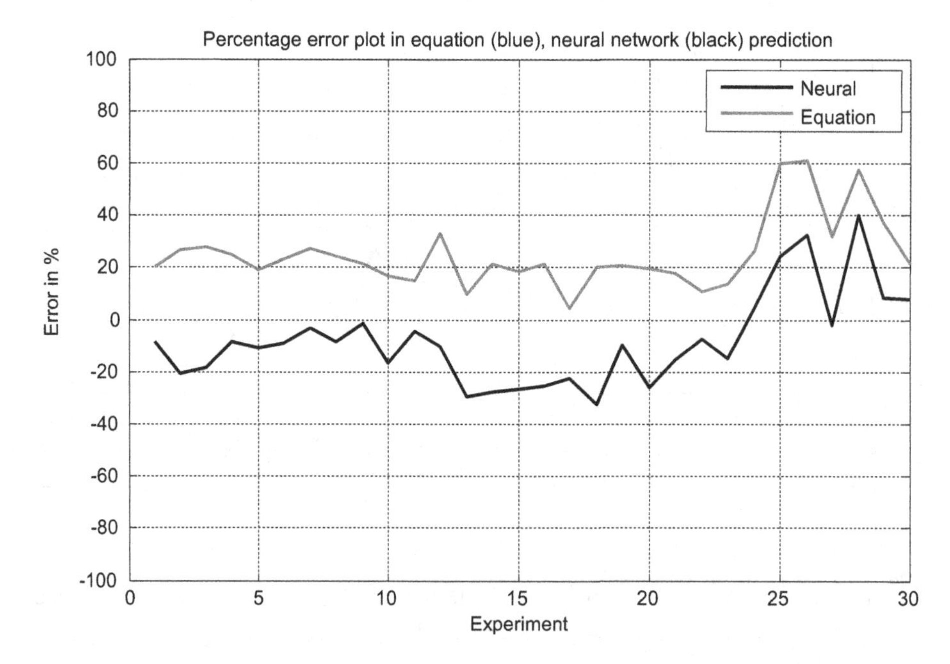

FIGURE 6.42 Graph of comparison of percentage error with mathematical equation-based and neural network prediction for the network for productivity of liner piston maintenance activity (z_{3P}).

Comparison of percentage error plot prediction for the network for productivity of liner piston overhauling.

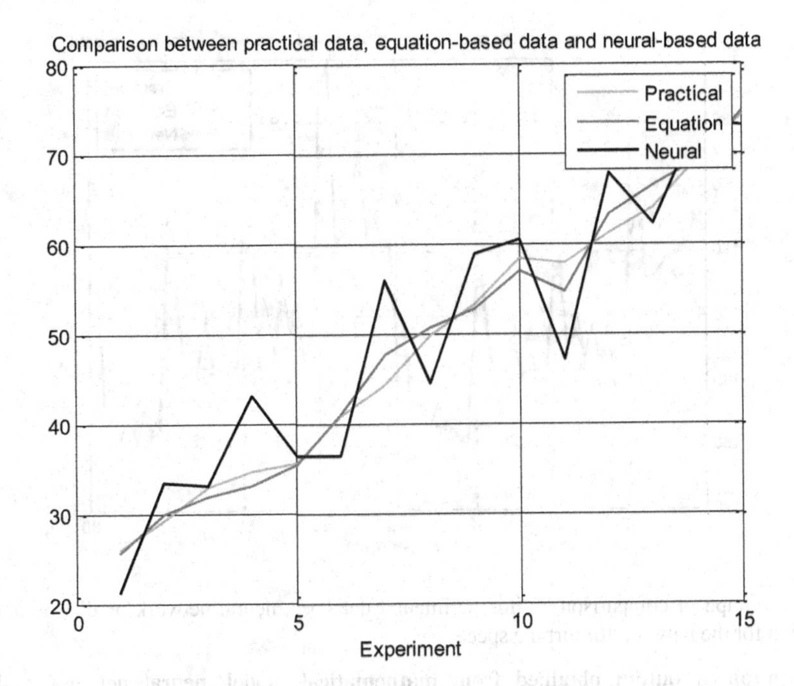

FIGURE 6.43 Graph of comparison with experimental database, neural network prediction and equation-based prediction for the network for brake thermal efficiency.

Comparison graph of output obtained from a mathematical model, neural network and recorded experimental reading for the dependent variable for brake thermal efficiency.

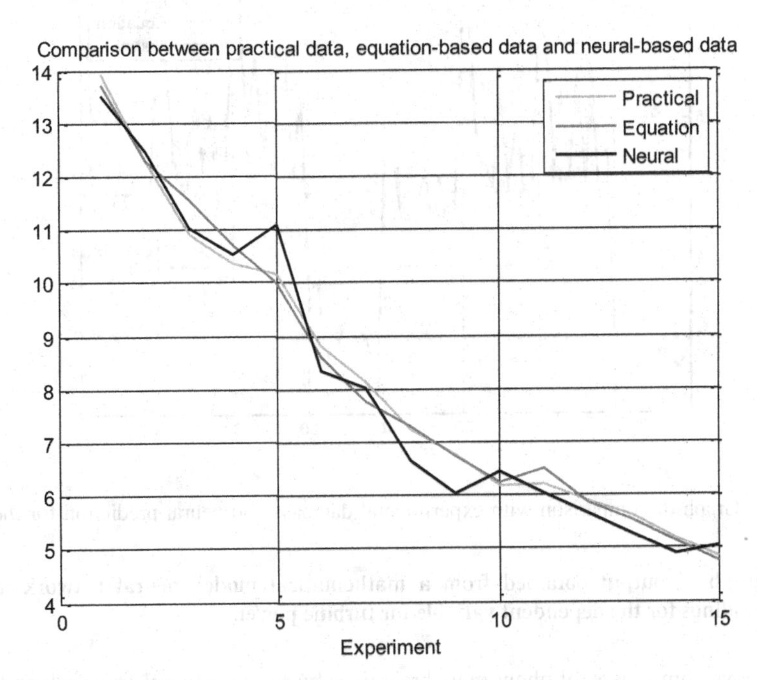

FIGURE 6.44 Graph of comparison with experimental database, neural network prediction and equation-based prediction for the network for brake-specific fuel consumption.

Comparison graph of output obtained from mathematical model, neural network and recorded experimental reading for the dependent variable for brake-specific fuel consumption.

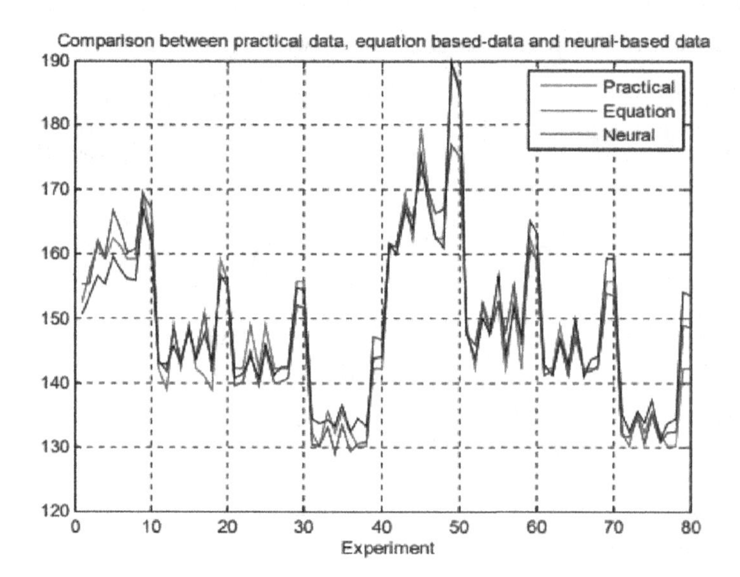

FIGURE 6.45 Graph of comparison with experimental database, neural network prediction and equation-based prediction for the network for turbine speed.

Comparison graph of output obtained from mathematical model, neural network and recorded experimental readings for the dependent variable for Turbine Speed.

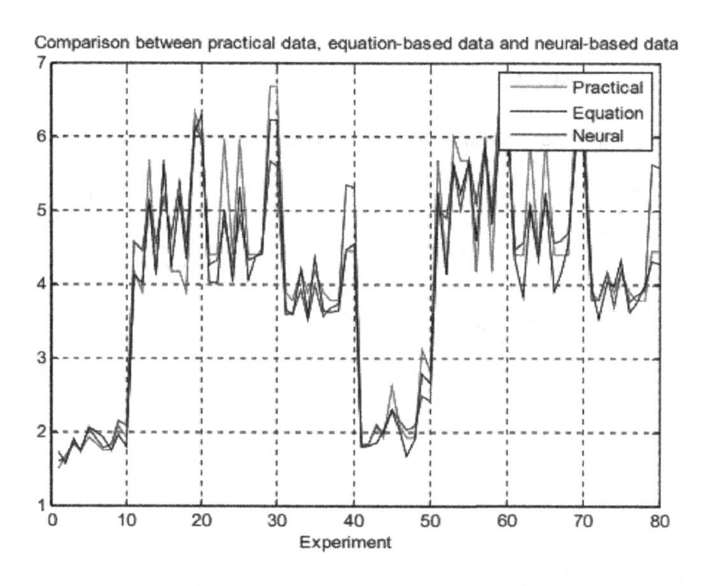

FIGURE 6.46 Graph of comparison with experimental database and neural prediction for the network for turbine power.

Comparison graph of output obtained from a mathematical model, neural network and recorded experimental readings for the dependent variable for turbine power.

From the above comparison of phenomenal response by a conventional approach and ANN simulation, it seems to be that the curve obtained by dependent pi terms for crankshaft maintenance activity for overhauling time (π_{D1}), human energy consumed (π_{D2}), and productivity (π_{D3}) of crankshaft overlap due to fewer errors, which is positive, and gives an accurate relationship between ANN simulation and Field Data.

ANN models have been developed for overhauling time, human energy consumed, and productivity of crankshaft maintenance activity and liner piston maintenance activities. These models proved to be successful in terms of agreement with actual values of experimentation. It can be concluded that an ANN model performs accurately to determine optimal values.

6.5 ANN PROGRAM FOR BRAKE THERMAL EFFICIENCY AND BRAKE-SPECIFIC FUEL CONSUMPTION

Based on experimental data for the dependent terms for brake thermal efficiency and brake-specific fuel consumption, experimental data-based modelling was obtained. Because the verification of experimental data-based models is not in close vicinity in such a complicated event, it became important to create ANN Simulations of the observational data.

6.5.1 ANN Program for Brake Thermal Efficiency

MATLAB software is selected for developing ANN. The program executed for brake thermal efficiency operation is as follows

```
clear all;
close all;
inputs3=[
   3.361134  1.546179688  79.409  0.01312683
   8.445761  1.529581151  80.542  0.01313494
  11.92674   1.49865104   82.3    0.01312683
  14.86954   1.443101042  85.93   0.01312683
  23.21995   1.42909625   88.34   0.01313494
   2.665727  0.773089844  79.409  0.01312683
   6.398304  0.764790575  80.542  0.01313494
   9.025641  0.74932552   82.3    0.01312683
  11.43811   0.721550521  85.93   0.01312683
  19.15646   0.714548125  88.34   0.01313494
   2.515056  0.515393229  79.409  0.01312683
   5.886439  0.509860384  80.542  0.01313494
   7.736264  0.499550347  82.3    0.01312683
   9.913029  0.481033681  85.93   0.01312683
  16.25397   0.476365417  88.34   0.01313494
]
 a1=inputs3
 a2=a1
input_data=a2;
output3=[
  25.8273
  29.2810
  32.9520
  34.7332
  35.4921
  40.8440
```

```
  44.2315
  49.7231
  53.2748
  58.3856
  57.8613
  61.1010
  63.9941
  69.4127
  74.2919
]
 y1=output3
 y2=y1
size(a2);
size(y2);
p=a2';
sizep=size(p);
t=y2';
sizet=size(t);
[S Q]=size(t)
[pn,meanp,stdp,tn,meant,stdt] = prestd(p,t);
 net = newff(minmax(pn),[15 1],{'logsig' 'purelin'},'trainlm');
net.performFcn='mse';
net.trainParam.goal=.99;
net.trainParam.show=200;
net.trainParam.epochs=50;
net.trainParam.mc=0.05;
net = train(net,pn,tn);
an = sim(net,pn);
[a] = poststd(an,meant,stdt);
error=t-a;
x1=1:15;
plot(x1,t,'rs-',x1,a,'b-')
legend('Experimental','Neural');
title('Output (Red) and Neural Network Prediction (Blue) Plot');
xlabel('Experiment No.');
ylabel('Output');
grid on;
figure
error_percentage=100×error./t
plot(x1,error_percentage)
legend('percentage error');
axis([0 15 -100 100]);
title('Percentage Error Plot in Neural Network Prediction');
xlabel('Experiment No.');
ylabel('Error in %');
grid on;
for ii=1:15
xx1=input_data(ii,1);
yy2=input_data(ii,2);
```

```
zz3=input_data(ii,3);
xx4=input_data(ii,4);
pause
yyy(1,ii)=85.1138xpower(xx1,0.1758)xpower(yy2,-0.738)xpower(zz3,
-0.695)xpower(xx4,-0.449);
yy_practical(ii)=(y2(ii,1));
yy_eqn(ii)=(yyy(1,ii))
yy_neur(ii)=(a(1,ii))
yy_practical_abs(ii)=(y2(ii,1));
yy_eqn_abs(ii)=(yyy(1,ii));
yy_neur_abs(ii)=(a(1,ii));
pause
end
figure;
plot(x1,yy_practical_abs,'r-',x1,yy_eqn_abs,'b-',x1,yy_neur_
abs,'k-');
legend('Practical','Equation','Neural');
title('Comparision between practical data, equation-based data and
neural based data');
xlabel('Experimental');
grid on;
figure;
plot(x1,yy_practical_abs,'r-',x1,yy_eqn_abs,'b-');
legend('Practical',' Equation');
title('Comparision between practical data, equation-based data and
neural based data');
xlabel('Experimental');
grid on;
figure;
plot(x1,yy_practical_abs,'r-',x1,yy_neur_abs,'k-');
legend('Practical','Neural');
title('Comparision between practical data, equation-based data and
neural based data');
xlabel('Experimental');
grid on;
error1=yy_practical_abs-yy_eqn_abs
figure
error_percentage1=100xerror1./yy_practical_abs;
plot(x1,error_percentage,'k-',x1,error_percentage1,'b-');
legend('Neural','Equation');
axis([0 15 -100 100]);
title('Percentage Error Plot in Equation (blue), Neural Network
(black) Prediction');
xlabel('Experiment No.');
ylabel('Error in %');
grid on;
meanexp=mean(output3)
meanann=mean(a)
meanmath=mean(yy_eqn_abs)
mean_absolute_error_performance_function = mae(error)
mean_squared_error_performance_function = mse(error)
```

From the above comparison of phenomenal response by a conventional approach and ANN simulation, it seems that the curve obtained by dependent pi terms for brake thermal efficiency π_D are overlapping due to fewer errors, which is positive.

6.5.2 ANN Program for Brake-specific Fuel Consumption

MATLAB software is selected for developing ANN. The program executed for brake-specific fuel consumption operation is as follows:

```
clear all;
close all;
inputs3=[
   3.361134  1.546179688  79.409  0.01312683
   8.445761  1.529581151  80.542  0.01313494
  11.92674   1.49865104   82.3    0.01312683
  14.86954   1.443101042  85.93   0.01312683
  23.21995   1.42909625   88.34   0.01313494
   2.665727  0.773089844  79.409  0.01312683
   6.398304  0.764790575  80.542  0.01313494
   9.025641  0.74932552   82.3    0.01312683
  11.43811   0.721550521  85.93   0.01312683
  19.15646   0.714548125  88.34   0.01313494
   2.515056  0.515393229  79.409  0.01312683
   5.886439  0.509860384  80.542  0.01313494
   7.736264  0.499550347  82.3    0.01312683
   9.913029  0.481033681  85.93   0.01312683
  16.25397   0.476365417  88.34   0.01313494
]
 a1=inputs3
 a2=a1
input_data=a2;
output3=[
  13.9388
  12.2947
  10.9250
  10.3647
  10.1431
   8.8140
   8.1390
   7.2401
   6.7574
   6.1659
   6.2218
   5.8919
```

```
    5.6255
    5.1864
    4.8458
]
 y1=output3
 y2=y1
size(a2);
size(y2);
p=a2';
sizep=size(p);
t=y2';
sizet=size(t);
[S Q]=size(t)
[pn,meanp,stdp,tn,meant,stdt] = prestd(p,t);
 net = newff(minmax(pn),[15 1],{'logsig' 'purelin'},'trainlm');
net.performFcn='mse';
net.trainParam.goal=.99;
net.trainParam.show=200;
net.trainParam.epochs=50;
net.trainParam.mc=0.05;
net = train(net,pn,tn);
an = sim(net,pn);
[a] = poststd(an,meant,stdt);
error=t-a;
x1=1:15;
plot(x1,t,'rs-',x1,a,'b-')
legend('Experimental','Neural');
title('Output (Red) and Neural Network Prediction (Blue) Plot');
xlabel('Experiment No.');
ylabel('Output');
grid on;
figure
error_percentage=100×error./t
plot(x1,error_percentage)
legend('percentage error');
axis([0 15 -100 100]);
title('Percentage Error Plot in Neural Network Prediction');
xlabel('Experiment No.');
ylabel('Error in %');
grid on;
for ii=1:15
xx1=input_data(ii,1);
yy2=input_data(ii,2);
zz3=input_data(ii,3);
xx4=input_data(ii,4);
pause
yyy(1,ii)=0.7348×power(xx1,-0.1016)×power(yy2,0.7070)×power(zz3,-0
.6336)×power(xx4,-1.2727);
yy_practical(ii)=(y2(ii,1));
```

```
yy_eqn(ii)=(yyy(1,ii))
yy_neur(ii)=(a(1,ii))
yy_practical_abs(ii)=(y2(ii,1));
yy_eqn_abs(ii)=(yyy(1,ii));
yy_neur_abs(ii)=(a(1,ii));
pause
end
figure;
plot(x1,yy_practical_abs,'r-',x1,yy_eqn_abs,'b-',x1,yy_neur_
abs,'k-');
legend('Practical','Equation','Neural');
title('Comparision between practical data, equation-based data and
neural based data');
xlabel('Experimental');
grid on;
figure;
plot(x1,yy_practical_abs,'r-',x1,yy_eqn_abs,'b-');
legend('Practical',' Equation');
title('Comparision between practical data, equation-based data and
neural based data');
xlabel('Experimental');
grid on;
figure;
plot(x1,yy_practical_abs,'r-',x1,yy_neur_abs,'k-');
legend('Practical','Neural');
title('Comparision between practical data, equation-based data and
neural based data');
xlabel('Experimental');
grid on;
error1=yy_practical_abs-yy_eqn_abs
figure
error_percentage1=100×error1./yy_practical_abs;
plot(x1,error_percentage,'k-',x1,error_percentage1,'b-');
legend('Neural','Equation');
axis([0 15 -100 100]);
title('Percentage Error Plot in Equation (blue), Neural Network
(black) Prediction');
xlabel('Experiment No.');
ylabel('Error in %');
grid on;
meanexp=mean(output3)
meanann=mean(a)
meanmath=mean(yy_eqn_abs)
mean_absolute_error_performance_function = mae(error)
mean_squared_error_performance_function = mse(error)
```

From the above comparison of phenomenal response by a conventional approach and ANN simulation, it seems that the curve obtained by dependent pi terms for brake-specific fuel consumption π_{D2} overlap due to the less percentage of error which is on the positive side.

6.6 ANN PROGRAM FOR SOLAR UPDRAFT TOWER

The experimental data-based modelling has been achieved based on experimental data for the two dependent π terms for turbine speed and power developed. In such complex phenomenon involving non-linear kinematics where in the validation of experimental data-based models is not in close proximity, it becomes necessary to formulate ANN Simulation of the observed data. Artificial Neural Network Simulation consists of three layers.

6.6.1 ANN PROGRAM FOR TURBINE SPEED

```
clear all; close all; inputs3=[
   2.48E-05  0.053  0.16  21.11  3.543  2981.25
   2.48E-05  0.053  0.17  21.37  3.543  3026.25
   2.48E-05  0.053  0.18  25.27  3.543  3093.75
   2.48E-05  0.053  0.16  24.12  3.543  3065.63
   2.48E-05  0.053  0.19  28.03  3.543  3121.88
   2.48E-05  0.053  0.17  26.65  3.543  3105
   2.48E-05  0.053  0.18  23.63  3.543  3037.5
   2.48E-05  0.053  0.19  24.12  3.543  3065.63
   2.48E-05  0.053  0.20  29.54  3.543  3150
   2.48E-05  0.053  0.20  28.03  3.543  3121.88
   1.24E-05  0.053  0.16  18.61  3.543  11497.5
   1.24E-05  0.053  0.17  17.4   3.543  11362.5
   1.24E-05  0.053  0.18  20.83  3.543  11700
   1.24E-05  0.053  0.16  18.61  3.543  11497.5
   1.24E-05  0.053  0.19  20.83  3.543  11700
   1.24E-05  0.053  0.17  18.61  3.543  11497.5
   1.24E-05  0.053  0.18  21.23  3.543  11362.5
   1.24E-05  0.053  0.19  17.4   3.543  11475
   1.24E-05  0.053  0.20  25.57  3.543  11925
   1.24E-05  0.053  0.20  24.21  3.543  11812.5
   1.24E-05  0.053  0.16  17.81  3.543  11842.1
   1.24E-05  0.053  0.17  17.59  3.543  11723.7
   1.24E-05  0.053  0.18  19.95  3.543  11889.5
   1.24E-05  0.053  0.16  17.81  3.543  11842.1
   1.24E-05  0.053  0.19  19.95  3.543  11889.5
   1.24E-05  0.053  0.17  17.81  3.543  11842.1
   1.24E-05  0.053  0.18  17.44  3.543  11723.7
   1.24E-05  0.053  0.19  17.59  3.543  11723.7
   1.24E-05  0.053  0.20  23.43  3.543  12031.6
   1.24E-05  0.053  0.20  23.23  3.543  12031.6
   1.24E-05  0.053  0.16  17     3.543  13218.8
   1.24E-05  0.053  0.17  16.64  3.543  13162.5
   1.24E-05  0.053  0.18  18.85  3.543  13640.6
   1.24E-05  0.053  0.16  17.03  3.543  13500
   1.24E-05  0.053  0.19  18.85  3.543  13640.6
```

1.24E-05	0.053	0.17	17	3.543	13500
1.24E-05	0.053	0.18	16.5	3.543	13078.1
1.24E-05	0.053	0.19	16.64	3.543	13162.5
1.24E-05	0.053	0.20	26.22	3.543	13781.3
1.24E-05	0.053	0.20	26	3.543	13781.3
2.48E-05	0.071	0.16	23.03	3.543	2981.25
2.48E-05	0.071	0.17	23.32	3.543	3026.25
2.48E-05	0.071	0.18	27.38	3.543	3093.75
2.48E-05	0.071	0.16	26.22	3.543	3065.63
2.48E-05	0.071	0.19	30.27	3.543	3121.88
2.48E-05	0.071	0.17	28.87	3.543	3105
2.48E-05	0.071	0.18	25.69	3.543	3037.5
2.48E-05	0.071	0.19	26.22	3.543	3065.63
2.48E-05	0.071	0.2	32.95	3.543	3150
2.48E-05	0.071	0.2	31.4	3.543	3121.88
1.24E-05	0.071	0.16	19.54	3.543	11497.5
1.24E-05	0.071	0.17	18.31	3.543	11362.5
1.24E-05	0.071	0.18	21.83	3.543	11700
1.24E-05	0.071	0.16	19.54	3.543	11497.5
1.24E-05	0.071	0.19	21.83	3.543	11700
1.24E-05	0.071	0.17	19.54	3.543	11497.5
1.24E-05	0.071	0.18	22.24	3.543	11362.5
1.24E-05	0.071	0.19	18.31	3.543	11475
1.24E-05	0.071	0.2	26.63	3.543	11925
1.24E-05	0.071	0.2	25.26	3.543	11812.5
1.24E-05	0.071	0.16	17.81	3.543	11842.1
1.24E-05	0.071	0.17	17.59	3.543	11723.7
1.24E-05	0.071	0.18	19.95	3.543	11889.5
1.24E-05	0.071	0.16	17.81	3.543	11842.1
1.24E-05	0.071	0.19	19.95	3.543	11889.5
1.24E-05	0.071	0.17	17.81	3.543	11842.1
1.24E-05	0.071	0.18	17.44	3.543	11723.7
1.24E-05	0.071	0.19	17.59	3.543	11723.7
1.24E-05	0.071	0.2	23.43	3.543	12031.6
1.24E-05	0.071	0.2	23.23	3.543	12031.6
1.24E-05	0.071	0.16	17	3.543	13218.8
1.24E-05	0.071	0.17	16.64	3.543	13162.5
1.24E-05	0.071	0.18	18.85	3.543	13640.6
1.24E-05	0.071	0.16	17.03	3.543	13500
1.24E-05	0.071	0.19	18.85	3.543	13640.6
1.24E-05	0.071	0.17	17	3.543	13500
1.24E-05	0.071	0.18	16.5	3.543	13078.1
1.24E-05	0.071	0.19	16.64	3.543	13162.5
1.24E-05	0.071	0.2	26.22	3.543	13781.3

```
  1.24E-05  0.071  0.2   26      3.543  13781.3
]
a1=inputs3
a2=a1
 input_data=a2;
output3=[
  152.4
  157.14
  161.2
  159.17
  162.56
  161.2
  159.17
  159.17
  169.33
  162.56
  142.24
  138.85
  149.01
  142.24
  149.01
  142.24
  140.88
  138.85
  159.17
  155.78
  142.24
  142.24
  149.01
  142.24
  149.01
  142.24
  142.24
  142.24
  155.78
  155.78
  132.08
  130.04
  135.46
  132.08
  135.46
  132.08
  130.04
  130.04
```

```
142.24
142.24
161.2
161.2
169.33
162.56
179.49
169.33
162.56
162.56
189.65
186.26
149.01
142.24
152.4
149.01
152.4
142.24
152.4
142.24
162.56
159.17
142.24
142.24
149.01
142.24
149.01
142.24
142.24
142.24
155.78
155.78
132.08
130.04
135.46
132.08
135.46
132.08
130.04
130.04
142.24
142.24
```

```
]
y1=output3
y2=y1
size(a2);
size(y2);
p=a2';
sizep=size(p);
t=y2';
sizet=size(t);
[S Q]=size(t)
[pn,meanp,stdp,tn,meant,stdt] = prestd(p,t);
net = newff(minmax(pn),[80 1],{'logsig' 'purelin'},'trainlm');
net.performFcn='mse';
net.trainParam.goal=.99;
net.trainParam.show=200;
net.trainParam.epochs=50;
net.trainParam.mc=0.05;
net = train(net,pn,tn);
an = sim(net,pn);
[a] = poststd(an,meant,stdt);
error=t-a;
x1=1:80;
plot(x1,t,'rs-',x1,a,'b-')
legend('Experimental','Neural');
title('Output (Red) and Neural Network Prediction (Blue) Plot');
xlabel('Experiment No.');
ylabel('Output'); grid on;
figure error_percentage=100×error./t
plot(x1,error_percentage)
legend('percentage error');
axis([0 80 -100 100]);
title('Percentage Error Plot in Neural Network Prediction');
xlabel('Experiment No.');
ylabel('Error in %'); grid on;
for ii=1:80
xx1=input_data(ii,1);
yy2=input_data(ii,2);
zz3=input_data(ii,3);
xx4=input_data(ii,4);
yy5=input_data(ii,5);
zz6=input_data(ii,6);
pause
yyy(1,ii)=0.5705×power(xx1,-0.9101)×power(yy2,0.0424)×power
(zz3,0.0474)
×power(xx4,0.3036)×power(yy5,-0.6156)×power(zz6,-0.4977);
yy_practical(ii)=(y2(ii,1));
yy_eqn(ii)=(yyy(1,ii))
yy_neur(ii)=(a(1,ii))
yy_practical_abs(ii)=(y2(ii,1));
yy_eqn_abs(ii)=(yyy(1,ii));
yy_neur_abs(ii)=(a(1,ii));
```

```
pause
end
 figure;
plot(x1,yy_practical_abs,'r-',x1,yy_eqn_abs,'b-',x1,yy_neur_
abs,'k-');
legend('Practical','Equation','Neural');
title('Comparision between practical data, equation-based data and
neural based data'); xlabel('Experimental');
grid on;
figure;
plot(x1,yy_practical_abs,'r-',x1,yy_eqn_abs,'b-');
legend('Practical',' Equation');
title('Comparision between practical data, equation-based data and
neural based data'); xlabel('Experimental');
grid on;
figure;
plot(x1,yy_practical_abs,'r-',x1,yy_neur_abs,'k-');
legend('Practical','Neural');
title('Comparision between practical data, equation-based data and
neural based data'); xlabel('Experimental');
grid on;
error1=yy_practical_abs-yy_eqn_abs
figure
error_percentage1=100×error1./yy_practical_abs;
plot(x1,error_percentage,'k-',x1,error_percentage1,'b-');
legend('Neural','Equation');
axis([0 80 -100 100]);
title('Percentage Error Plot in Equation (blue), Neural Network
(black) Prediction');
xlabel('Experiment No.');
ylabel('Error in %');
grid on;
meanexp=mean(output3)
meanann=mean(a)
meanmath=mean(yy_eqn_abs)
```

6.6.2 ANN Program for Turbine Power

MATLAB program for the Turbine power is the same as that for Turbine speed, except for the output for the turbine power and the model equation, which is different, as shown below:

```
Output for the Turbine Power
output3=[
  1.5
  1.6667
  1.8333
  1.75
  1.9167
  1.8333
```

```
1.75
1.75
2.0833
1.9167
4.1667
3.8889
5.6667
4.1667
5.6667
4.1667
4.1667
3.8889
6.3333
6
4.386
4.386
5.9649
4.386
5.9649
4.386
4.386
4.386
6.6667
6.6667
3.8889
3.7778
4.1667
3.8889
4.1667
3.8889
3.7778
3.7778
4.4444
4.4444
1.8333
1.8333
2.0833
1.9167
2.625
2.0833
1.9167
1.9167
3.1111
2.8194
5.6667
```

```
    4.1667
    6
    5.6667
    5.6667
    4.1667
    6
    4.1667
    6.6667
    6.3333
    4.386
    4.386
    5.9649
    4.386
    5.9649
    4.386
    4.386
    4.386
    6.6667
    6.6667
    3.8889
    3.7778
    4.1667
    3.8889
    4.1667
    3.8889
    3.7778
    3.7778
    4.4444
    4.4444
]
```

Equation for the turbine power

```
yyy(1,ii) = 7.1796 x 10-19xpower(xx1,-3.9624)xpower(yy2,0.1755)
            xpower(zz3,0.2861)xpower(xx4,0.8756)
            xpower(yy5,6.6344)xpower(zz6,-1.1456) ;
```

The above output and equation of turbine power is placed instead of turbine speed output and equation in MATLAB program. The various output graphs obtained for turbine power are shown as below by ANN program.

From the figure comparison of phenomenal response by a conventional approach and ANN simulation, it seems that the curve obtained by dependent pi terms for turbine speed π_{D1} and Power developed π_{D2} overlap due to fewer errors, which is positive, and gives an accurate relationship between ANN simulation and Experimental data.

7 Sensitivity Analysis

7.1 SENSITIVITY ANALYSIS OF CRANKSHAFT MAINTENANCE ACTIVITY

The influences of various independent π terms of crankshaft maintenance activity have been studied by analyzing the indices of the various π terms in the models. Through the technique of sensitivity analysis, the change in the value of a dependent π term caused due to an introduced change in the value of individual independent pi term is evaluated.

In this case the percentage change of ±10% is introduced in individual pi terms independently (one at time). Therefore, total range of introduced change is 20%. The effect of this introduced change on percentage change values of the dependent pi term is evaluated. The average values of the change in the dependent pi terms (z_{1C}, z_{2C}, and z_{3C}) due to the introduced change of 20% in each independent pi term (π_1, π_2, π_3, π_4,π_5, π_6, π_7, π_8, π_9, π_{10}, and π_{11}) is shown in the below Tables 7.1–7.4. These values are plotted on the graphs shown in Figures 7.1–7.3.

The model equation of overhauling time of crankshaft maintenance activity (z_{1C}) is given by:

$$
\begin{aligned}
z_{1C} &= \pi_{D1} \\
&= 1.7198(\pi_1)^{0.0873}(\pi_2)^{0.1026}(\pi_3)^{2.8137}(\pi_4)^{0.2115}(\pi_5)^{-0.2101}(\pi_6)^{-0.0025}(\pi_7)^{3.5409}(\pi_8)^{-0.4451} \\
&\quad (\pi_9)^{-0.125}(\pi_{10})^{-0.0609}(\pi_{11})^{0.1131}
\end{aligned}
$$

Model equation of human energy consumed in crankshaft maintenance activity (z_{2C}) is given by:

$$
\begin{aligned}
z_{2C} &= \pi_{D2} \\
&= 1.2203(\pi_1)^{0.2769}(\pi_2)^{0.018}(\pi_3)^{-0.2583}(\pi_4)^{3.3045}(\pi_5)^{0.0597}(\pi_6)^{-0.0314}(\pi_7)^{-1.2341}(\pi_8)^{0.0415} \\
&\quad (\pi_9)^{0.0103}(\pi_{10})^{0.0002}(\pi_{11})^{-0.3935}
\end{aligned}
$$

Model equation of productivity of crankshaft maintenance activity (z_{3C}) is given by:

$$
\begin{aligned}
z_{3C} &= \pi_{D3} \\
&= 1.6837(\pi_1)^{-0.0873}(\pi_2)^{-0.1026}(\pi_3)^{-0.1837}(\pi_4)^{-1.2275}(\pi_5)^{0.2101}(\pi_6)^{0.0025}(\pi_7)^{0.9916}(\pi_8)^{0.4451} \\
&\quad (\pi_9)^{0.125}(\pi_{10})^{0.0609}(\pi_{11})^{-0.1131}
\end{aligned}
$$

7.1.1 EFFECT OF INTRODUCED CHANGE ON THE DEPENDENT π TERM – OVERHAULING TIME OF A MAINTENANCE ACTIVITY OF CRANKSHAFT MAINTENANCE ACTIVITY (z_{1C})

In this model, when a total 20% i.e. ±10% limit of the change is introduced in the first dependent pi term (π_1), such as anthropometric data of a worker, then 1.75% change occurs in the value of z_{1C}, i.e. overhauling time of crankshaft maintenance activity. Again, change of 2.06% brought in the value of dependent pi terms (z_{1C}), due to ±10% change introduced in the value of second next pi term (π_2), i.e. workers' data for crankshaft maintenance activity.

Similarly, the change of about 12.855%, 4.24%, –4.221%, –0.05%, 30.81%, –8.96%, –2.51%, –1.22%, and 2.27% takes place due to change in the pi term values of π_3, π_4, π_5, π_6, π_7, π_8, π_9, π_{10}, and π_{11} respectively.

DOI: 10.1201/9781003318699-7

TABLE 7.1
Sensitivity Analysis of Overhauling Time of Crankshaft Maintenance Activity (z_{1C})

π_1	π_2	π_3	π_4	π_5	π_6	π_7	π_8	π_9	π_{10}	π_{11}	Computed z_{1C}	% Change
1.0879	3.684	1898.906	54.739	131.535	0.1587	4.2872	0.3273	48.067	252419.9	77.338	1252.01	
1.1967	3.684	1898.906	54.739	131.535	0.1587	4.2872	0.3273	48.067	252419.9	77.338	1262.471	1.75%
0.9791	3.684	1898.906	54.739	131.535	0.1587	4.2872	0.3273	48.067	252419.9	77.338	1240.547	
1.0879	3.684	1898.906	54.739	131.535	0.1587	4.2872	0.3273	48.067	252419.9	77.338	1252.01	
1.0879	4.053	1898.906	54.739	131.535	0.1587	4.2872	0.3273	48.067	252419.9	77.338	1264.313	2.06%
1.0879	3.316	1898.906	54.739	131.535	0.1587	4.2872	0.3273	48.067	252419.9	77.338	1238.549	
1.0879	3.684	1898.906	54.739	131.535	0.1587	4.2872	0.3273	48.067	252419.9	77.338	1252.01	
1.0879	3.684	2088.797	54.739	131.535	0.1587	4.2872	0.3273	48.067	252419.9	77.338	1327.427	12.285%
1.0879	3.684	1709.015	54.739	131.535	0.1587	4.2872	0.3273	48.067	252419.9	77.338	1173.617	
1.0879	3.684	1898.906	54.739	131.535	0.1587	4.2872	0.3273	48.067	252419.9	77.338	1252.01	
1.0879	3.684	1898.906	60.213	131.535	0.1587	4.2872	0.3273	48.067	252419.9	77.338	1277.504	4.240%
1.0879	3.684	1898.906	49.265	131.535	0.1587	4.2872	0.3273	48.067	252419.9	77.338	1224.419	
1.0879	3.684	1898.906	54.739	131.535	0.1587	4.2872	0.3273	48.067	252419.9	77.338	1252.01	
1.0879	3.684	1898.906	54.739	144.689	0.1587	4.2872	0.3273	48.067	252419.9	77.338	1227.188	–4.221%
1.0879	3.684	1898.906	54.739	118.382	0.1587	4.2872	0.3273	48.067	252419.9	77.338	1280.034	
1.0879	3.684	1898.906	54.739	131.535	0.1587	4.2872	0.3273	48.067	252419.9	77.338	1252.01	
1.0879	3.684	1898.906	54.739	131.535	0.1746	4.2872	0.3273	48.067	252419.9	77.338	1251.712	–0.05016%
1.0879	3.684	1898.906	54.739	131.535	0.1429	4.2872	0.3273	48.067	252419.9	77.338	1252.34	
1.0879	3.684	1898.906	54.739	131.535	0.1587	4.2872	0.3273	48.067	252419.9	77.338	1252.01	
1.0879	3.684	1898.906	54.739	131.535	0.1587	4.7159	0.3273	48.067	252419.9	77.338	1450.073	30.81%
1.0879	3.684	1898.906	54.739	131.535	0.1587	3.8585	0.3273	48.067	252419.9	77.338	1064.388	
1.0879	3.684	1898.906	54.739	131.535	0.1587	4.2872	0.3273	48.067	252419.9	77.338	1252.01	
1.0879	3.684	1898.906	54.739	131.535	0.1587	4.2872	0.36	48.067	252419.9	77.338	1200.007	–8.96%
1.0879	3.684	1898.906	54.739	131.535	0.1587	4.2872	0.2946	48.067	252419.9	77.338	1312.123	
1.0879	3.684	1898.906	54.739	131.535	0.1587	4.2872	0.3273	48.067	252419.9	77.338	1252.01	
1.0879	3.684	1898.906	54.739	131.535	0.1587	4.2872	0.3273	52.873	252419.9	77.338	1237.182	–2.51%
1.0879	3.684	1898.906	54.739	131.535	0.1587	4.2872	0.3273	43.26	252419.9	77.338	1268.608	
1.0879	3.684	1898.906	54.739	131.535	0.1587	4.2872	0.3273	48.067	252419.9	77.338	1252.01	
1.0879	3.684	1898.906	54.739	131.535	0.1587	4.2872	0.3273	48.067	277661.9	77.338	1244.764	–1.22%
1.0879	3.684	1898.906	54.739	131.535	0.1587	4.2872	0.3273	48.067	227177.9	77.338	1260.069	
1.0879	3.684	1898.906	54.739	131.535	0.1587	4.2872	0.3273	48.067	252419.9	77.338	1252.01	
1.0879	3.684	1898.906	54.739	131.535	0.1587	4.2872	0.3273	48.067	252419.9	85.072	1265.579	2.27%
1.0879	3.684	1898.906	54.739	131.535	0.1587	4.2872	0.3273	48.067	252419.9	69.604	1237.179	

TABLE 7.2
Sensitivity Analysis of Human Energy Consumed in Crankshaft Maintenance Activity (z_{2C})

π_1	π_2	π_3	π_4	π_5	π_6	π_7	π_8	π_9	π_{10}	π_{11}	Computed z_{2C}	% Change
1.08791	3.684437	1898.906	54.73918	131.5352	0.15874	4.28719	0.3273	48.06667	252419.9	77.33833	3694.691	
1.19671	3.684437	1898.906	54.73918	131.5352	0.15874	4.28719	0.3273	48.06667	252419.9	77.33833	3793.497	5.55%
0.97912	3.684437	1898.906	54.73918	131.5352	0.15874	4.28719	0.3273	48.06667	252419.9	77.33833	3588.458	
1.08791	3.684437	1898.906	54.73918	131.5352	0.15874	4.28719	0.3273	48.06667	252419.9	77.33833	3694.691	
1.08791	4.0528807	1898.906	54.73918	131.5352	0.15874	4.28719	0.3273	48.06667	252419.9	77.33833	3701.035	0.36%
1.08791	3.3159933	1898.906	54.73918	131.5352	0.15874	4.28719	0.3273	48.06667	252419.9	77.33833	3687.69	
1.08791	3.684437	1898.906	54.73918	131.5352	0.15874	4.28719	0.3273	48.06667	252419.9	77.33833	3694.691	
1.08791	3.684437	2088.7966	54.73918	131.5352	0.15874	4.28719	0.3273	48.06667	252419.9	77.33833	3604.843	–5.19%
1.08791	3.684437	1709.0154	54.73918	131.5352	0.15874	4.28719	0.3273	48.06667	252419.9	77.33833	3796.621	
1.08791	3.684437	1898.906	54.73918	131.5352	0.15874	4.28719	0.3273	48.06667	252419.9	77.33833	3694.691	
1.08791	3.684437	1898.906	60.21309	131.5352	0.15874	4.28719	0.3273	48.06667	252419.9	77.33833	5062.444	66.42%
1.08791	3.684437	1898.906	49.26526	131.5352	0.15874	4.28719	0.3273	48.06667	252419.9	77.33833	2608.389	
1.08791	3.684437	1898.906	54.73918	131.5352	0.15874	4.28719	0.3273	48.06667	252419.9	77.33833	3694.691	
1.08791	3.684437	1898.906	54.73918	144.6887	0.15874	4.28719	0.3273	48.06667	252419.9	77.33833	3715.773	1.20%
1.08791	3.684437	1898.906	54.73918	118.3816	0.15874	4.28719	0.3273	48.06667	252419.9	77.33833	3671.524	
1.08791	3.684437	1898.906	54.73918	131.5352	0.15874	4.28719	0.3273	48.06667	252419.9	77.33833	3694.691	
1.08791	3.684437	1898.906	54.73918	131.5352	0.17461	4.28719	0.3273	48.06667	252419.9	77.33833	3683.65	–0.63%
1.08791	3.684437	1898.906	54.73918	131.5352	0.14286	4.28719	0.3273	48.06667	252419.9	77.33833	3706.934	
1.08791	3.684437	1898.906	54.73918	131.5352	0.15874	4.28719	0.3273	48.06667	252419.9	77.33833	3694.691	
1.08791	3.684437	1898.906	54.73918	131.5352	0.15874	4.71590	0.3273	48.06667	252419.9	77.33833	3253.54	–27.03%
1.08791	3.684437	1898.906	54.73918	131.5352	0.15874	3.85847	0.3273	48.06667	252419.9	77.33833	4252.293	
1.08791	3.684437	1898.906	54.73918	131.5352	0.15874	4.28719	0.3273	48.06667	252419.9	77.33833	3694.691	
1.08791	3.684437	1898.906	54.73918	131.5352	0.15874	4.28719	0.3600	48.06667	252419.9	77.33833	3709.333	0.83%
1.08791	3.684437	1898.906	54.73918	131.5352	0.15874	4.28719	0.2945	48.06667	252419.9	77.33833	3678.571	
1.08791	3.684437	1898.906	54.73918	131.5352	0.15874	4.28719	0.3273	48.06667	252419.9	77.33833	3694.691	
1.08791	3.684437	1898.906	54.73918	131.5352	0.15874	4.28719	0.3273	52.87333	252419.9	77.33833	3698.319	0.21%
1.08791	3.684437	1898.906	54.73918	131.5352	0.15874	4.28719	0.3273	43.26000	252419.9	77.33833	3690.683	
1.08791	3.684437	1898.906	54.73918	131.5352	0.15874	4.28719	0.3273	48.06667	252419.9	77.33833	3694.691	
1.08791	3.684437	1898.906	54.73918	131.5352	0.15874	4.28719	0.3273	48.06667	277661.8	77.33833	3694.761	0.004%
1.08791	3.684437	1898.906	54.73918	131.5352	0.15874	4.28719	0.3273	48.06667	227177.9	77.33833	3694.613	
1.08791	3.684437	1898.906	54.73918	131.5352	0.15874	4.28719	0.3273	48.06667	252419.9	77.33833	3694.691	
1.08791	3.684437	1898.906	54.73918	131.5352	0.15874	4.28719	0.3273	48.06667	252419.9	85.07216	3558.689	–7.91%
1.08799	3.684437	1898.906	54.73918	131.5352	0.15874	4.28719	0.3273	48.06667	252419.9	69.60449	3851.09	

TABLE 7.3
Sensitivity Analysis of Productivity of Crankshaft Maintenance Activity (z_{3C})

π_1	π_2	π_3	π_4	π_5	π_6	π_7	π_8	π_9	π_{10}	π_{11}	Computed z_{3C}	% Change
1.087919	3.684437	1898.906	54.7391	131.5352	0.15874	4.28719	0.3273	48.06667	252419.9	77.3383	0.040997	
1.196710	3.684437	1898.906	54.7391	131.5352	0.15874	4.28719	0.3273	48.06667	252419.9	77.3383	0.040658	−1.751%
0.979127	3.684437	1898.906	54.7391	131.5352	0.15874	4.28719	0.3273	48.06667	252419.9	77.3383	0.041376	
1.087919	3.684437	1898.906	54.7391	131.5352	0.15874	4.28719	0.3273	48.06667	252419.9	77.3383	0.040997	
1.087919	4.052880	1898.906	54.7391	131.5352	0.15874	4.28719	0.3273	48.06667	252419.9	77.3383	0.040598	−2.06%
1.087919	3.315993	1898.906	54.7391	131.5352	0.15874	4.28719	0.3273	48.06667	252419.9	77.3383	0.041443	
1.087919	3.684437	1898.906	54.7391	131.5352	0.15874	4.28719	0.3273	48.06667	252419.9	77.3383	0.040997	
1.087919	3.684437	2088.796	54.7391	131.5352	0.15874	4.28719	0.3273	48.06667	252419.9	77.3383	0.040286	−3.69%
1.087919	3.684437	1709.015	54.7391	131.5352	0.15874	4.28719	0.3273	48.06667	252419.9	77.3383	0.041798	
1.087919	3.684437	1898.906	54.7391	131.5352	0.15874	4.28719	0.3273	48.06667	252419.9	77.3383	0.040997	
1.087919	3.684437	1898.906	60.2130	131.5352	0.15874	4.28719	0.3273	48.06667	252419.9	77.3383	0.036471	−24.85%
1.087919	3.684437	1898.906	49.2652	131.5352	0.15874	4.28719	0.3273	48.06667	252419.9	77.3383	0.046658	
1.087919	3.684437	1898.906	54.7391	131.5352	0.15874	4.28719	0.3273	48.06667	252419.9	77.3383	0.040997	
1.087919	3.684437	1898.906	54.7391	144.6887	0.15874	4.28719	0.3273	48.06667	252419.9	77.3383	0.041826	4.21%
1.087919	3.684437	1898.906	54.7391	118.3816	0.15874	4.28719	0.3273	48.06667	252419.9	77.3383	0.0401	
1.087919	3.684437	1898.906	54.7391	131.5352	0.15874	4.28719	0.3273	48.06667	252419.9	77.3383	0.040997	
1.087919	3.684437	1898.906	54.7391	131.5352	0.17461	4.28719	0.3273	48.06667	252419.9	77.3383	0.040987	−0.05%
1.087919	3.684437	1898.906	54.7391	131.5352	0.14286	4.28719	0.3273	48.06667	252419.9	77.3383	0.041008	
1.087919	3.684437	1898.906	54.7391	131.5352	0.15874	4.28719	0.3273	48.06667	252419.9	77.3383	0.040997	
1.087919	3.684437	1898.906	54.7391	131.5352	0.15874	4.71590	0.3273	48.06667	252419.9	77.3383	0.045061	19.83%
1.087919	3.684437	1898.906	54.7391	131.5352	0.15874	3.85847	0.3273	48.06667	252419.9	77.3383	0.03693	
1.087919	3.684437	1898.906	54.7391	131.5352	0.15874	4.28719	0.3273	48.06667	252419.9	77.3383	0.040997	
1.087919	3.684437	1898.906	54.7391	131.5352	0.15874	4.28719	0.3600	48.06667	252419.9	77.3383	0.042774	8.92%
1.087919	3.684437	1898.906	54.7391	131.5352	0.15874	4.28719	0.2945	48.06667	252419.9	77.3383	0.039119	
1.087919	3.684437	1898.906	54.7391	131.5352	0.15874	4.28719	0.3273	48.06667	252419.9	77.3383	0.040997	
1.087919	3.684437	1898.906	54.7391	131.5352	0.15874	4.28719	0.3273	52.87334	252419.9	77.3383	0.041489	2.51%
1.087919	3.684437	1898.906	54.7391	131.5352	0.15874	4.28719	0.3273	43.26	252419.9	77.3383	0.040461	
1.087919	3.684437	1898.906	54.7391	131.5352	0.15874	4.28719	0.3273	48.06667	252419.9	77.3383	0.040997	
1.087919	3.684437	1898.906	54.7391	131.5352	0.15874	4.28719	0.3273	48.06667	277661.8	77.3383	0.041236	1.22%
1.087919	3.684437	1898.906	54.7391	131.5352	0.15874	4.28719	0.3273	48.06667	227177.9	77.3383	0.040735	
1.087919	3.684437	1898.906	54.7391	131.5352	0.15874	4.28719	0.3273	48.06667	252419.9	77.3383	0.040997	
1.087919	3.684437	1898.906	54.7391	131.5352	0.15874	4.28719	0.3273	48.06667	252419.9	85.0721	0.040558	−2.27%
									[illegible]	69.6044	0.041489	

TABLE 7.4
Percentage Change in z_{1C}, z_{2C} and z_{3C} with a ±10% Variation of Mean Value of Each π Term at a Time – Crankshaft Maintenance Activity

Pi terms	z_{1C}	z_{2C}	z_{3C}
π_1	1.75%	5.55%	1.75%
π_2	2.06%	0.36%	2.06%
π_3	12.29%	5.19%	3.69%
π_4	4.24%	66.42%	24.85%
π_5	4.22%	1.20%	4.21%
π_6	0.05%	0.63%	0.05%
π_7	30.81%	27.03%	19.83%
π_8	8.96%	0.83%	8.92%
π_9	2.51%	0.21%	2.51%
π_{10}	1.22%	0.004%	1.22%
π_{11}	2.27%	7.91%	2.27%

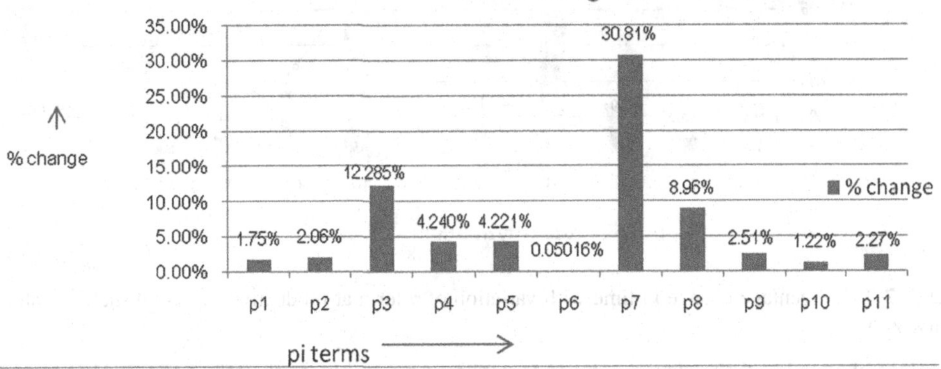

FIGURE 7.1 Percentage change in time with variation of π term at an overhauling time of crankshaft maintenance activity (z_{1C}).

Percentage change in all the eleven independent Pi terms to evaluate the effect of this changes on the dependent variable Time.

From Figure 7.1, it can be seen that π_7 is most sensitive among all pi terms, and highest change takes place in response variables because of independent pi term π_7 (workstation data), where π_6 (axial clearance of crankpin bearing) is least sensitive among all pi terms, and least change takes place in the response variable.

The sequence of influence of various pi terms on response variable in descending order is as follows: π_7, π_3, π_8, π_4, π_5, π_9, π_{11}, π_2, π_1, π_{10}, and π_6.

7.1.2 Effect of Introduced Change on the Dependent π Term – Human Energy Consumed in Crankshaft Maintenance Activity (z_{2C})

In this model, when a total 20% (i.e. ±10%) limit of the change is introduced in the first dependent pi term (π_1) i.e. anthropometric data then 5.55% change occurs in the value of z_{2C} (i.e. human energy consumed in crankshaft maintenance activity). Again, change of 0.36% brought in the value of

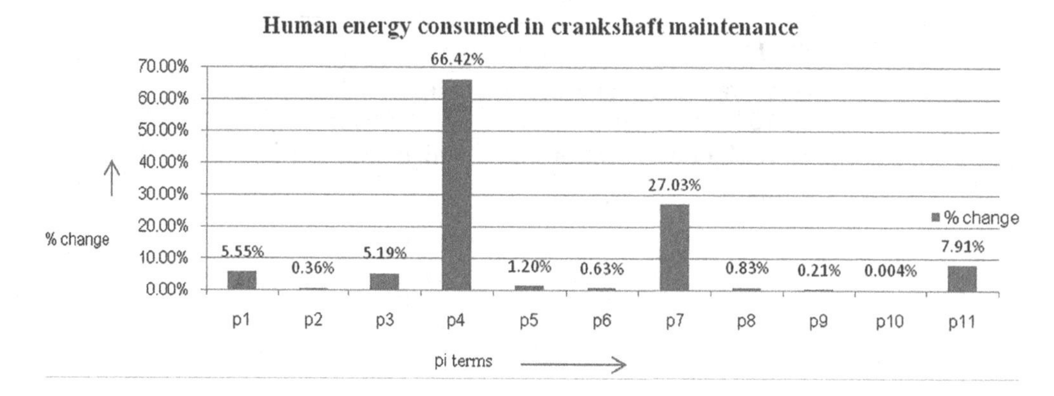

FIGURE 7.2 Percentage change in time with variation of π term at a human energy consumed in crankshaft maintenance activity (z_{1C}).

Percentage change in all the eleven independent pi terms to evaluate the effect of this changes on the dependent variable human energy consumption.

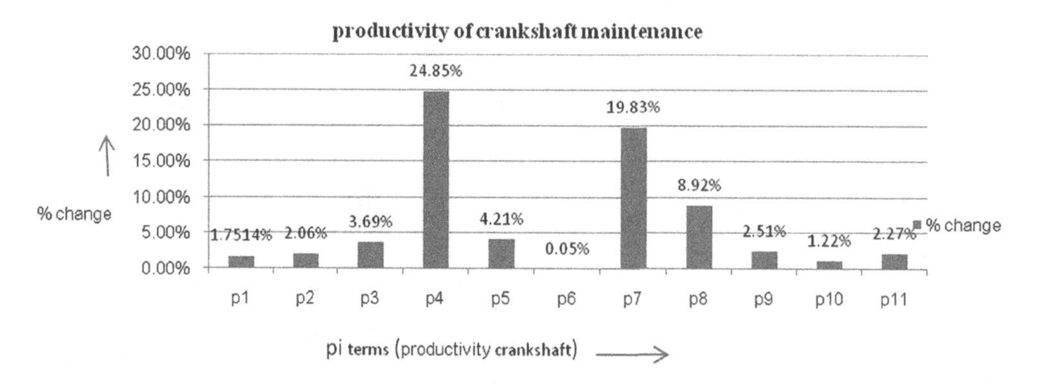

FIGURE 7.3 Percentage change in time with variation of π term at productivity of crankshaft maintenance activity (z_{1C}).

Percentage change in all the eleven independent pi terms to evaluate the effect of this changes on the dependent variable productivity.

dependent pi terms (z_{2C}) i.e. human energy consumed in liner piston maintenance activity due to ±10% change introduced in the value of second next pi term (π_2) (i.e. workers data).

Similarly, the change of about −5.19%, 66.42%, 1.20%, −0.63%, −27.03%, 0.83%, 0.21%, 0.004%, and −7.91% takes place due to change in the independent pi term values of π_3, π_4, π_5, π_6, π_7, π_8, π_9, π_{10}, and π_{11} respectively.

From Figure 7.2, it can be seen that π_4 is most sensitive among all pi terms, and highest change takes place in the response variable because of independent pi term π_4 (specification of tools) where π_{10} (illumination at workplace) is least sensitive among all pi terms, and least change take place in the response variable. The sequence of influence of various pi terms on response variable in descending order is as follows: π_4, π_7, π_{11}, π_1, π_3, π_5, π_8, π_6, π_2, π_9, and π_{10}.

7.1.3 Effect of Introduced Change on the Dependent π term – Productivity of Crankshaft Maintenance Activity (z_{3C})

In this model, when a total of 20% (i.e. ±10%) limit of the change is introduced in first dependent pi term (π_1) i.e. anthropometric data then −1.7514% change occurs in the value of z_{3C} i.e. productivity of crankshaft maintenance activity. Again change of −2.06% brought in the value of dependent pi terms (z_{3C}), due to ±10% change introduced in the value of second next pi term (π_2) i.e. workers data.

Similarly, the change of about 20% is introduced in other pi terms and −3.69%, −24.85%, 4.21%, −0.05%, 19.83%, 8.92%, 2.51%, 1.22%, and −2.27% takes place due to change in the independent pi term values of π_3, π_4, π_5, π_6, π_7, π_8, π_9, π_{10}, and π_{11} respectively.

From Figure 7.3, it can be seen that π_4 is most sensitive among all pi terms, and highest change takes place in the response variable because of independent pi term π_4 (specification of Tools) where π_6 (axial clearance of crankpin bearing) is least sensitive among all pi terms, and least change take place in the response variable.

The sequence of influence of various pi terms on the response variable in descending order is as follows: π_4, π_7, π_8, π_5, π_3, π_9, π_{11}, π_2, π_1, π_{10}, and π_6.

7.2 SENSITIVITY ANALYSIS OF LINER PISTON MAINTENANCE ACTIVITY

The influences of various independent π terms of liner piston maintenance activity have been studied by analyzing the indices of the various π terms in the models. Through the technique of sensitivity analysis the change in the value of a dependent pi term caused due to an introduced change in the value of individual independent pi term is evaluated.

In this case, the change of ±10% is introduced in individual pi terms independently (one at time). Therefore, total range of introduced change is 20%. The effect of this introduced change on percentage change of dependent pi term is evaluated. The average value of the change in the dependent pi terms (z_{1P}, z_{2P}, z_{3P}) due to the introduced change of 20% in each independent pi term (π_1, π_2, π_3, π_4,π_5, π_6, π_7, π_8, π_9, π_{10}, and π_{11}) is shown in Tables 7.5–7.8. These values are plotted on the graphs as shown in Figures 7.4–7.6.

Model equation of overhauling time of liner piston maintenance activity (z_{1P}) is given by:

$$\begin{aligned} Z_{1P} &= \pi_{D1} \\ &= 1.0122(\pi_1)^{-0.0973}(\pi_2)^{-0.1917}(\pi_3)^{0.8772}(\pi_4)^{-6.0098}(\pi_5)^{0.0896}(\pi_6)^{0.378}(\pi_7)^{0.6869}(\pi_8)^{-0.5615} \\ &\quad (\pi_9)^{-0.2282}(\pi_{10})^{0.0887}(\pi_{11})^{0.0838} \end{aligned}$$

Model equation of human energy consumed in liner piston maintenance activity (z_{2P}) is given by

$$\begin{aligned} Z_{2P} &= \pi_{D2} \\ &= 1.7874(\pi_1)^{0.355}(\pi_2)^{0.3448}(\pi_3)^{0.1681}(\pi_4)^{4.4021}(\pi_5)^{-0.0249}(\pi_6)^{-0.131}(\pi_7)^{-0.3313}(\pi_8)^{-0.1018} \\ &\quad (\pi_9)^{-0.1479}(\pi_{10})^{0.0643}(\pi_{11})^{0.2221} \end{aligned}$$

Model equation of productivity of liner piston maintenance activity (z_{3P}) is given by:

$$\begin{aligned} Z_{3P} &= \pi_{D3} \\ &= 1.1148(\pi_1)^{0.0973}(\pi_2)^{0.1917}(\pi_3)^{0.0188}(\pi_4)^{-2.5291}(\pi_5)^{-0.0896}(\pi_6)^{-0.378}(\pi_7)^{2.9719}(\pi_8)^{0.5615} \\ &\quad (\pi_9)^{0.2282}(\pi_{10})^{-0.0887}(\pi_{11})^{-0.0838} \end{aligned}$$

7.2.1 EFFECT OF INTRODUCED CHANGE ON THE DEPENDENT π TERM – OVERHAULING TIME OF A OF LINER PISTON MAINTENANCE ACTIVITY (Z_{1P})

In this model, when a total 20% (i.e. ±10%) limit of the change is introduced in first dependent pi term (π_1) (e.g. anthropometric data) then −1.95% change occurs in the value of z_{1P} (i.e. overhauling time of liner piston maintenance activity). Again, change of −3.85% brought in the value of

TABLE 7.5
Sensitivity Analysis for Overhauling Time of Liner Piston Maintenance Activity (z_{1P})

π_1	π_2	π_3	π_4	π_5	π_6	π_7	π_8	π_9	π_{10}	π_{11}	Computed z_{1P}	% Change
1.08791	2.834828	2.29E+10	0.459878	76.22086	0.140796	0.21384	0.331667	50.6667	251024.3	79.08	5069.086	
1.196710	2.834828	2.29E+10	0.459878	76.22086	0.140796	0.21384	0.331667	50.6667	251024.3	79.08	5022.294	–1.95%
0.979127	2.834828	2.29E+10	0.459878	76.22086	0.140796	0.21384	0.331667	50.6667	251024.3	79.08	5121.319	
1.08791	2.834828	2.29E+10	0.459878	76.22086	0.140796	0.21384	0.331667	50.6667	251024.3	79.08	5069.086	
1.08791	3.118310	2.29E+10	0.459878	76.22086	0.140796	0.21384	0.331667	50.6667	251024.3	79.08	4977.31	–3.85%
1.08791	2.551345	2.29E+10	0.459878	76.22086	0.140796	0.21384	0.331667	50.6667	251024.3	79.08	5172.51	
1.08791	2.834828	2.29E+10	0.459878	76.22086	0.140796	0.21384	0.331667	50.6667	251024.3	79.08	5069.086	
1.08791	2.834828	2.52E+10	0.459878	76.22086	0.140796	0.21384	0.331667	50.6667	251024.3	79.08	5126.027	2.35%
1.08791	2.834828	2.06E+10	0.459878	76.22086	0.140796	0.21384	0.331667	50.6667	251024.3	79.08	5006.876	
1.08791	2.834828	2.29E+10	0.459878	76.22086	0.140796	0.21384	0.331667	50.6667	251024.3	79.08	5069.086	
1.08791	2.834828	2.29E+10	0.505865	76.22086	0.140796	0.21384	0.331667	50.6667	251024.3	79.08	2858.695	–131.9%
1.08791	2.834828	2.29E+10	0.413890	76.22086	0.140796	0.21384	0.331667	50.6667	251024.3	79.08	9548.233	
1.08791	2.834828	2.29E+10	0.459878	76.22086	0.140796	0.21384	0.331667	50.6667	251024.3	79.08	5069.086	
1.08791	2.834828	2.29E+10	0.459878	83.84294	0.140796	0.21384	0.331667	50.6667	251024.3	79.08	5112.56	1.80%
1.08791	2.834828	2.29E+10	0.459878	68.59877	0.140796	0.21384	0.331667	50.6667	251024.3	79.08	5021.457	
1.08791	2.834828	2.29E+10	0.459878	76.22086	0.140796	0.21384	0.331667	50.6667	251024.3	79.08	5069.086	
1.08791	2.834828	2.29E+10	0.459878	76.22086	0.1548756	0.21384	0.331667	50.6667	251024.3	79.08	4889.711	–7.60%
1.08791	2.834828	2.29E+10	0.459878	76.22086	0.1267164	0.21384	0.331667	50.6667	251024.3	79.08	5275.043	
1.08791	2.834828	2.29E+10	0.459878	76.22086	0.140796	0.21384	0.331667	50.6667	251024.3	79.08	5069.086	
1.08791	2.834828	2.29E+10	0.459878	76.22086	0.140796	0.23523	0.331667	50.6667	251024.3	79.08	5412.056	13.75%
1.08791	2.834828	2.29E+10	0.459878	76.22086	0.140796	0.19246	0.331667	50.6667	251024.3	79.08	4715.186	
1.08791	2.834828	2.29E+10	0.459878	76.22086	0.140796	0.21384	0.331667	50.6667	251024.3	79.08	5069.086	
1.08791	2.834828	2.29E+10	0.459878	76.22086	0.140796	0.21384	0.364833	50.6667	251024.3	79.08	4804.937	–11.31%
1.08791	2.834828	2.29E+10	0.459878	76.22086	0.140796	0.21384	0.298500	50.6667	251024.3	79.08	5378.021	
1.08791	2.834828	2.29E+10	0.459878	76.22086	0.140796	0.21384	0.331667	50.6667	251024.3	79.08	5069.086	
1.08791	2.834828	2.29E+10	0.459878	76.22086	0.140796	0.21384	0.331667	55.7333	251024.3	79.08	4960.025	–4.58%
1.08791	2.834828	2.29E+10	0.459878	76.22086	0.140796	0.21384	0.331667	45.6000	251024.3	79.08	5192.44	
1.08791	2.834828	2.29E+10	0.459878	76.22086	0.140796	0.21384	0.331667	50.6667	251024.3	79.08	5069.086	
1.08791	2.834828	2.29E+10	0.459878	76.22086	0.140796	0.21384	0.331667	50.6667	276126.7	79.08	5112.122	1.78%
1.08791	2.834828	2.29E+10	0.459878	76.22086	0.140796	0.21384	0.331667	50.6667	225921.8	79.08	5021.934	
1.08791	2.834828	2.29E+10	0.459878	76.22086	0.140796	0.21384	0.331667	50.6667	251024.3	79.08	5069.086	
1.08791	2.834828	2.29E+10	0.459878	76.22086	0.140796	0.21384	0.331667	50.6667	251024.3	86.98	5109.735	1.68%
1.08791	2.834828	2.29E+10	0.459878	76.22086	0.140796	0.21384	0.331667	50.6667	251024.3	71.12	5024.527	

TABLE 7.6

Sensitivity Analysis for Human Energy Consumed in Liner Piston Maintenance Activity (z_{2P})

π_1	π_2	π_3	π_4	π_5	π_6	π_7	π_8	π_9	π_{10}	π_{11}	Computed z_{2P}	% Change
1.087919	2.834828	2.29E+10	0.459878	76.22086	0.140796	0.213847754	0.331667	50.6667	251024.3	79.08	3461.582	
1.196710	2.834828	2.29E+10	0.459878	76.22086	0.140796	0.213847754	0.331667	50.6667	251024.3	79.08	3580.709	7.11%
0.979127	2.834828	2.29E+10	0.459878	76.22086	0.140796	0.213847754	0.331667	50.6667	251024.3	79.08	3334.5	
1.087919	2.834828	2.29E+10	0.459878	76.22086	0.140796	0.213847754	0.331667	50.6667	251024.3	79.08	3461.582	
1.087919	3.118310	2.29E+10	0.459878	76.22086	0.140796	0.213847754	0.331667	50.6667	251024.3	79.08	3577.23	6.91%
1.087919	2.551345	2.29E+10	0.459878	76.22086	0.140796	0.213847754	0.331667	50.6667	251024.3	79.08	3338.086	
1.087919	2.834828	2.29E+10	0.459878	76.22086	0.140796	0.213847754	0.331667	50.6667	251024.3	79.08	3461.582	
1.087919	2.834828	2.52E+10	0.459878	76.22086	0.140796	0.213847754	0.331667	50.6667	251024.3	79.08	3517.489	3.37%
1.087919	2.834828	2.06E+10	0.459878	76.22086	0.140796	0.213847754	0.331667	50.6667	251024.3	79.08	3400.814	
1.087919	2.834828	2.29E+10	0.459878	76.22086	0.140796	0.213847754	0.331667	50.6667	251024.3	79.08	3461.582	
1.087919	2.834828	2.29E+10	0.505865	76.22086	0.140796	0.213847754	0.331667	50.6667	251024.3	79.08	5266.104	89.24%
1.087919	2.834828	2.29E+10	0.413890	76.22086	0.140796	0.213847754	0.331667	50.6667	251024.3	79.08	2176.936	
1.087919	2.834828	2.29E+10	0.459878	76.22086	0.140796	0.213847754	0.331667	50.6667	251024.3	79.08	3461.582	
1.087919	2.834828	2.29E+10	0.459878	83.84294	0.140796	0.213847754	0.331667	50.6667	251024.3	79.08	3453.377	−0.50%
1.087919	2.834828	2.29E+10	0.459878	68.59877	0.140796	0.213847754	0.331667	50.6667	251024.3	79.08	3470.676	
1.087919	2.834828	2.29E+10	0.459878	76.22086	0.140796	0.213847754	0.331667	50.6667	251024.3	79.08	3461.582	
1.087919	2.834828	2.29E+10	0.459878	76.22086	0.154875	0.213847754	0.331667	50.6667	251024.3	79.08	3418.631	−2.63%
1.087919	2.834828	2.29E+10	0.459878	76.22086	0.126716	0.213847754	0.331667	50.6667	251024.3	79.08	3509.691	
1.087919	2.834828	2.29E+10	0.459878	76.22086	0.140796	0.213847754	0.331667	50.6667	251024.3	79.08	3461.582	
1.087919	2.834828	2.29E+10	0.459878	76.22086	0.140796	0.235232529	0.331667	50.6667	251024.3	79.08	2519.899	−69.25%
1.087919	2.834828	2.29E+10	0.459878	76.22086	0.140796	0.192462979	0.331667	50.6667	251024.3	79.08	4917.072	
1.087919	2.834828	2.29E+10	0.459878	76.22086	0.140796	0.213847754	0.331667	50.6667	251024.3	79.08	3461.582	
1.087919	2.834828	2.29E+10	0.459878	76.22086	0.140796	0.213847754	0.364833	50.6667	251024.3	79.08	3428.158	−2.04%
1.087919	2.834828	2.29E+10	0.459878	76.22086	0.140796	0.213847754	0.298500	50.6667	251024.3	79.08	3498.91	
1.087919	2.834828	2.29E+10	0.459878	76.22086	0.140796	0.213847754	0.331667	50.6667	251024.3	79.08	3461.582	
1.087919	2.834828	2.29E+10	0.459878	76.22086	0.140796	0.213847754	0.331667	55.7333	251024.3	79.08	3413.129	−2.97%
1.087919	2.834828	2.29E+10	0.459878	76.22086	0.140796	0.213847754	0.331667	45.6000	251024.3	79.08	3515.946	
1.087919	2.834828	2.29E+10	0.459878	76.22086	0.140796	0.213847754	0.331667	50.6667	251024.3	79.08	3461.582	
1.087919	2.834828	2.29E+10	0.459878	76.22086	0.140796	0.213847754	0.331667	50.6667	276126.73	79.08	3482.861	1.29%
1.087919	2.834828	2.29E+10	0.459878	76.22086	0.140796	0.213847754	0.331667	50.6667	225921.87	79.08	3438.21	
1.087919	2.834828	2.29E+10	0.459878	76.22086	0.140796	0.213847754	0.331667	50.6667	251024.3	79.08	3461.582	
1.087919	2.834828	2.29E+10	0.459878	76.22086	0.140796	0.213847754	0.331667	50.6667	251024.3	86.98	3535.639	4.45%
1.087919	2.834828	2.29E+10	0.459878	76.22086	0.140796	0.213847754	0.331667	50.6667	251024.3	71.17	3381.52	

TABLE 7.7
Sensitivity Analysis for Productivity of Liner Piston Maintenance Activity (z_{3P})

π_1	π_2	π_3	π_4	π_5	π_6	π_7	π_8	π_9	π_{10}	π_{11}	Computed z_{3P}	% Change
1.08791	2.834828	2.29E+10	0.459878	76.2208	0.140796	0.213847	0.331667	50.6667	251024.3	79.08	0.35127	
1.19671	2.834828	2.29E+10	0.459878	76.2208	0.140796	0.213847	0.331667	50.6667	251024.3	79.08	0.354543	1.95149
0.97911	2.834828	2.29E+10	0.459878	76.2208	0.140796	0.213847	0.331667	50.6667	251024.3	79.08	0.347688	
1.08799	2.834828	2.29E+10	0.459878	76.2208	0.140796	0.213847	0.331667	50.6667	251024.3	79.08	0.35127	
1.08791	3.118310	2.29E+10	0.459878	76.2208	0.140796	0.213847	0.331667	50.6667	251024.3	79.08	0.357747	3.843198
1.08799	2.551345	2.29E+10	0.459878	76.2208	0.140796	0.213847	0.331667	50.6667	251024.3	79.08	0.344247	
1.08799	2.834828	2.29E+10	0.459878	76.2208	0.140796	0.213847	0.331667	50.6667	251024.3	79.08	0.35127	
1.08799	2.834828	2.52E+10	0.459878	76.2208	0.140796	0.213847	0.331667	50.6667	251024.3	79.08	0.354226	1.763316
1.08799	2.834828	2.06E+10	0.459878	76.2208	0.140796	0.213847	0.331667	50.6667	251024.3	79.08	0.348032	
1.08799	2.834828	2.29E+10	0.459878	76.2208	0.140796	0.213847	0.331667	50.6667	251024.3	79.08	0.35127	
1.08799	2.834828	2.29E+10	0.505865	76.2208	0.140796	0.213847	0.331667	50.6667	251024.3	79.08	0.276029	−51.9543
1.08799	2.834828	2.29E+10	0.413890	76.2208	0.140796	0.213847	0.331667	50.6667	251024.3	79.08	0.458529	
1.08799	2.834828	2.29E+10	0.459878	76.2208	0.140796	0.213847	0.331667	50.6667	251024.3	79.08	0.35127	
1.08799	2.834828	2.29E+10	0.459878	83.8429	0.140796	0.213847	0.331667	50.6667	251024.3	79.08	0.348283	−1.7989
1.08799	2.834828	2.29E+10	0.459878	68.5987	0.140796	0.213847	0.331667	50.6667	251024.3	79.08	0.354602	
1.08799	2.834828	2.29E+10	0.459878	76.2208	0.140796	0.213847	0.331667	50.6667	251024.3	79.08	0.35127	
1.08799	2.834828	2.29E+10	0.459878	76.2208	0.154875	0.213847	0.331667	50.6667	251024.3	79.08	0.33884	−7.60184
1.08799	2.834828	2.29E+10	0.459878	76.2208	0.126716	0.213847	0.331667	50.6667	251024.3	79.08	0.365543	
1.08799	2.834828	2.29E+10	0.459878	76.2208	0.140796	0.213847	0.331667	50.6667	251024.3	79.08	0.35127	
1.08799	2.834828	2.29E+10	0.459878	76.2208	0.140796	0.235232	0.331667	50.6667	251024.3	79.08	0.46629	59.62792
1.08799	2.834828	2.29E+10	0.459878	76.2208	0.140796	0.192462	0.331667	50.6667	251024.3	79.08	0.256835	
1.08799	2.834828	2.29E+10	0.459878	76.2208	0.140796	0.213847	0.331667	50.6667	251024.3	79.08	0.35127	
1.08799	2.834828	2.29E+10	0.459878	76.2208	0.140796	0.213847	0.364833	50.6667	251024.3	79.08	0.370581	11.24178
1.08799	2.834828	2.29E+10	0.459878	76.2208	0.140796	0.213847	0.298500	50.6667	251024.3	79.08	0.331092	
1.08799	2.834828	2.29E+10	0.459878	76.2208	0.140796	0.213847	0.331667	50.6667	251024.3	79.08	0.35127	
1.08799	2.834828	2.29E+10	0.459878	76.2208	0.140796	0.213847	0.331667	55.7333	251024.3	79.08	0.358994	4.574544
1.08799	2.834828	2.29E+10	0.459878	76.2208	0.140796	0.213847	0.331667	45.600	251024.3	79.08	0.342925	
1.08799	2.834828	2.29E+10	0.459878	76.2208	0.140796	0.213847	0.331667	50.6667	251024.3	79.08	0.35127	
1.08799	2.834828	2.29E+10	0.459878	76.2208	0.140796	0.213847	0.331667	50.6667	276126.73	79.08	0.348313	−1.7810
1.08799	2.834828	2.29E+10	0.459878	76.2208	0.140796	0.213847	0.331667	50.6667	225921.87	79.08	0.354569	
1.08799	2.834828	2.29E+10	0.459878	76.2208	0.140796	0.213847	0.331667	50.6667	251024.3	79.08	0.35127	
1.08799	2.834828	2.29E+10	0.459878	76.2208	0.140796	0.213847	0.331667	50.6667	251024.3	86.988	0.348476	−1.6825
1.08799	2.834828	2.29E+10	0.459878	76.2208	0.140796	0.213847	0.331667	50.6667	251024.3	71.172	0.354386	

TABLE 7.8
Percentage Change in z_{1P}, z_{2P} and z_{3P} with a ±10% Variation of Mean Value of Each π Term at a Time – Liner Piston Maintenance Activity

Pi terms	z_{1P}	z_{2P}	z_{3P}
π_1	1.95%	7.11%	1.95%
π_2	3.85%	6.91%	3.84%
π_3	2.35%	3.37%	1.76%
π_4	131.97%	89.24%	51.95%
π_5	1.80%	0.50%	1.80%
π_6	7.60%	2.63%	7.60%
π_7	13.75%	69.25%	59.63%
π_8	11.31%	2.04%	11.24%
π_9	4.58%	2.97%	4.58%
π_{10}	1.78%	1.29%	1.78%
π_{11}	1.68%	4.45%	1.68%

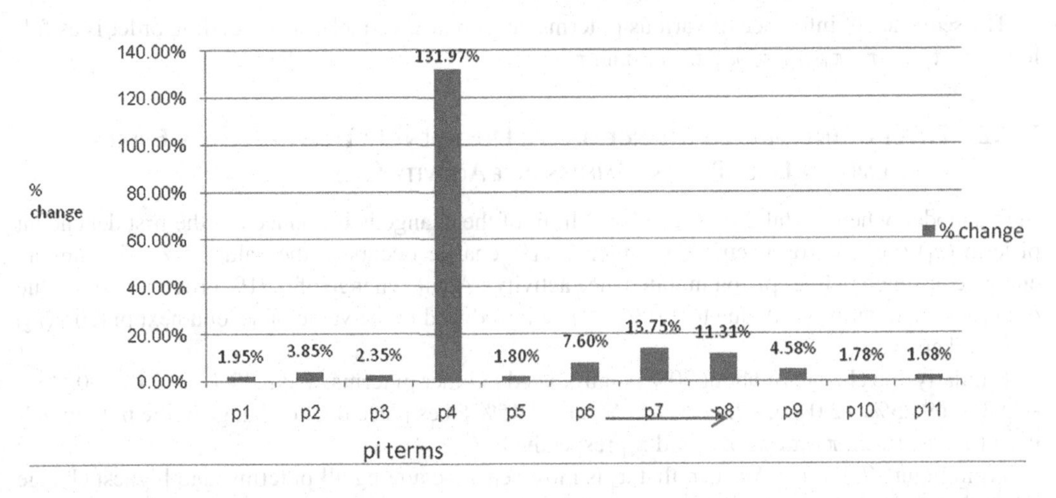

FIGURE 7.4 Percentage change in time with variation of π term at an overhauling time of liner piston maintenance activity (z_{1P}).

Percentage change in all the eleven independent Pi terms to evaluate the effect of these changes on the dependent variable overhauling time of liner piston maintenance activity.

dependent pi terms (z_{1P}), due to ±10% change introduced in the value of second next pi term (π_2) (i.e. workers data of liner piston maintenance activity).

Similarly, the change of about 20% is introduced in other pi terms, and 2.35%, 131.97%, 1.80%, −7.60%, 13.75%, −11.31%, −4.58%, 1.78%, and 1.68% takes place due to change in the independent pi term values of π_3, π_4, π_5, π_6, π_7, π_8, π_9, π_{10}, and π_{11} respectively.

From Figure 7.5, it can be seen that π_4 is most sensitive among all pi terms, and highest change takes place in the response variable because of independent pi term π_4 (specification of tools) where π_{11} (noise at workplace) is least sensitive among all pi terms, and least change takes place in the response variable.

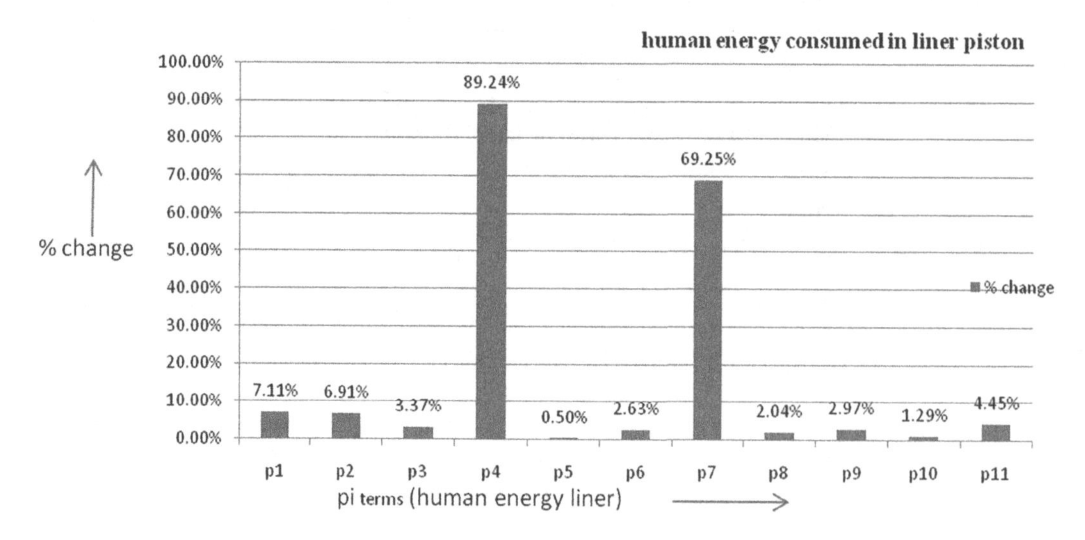

FIGURE 7.5 Percentage change in time with variation of π term at human energy consumed in liner piston maintenance activity (z_{2P}).

Percentage change in all the eleven independent Pi terms to evaluate the effect of this changes on the dependent variable human energy consumed in liner piston maintenance activity.

The sequence of influence of various pi terms on response variable in descending order is as follows: π_4, π_7, π_8, π_6, π_9, π_2, π_3, π_1, π_5, π_{10}, and π_{11}.

7.2.2 Effect of Introduced Change on the Dependent Pi Terms – Human Energy Consumed in Liner Piston Maintenance Activity (z_{2P})

In this model, when a total 20% (i.e. ±10%) limit of the change is introduced in the first dependent pi term (π_1) (i.e. anthropometric data) then 7.11% change occurs in the value of z_{2P} (i.e. human energy consumed in liner piston maintenance activity). Again, change of 6.91% brought in the value of dependent pi terms (z_{2P}), due to ±10% change introduced in the value of second next pi term (π_2) (i.e. workers data).

Similarly, the change of about 20% is introduced in other pi terms, and 3.37%, 89.24%, −0.50%, −2.63%, 69.25%, −2.04%, −2.97%, 1.29%, and 4.45% takes place due to change in the pi term values of π_3, π_4, π_5, π_6, π_7, π_8, π_9, π_{10}, and π_{11} respectively.

From Figure 7.5, it can be seen that π_4 is most sensitive among all pi terms, and highest change takes place in the response variable because of independent pi term π_4 (specification of tools) where π_5 (specification of solvents) is least sensitive among all pi terms, and least change take place in the response variable. The sequence of influence of various pi terms on response variable in descending order is as follows: π_4, π_7, π_1, π_2, π_{11}, π_3, π_9, π_6, π_8, π_{10}, and π_5.

7.2.3 Effect of Introduced Change on the Dependent π Term – Productivity of Liner Piston Maintenance Activity (z_{3P})

In this model, when a total 20% (i.e. ±10%) limit of the change is introduced in the first dependent pi term, π_1 (i.e. anthropometric data), then 1.95% change occurs in the value of z_{3P} (i.e. productivity of liner piston maintenance activity). Again, change of 3.84% brought in the value of dependent pi terms (z_{3P}) (i.e. productivity) due to ±10% change introduced in the value of second next pi term, π_2 (i.e. workers data).

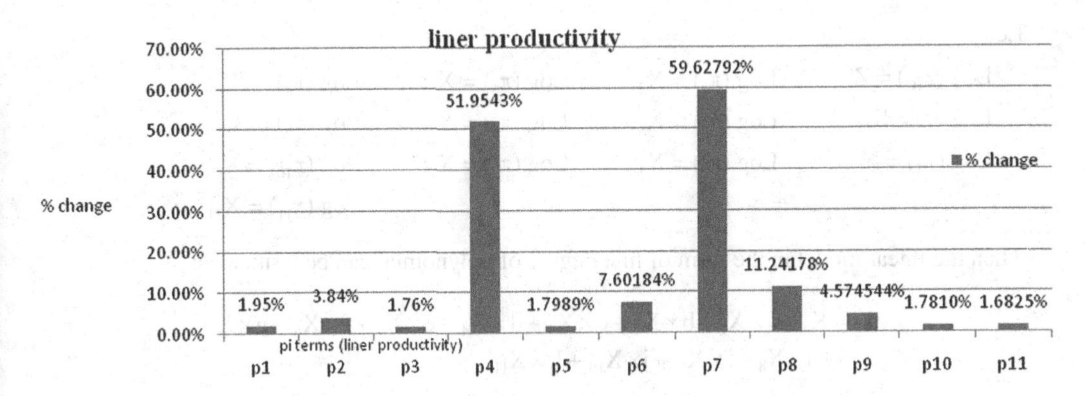

FIGURE 7.6 Percentage change in time with variation of π term at productivity of liner piston maintenance activity (z_{3P}).

Percentage change in all the eleven independent pi terms to evaluate the effect of this changes on the dependent variable productivity of liner piston.

Similarly, the change of about 20% is introduced in other pi terms, and 1.76%, −51.9543%, −1.7989%, −7.60184%, 59.62792%, 11.24178%, 4.574544%, −1.7810%, and −1.6825% takes place due to change in the pi term values of π_3, π_4, π_5, π_6, π_7, π_8, π_9, π_{10}, and π_{11} respectively.

From Figure 7.6, it can be seen that π_4 is most sensitive among all pi terms and the highest change takes place in the response variable because of independent pi term π_7 (workstation data) where π_{11} is least sensitive among all pi terms, and least change takes place in response variable. The sequence of influence of various pi terms on response variable in descending order is as follows: π_7, π_4, π_8, π_6, π_9, π_2, π_1, π_5, π_{10}, π_3, and π_{11}.

7.3 OPTIMIZATION OF MODELS FOR CRANKSHAFT MAINTENANCE ACTIVITY

Three models have been developed corresponding to the overhauling time of crankshaft maintenance activity (z_{1C}), human energy consumed in crankshaft maintenance activity (z_{2C}), and productivity of crankshaft maintenance activity (z_{3C}). The ultimate objective of this work is not merely developing models, but to find the best set of independent variables that will result in maximization/minimization of the objective function. There will be three objective functions corresponding to this model. The model for the productivity of crankshaft maintenance activity (z_{3C}) needs to be maximized, whereas models for overhauling time of crankshaft maintenance activity (z_{1C}) and human energy consumed in crankshaft maintenance activity (z_{2C}) need to be minimized. The developed models are in non-linear form. Hence, for the purpose of optimization of model, this is to be converted into linear form. This is done by taking log on both the sides of the equation. To maximize the linear function, linear programming technique is used as shown below.

For the dependent pi term (z_{1C}), we have:

$$z_{1C} = K_1 \times (\pi_1)^a \times (\pi_2)^b \times (\pi_3)^c \times (\pi_4)^d \times (\pi_5)^e \times (\pi_6)^f \times (\pi_7)^g \times (\pi_8)^h \times (\pi_9)^i (\pi_{10})^j \times (\pi_{11})^k$$

Taking Log on both the side of this equation,

$$\log(z_{1C}) = \log K_1 + a\mathrm{Log}(\pi_1) + b\mathrm{Log}(\pi_2) + c\mathrm{Log}(\pi_3) + d\mathrm{Log}(\pi_4) + e\mathrm{Log}(\pi_5) + f\mathrm{Log}(\pi_6)$$
$$+ g\mathrm{Log}(\pi_7) + h\mathrm{Log}(\pi_8) + i\mathrm{Log}(\pi_9) + j\mathrm{Log}(\pi_{10}) + k\mathrm{Log}(\pi_{11})$$

Let:

$\text{Log}(z_{1C}) = Z_1$; $\text{Log}(\pi_2) = X_2$; $\text{Log}(\pi_5) = X_5$; $\text{Log}(\pi_8) = X_8$;
$\text{Log}\,K_1 = K_1$; $\text{Log}(\pi_3) = X_3$; $\text{Log}(\pi_6) = X_6$; $\text{Log}(\pi_9) = X_9$;
$\text{Log}(\pi_1) = X_1$; $\text{Log}(\pi_4) = X_4$; $\text{Log}(\pi_7) = X_7$; $\text{Log}(\pi_{10}) = X_{10}$;
$\text{Log}(\pi_{11}) = X_{11}$

Then the linear model in the form of first degree of polynomial can be written as:

$$z_{1C} = K_1 + a\times X_1 + b\times X_2 + c\times X_3 + d\times X_4 + e\times X_5 + f\times X_6 + g\times X_7 + h\times X_8 + i\times X_9 + j\times X_{10} + k\times X_{11}$$

Thus the equation will be objective function for the optimization or to be very specific for minimization for the purpose of formulation of the linear programming problems. The next task is to define the constraints for the problem. The constraints can be boundaries defined for the various independent pi terms involved in the function. During the collection of data, certain ranges of independent pi term have been observed. In fact this range has a minimum and maximum value. Therefore, there are two constraints for each independent variable.

If we denote the maximum and minimum value of dependent pi term z_1 is π_{1max} and π_{1min}, by then, the first two constraints for the problem will be obtained by taking Log of these quantities, and by substituting the value of multipliers of all other variables except the one under consideration equal to zero. Let the log limits be defined as C_1 (i.e. $C_1 = \text{Log}\,\pi_{1\,max}$) and C_2, (i.e. $C_2 = \text{Log}\,\pi_{1\,min}$).

Hence the equations of constraints will be as follows::

$$1X_1 + 0X_2 + 0X_3 + 0X_4 + 0X_5 + 0X_6 + 0X_7 + 0X_8 + 0X_9 + 0X_{10} + 0X_{11} \le C_1$$

$$1X_1 + 0X_2 + 0X_3 + 0X_4 + 0X_5 + 0X_6 + 0X_7 + 0X_8 + 0X_9 + 0X_{10} + 0X_{11} \ge C_2$$

The other constraints can be likewise found as follows::

$$0X_1 + 1X_2 + 0X_3 + 0X_4 + 0X_5 + 0X_6 + 0X_7 + 0X_8 + 0X_9 + 0X_{10} + 0X_{11} \le C_3$$

$$0X_1 + 1X_2 + 0X_3 + 0X_4 + 0X_5 + 0X_6 + 0X_7 + 0X_8 + 0X_9 + 0X_{10} + 0X_{11} \ge C_4$$

$$0X_1 + 0X_2 + 1X_3 + 0X_4 + 0X_5 + 0X_6 + 0X_7 + 0X_8 + 0X_9 + 0X_{10} + 0X_{11} \le C_5$$

$$0X_1 + 0X_2 + 1X_3 + 0X_4 + 0X_5 + 0X_6 + 0X_7 + 0X_8 + 0X_9 + 0X_{10} + 0X_{11} \ge C_6$$

$$0X_1 + 0X_2 + 0X_3 + 1X_4 + 0X_5 + 0X_6 + 0X_7 + 0X_8 + 0X_9 + 0X_{10} + 0X_{11} \le C_7$$

$$0X_1 + 0X_2 + 0X_3 + 1X_4 + 0X_5 + 0X_6 + 0X_7 + 0X_8 + 0X_9 + 0X_{10} + 0X_{11} \ge C_8$$

$$0X_1 + 0X_2 + 0X_3 + 0X_4 + 1X_5 + 0X_6 + 0X_7 + 0X_8 + 0X_9 + 0X_{10} + 0X_{11} \le C_9$$

$$0X_1 + 0X_2 + 0X_3 + 0X_4 + 1X_5 + 0X_6 + 0X_7 + 0X_8 + 0X_9 + 0X_{10} + 0X_{11} \ge C_{10}$$

$$0X_1 + 0X_2 + 0X_3 + 0X_4 + 0X_5 + 1X_6 + 0X_7 + 0X_8 + 0X_9 + 0X_{10} + 0X_{11} \le C_{11}$$

$$0X_1 + 0X_2 + 0X_3 + 0X_4 + 0X_5 + 1X_6 + 0X_7 + 0X_8 + 0X_9 + 0X_{10} + 0X_{11} \ge C_{12}$$

$$0X_1 + 0X_2 + 0X_3 + 0X_4 + 0X_5 + 0X_6 + 1X_7 + 0X_8 + 0X_9 + 0X_{10} + 0X_{11} \le C_{13}$$

$$0X_1 + 0X_2 + 0X_3 + 0X_4 + 0X_5 + 0X_6 + 1X_7 + 0X_8 + 0X_9 + 0X_{10} + 0X_{11} \geq C_{14}$$

$$0X_1 + 0X_2 + 0X_3 + 0X_4 + 0X_5 + 0X_6 + 0X_7 + 1X_8 + 0X_9 + 0X_{10} + 0X_{11} \leq C_{15}$$

$$0X_1 + 0X_2 + 0X_3 + 0X_4 + 0X_5 + 0X_6 + 0X_7 + 1X_8 + 0X_9 + 0X_{10} + 0X_{11} \geq C_{16}$$

$$0X_1 + 0X_2 + 0X_3 + 0X_4 + 0X_5 + 0X_6 + 0X_7 + 0X_8 + 1X_9 + 0X_{10} + 0X_{11} \leq C_{17}$$

$$0X_1 + 0X_2 + 0X_3 + 0X_4 + 0X_5 + 0X_6 + 0X_7 + 0X_8 + 1X_9 + 0X_{10} + 0X_{11} \geq C_{18}$$

$$0X_1 + 0X_2 + 0X_3 + 0X_4 + 0X_5 + 0X_6 + 0X_7 + 0X_8 + 0X_9 + 1X_{10} + 0X_{11} \leq C_{19}$$

$$0X_1 + 0X_2 + 0X_3 + 0X_4 + 0X_5 + 0X_6 + 0X_7 + 0X_8 + 0X_9 + 1X_{10} + 0X_{11} \geq C_{20}$$

$$0X_1 + 0X_2 + 1X_3 + 0X_4 + 0X_5 + 0X_6 + 0X_7 + 0X_8 + 0X_9 + 0X_{10} + 1X_{11} \leq C_{21}$$

$$0X_1 + 0X_2 + 1X_3 + 0X_4 + 0X_5 + 0X_6 + 0X_7 + 0X_8 + 0X_9 + 0X_{10} + 1X_{11} \geq C_{22}$$

After solving this linear programming problem, we get the minimum value of the z_{1C} and the set of values of the variables to achieve this minimum value. The values of the independent pi terms can then be obtained by finding the antilog of the values of z_{1C}, X_1, X_2, X_3, X_4, X_5, X_6, X_7, X_8, X_9, X_{10}, and X11.The actual vales of the multipliers and the variables are found and substituted in the above equations and the actual problem in this case can be stated below. This can now be solved as a linear programming problem using matrix analysis. Thus, the actual problem is to minimize z_{1C}, where:

$$z_{1C} = K_1 + a \times X_1 + b \times X_2 + c \times X_3 + d \times X_4 + e \times X_5 + f \times X_6 + g \times X_7$$
$$+ h \times X_8 + i \times X_9 + j \times X_{10} + k \times X_{11}$$

$$z_{1C} = \log(1.7198) + (0.0873 \times \log \pi_1) + (0.1026 \times \log \pi_2) + (2.8137 \times \log \pi_3) + (0.2115 \times \log \pi_4)$$
$$-(0.2101 \times \log \pi_5) - (0.0025 \times \log \pi_6) + (3.5409 \times \log \pi_7) - (0.4451 \times \log \pi_8)$$
$$-(0.125 \times \log \pi_9) - (0.0609 \times \log \pi_{10}) + (0.1131 \times \log \pi_{11})$$

$$z_{1C} = \log(1.7198) + 0.0873 \times 1 + 0.1026 \times 2 + 2.8137 \times 3 + (0.2115 \times 4)$$
$$-(0.2101 \times 5) - (0.0025 \times 6) + (3.5409 \times 7) - (0.4451 \times 8) - (0.125 \times 9)$$
$$-(0.0609 \times 10) + (0.1131 \times 11)$$

Subject to following constraints:

$$1X_1 + 0X_2 + 0X_3 + 0X_4 + 0X_5 + 0X_6 + 0X_7 + 0X_8 + 0X_9 + 0X_{10} + 0X_{11} \leq 1.65875$$

$$1X_1 + 0X_2 + 0X_3 + 0X_4 + 0X_5 + 0X_6 + 0X_7 + 0X_8 + 0X_9 + 0X_{10} + 0X_{11} \geq 0.531259$$

The other constraints can be likewise found as follows::

$$0X_1 + 1X_2 + 0X_3 + 0X_4 + 0X_5 + 0X_6 + 0X_7 + 0X_8 + 0X_9 + 0X_{10} + 0X_{11} \leq 3.8444018$$

$$0X_1 + 1X_2 + 0X_3 + 0X_4 + 0X_5 + 0X_6 + 0X_7 + 0X_8 + 0X_9 + 0X_{10} + 0X_{11} \geq 2.090833$$

$$0X_1 + 0X_2 + 1X_3 + 0X_4 + 0X_5 + 0X_6 + 0X_7 + 0X_8 + 0X_9 + 0X_{10} + 0X_{11} \leq 1898.9064$$

$$0X_1 + 0X_2 + 1X_3 + 0X_4 + 0X_5 + 0X_6 + 0X_7 + 0X_8 + 0X_9 + 0X_{10} + 0X_{11} \geq 1898.9064$$

$$0X_1 + 0X_2 + 0X_3 + 1X_4 + 0X_5 + 0X_6 + 0X_7 + 0X_8 + 0X_9 + 0X_{10} + 0X_{11} \leq 54.73918$$

$$0X_1 + 0X_2 + 0X_3 + 1X_4 + 0X_5 + 0X_6 + 0X_7 + 0X_8 + 0X_9 + 0X_{10} + 0X_{11} \geq 54.73918$$

$$0X_1 + 0X_2 + 0X_3 + 0X_4 + 1X_5 + 0X_6 + 0X_7 + 0X_8 + 0X_9 + 0X_{10} + 0X_{11} \leq 235.2941$$

$$0X_1 + 0X_2 + 0X_3 + 0X_4 + 1X_5 + 0X_6 + 0X_7 + 0X_8 + 0X_9 + 0X_{10} + 0X_{11} \geq 62.5$$

$$0X_1 + 0X_2 + 0X_3 + 0X_4 + 0X_5 + 1X_6 + 0X_7 + 0X_8 + 0X_9 + 0X_{10} + 0X_{11} \leq 0.561728$$

$$0X_1 + 0X_2 + 0X_3 + 0X_4 + 0X_5 + 1X_6 + 0X_7 + 0X_8 + 0X_9 + 0X_{10} + 0X_{11} \geq 0.12963$$

$$0X_1 + 0X_2 + 0X_3 + 0X_4 + 0X_5 + 0X_6 + 1X_7 + 0X_8 + 0X_9 + 0X_{10} + 0X_{11} \leq 4.287195$$

$$0X_1 + 0X_2 + 0X_3 + 0X_4 + 0X_5 + 0X_6 + 1X_7 + 0X_8 + 0X_9 + 0X_{10} + 0X_{11} \geq 4.281795$$

$$0X_1 + 0X_2 + 0X_3 + 0X_4 + 0X_5 + 0X_6 + 0X_7 + 1X_8 + 0X_9 + 0X_{10} + 0X_{11} \leq 0.455$$

$$0X_1 + 0X_2 + 0X_3 + 0X_4 + 0X_5 + 0X_6 + 0X_7 + 1X_8 + 0X_9 + 0X_{10} + 0X_{11} \geq 0.26$$

$$0X_1 + 0X_2 + 0X_3 + 0X_4 + 0X_5 + 0X_6 + 0X_7 + 0X_8 + 1X_9 + 0X_{10} + 0X_{11} \leq 80$$

$$0X_1 + 0X_2 + 0X_3 + 0X_4 + 0X_5 + 0X_6 + 0X_7 + 0X_8 + 1X_9 + 0X_{10} + 0X_{11} \geq 27$$

$$0X_1 + 0X_2 + 0X_3 + 0X_4 + 0X_5 + 0X_6 + 0X_7 + 0X_8 + 0X_9 + 1X_{10} + 0X_{11} \leq 353501$$

$$0X_1 + 0X_2 + 0X_3 + 0X_4 + 0X_5 + 0X_6 + 0X_7 + 0X_8 + 0X_9 + 1X_{10} + 0X_{11} \geq 195157.7$$

$$0X_1 + 0X_2 + 1X_3 + 0X_4 + 0X_5 + 0X_6 + 0X_7 + 0X_8 + 0X_9 + 0X_{10} + 1X_{11} \leq 85$$

$$0X_1 + 0X_2 + 1X_3 + 0X_4 + 0X_5 + 0X_6 + 0X_7 + 0X_8 + 0X_9 + 0X_{10} + 1X_{11} \geq 62.75$$

On solving the above problem, we get:

$X_1 = -0.2746$	$X_4 = 1.738$	$X_7 = 0.6321$	$X_{10} = 5.2903$
$X_2 = 0.3203$	$X_5 = 1.79588$	$X_8 = -0.5850$	$X_{11} = 1.7976$
$X_3 = 3.27850$	$X_6 = -0.8872$	$X_9 = 1.4313$	

Thus substituting these values in objective function equation we get min value of z_{1C}

$$\begin{aligned} z_{1C} = {} & \log(1.7198) + (0.0873 \times -0.2746) + (0.1026 \times 0.3203) + (2.8137 \times 3.27850) \\ & + (0.2115 \times 1.738) - (0.2101 \times 1.79588) - (0.0025 \times -0.8872) + (3.5409 \times 0.6321) \\ & - (0.4451 \times -0.5850) - (0.125 \times 1.4313) - (0.0609 \times 5.2903) + (0.1131 \times 1.7976) \end{aligned}$$

$$z_{1C} = 11.06237738$$

The minimum value of z can be obtained by taking antilog of z_{1C}

Hence:

z_{1Cmin} = antilog(11.06237738) = $1.154455986 \times 10^{11}$ and corresponding to this, the values of z_{1Cmin}, the independent pi terms are obtained by taking the antilog of X_1, X_2, X_3, X_4, X_5, X_6, X_7, X_8,

X_9, X_{10}, and X_{11}. These values are 0.531259, 2.0090833, 1898.9064, 54.73, 62.5, 0.12963, 4.2864, 0.260, 27, 195157.7, and 62.75 respectively.

Similar procedure has been adopted to optimize the models for z_{2C} and z_{3C}

z_{2Cmin} = antilog(3.4310) = 2697.7394 and corresponding to this, the values of z_{2Cmin}, the independent pi terms are obtained by taking the antilog of X_1, X_2, X_3, X_4, X_5, X_6, X_7, X_8, X_9, X_{10}, and X_{11}. These values 0.531259, 2.0090833, 1898.9064, 54.73, 62.5, 0.12963, 4.2864, 0.260, 27, 195157.7, and 62.75 respectively.

z_{3Cmax} = antilog(−1.259121) = 0.05506 and corresponding to this, the values of z_{3Cmax}, the independent pi terms are obtained by taking the antilog of X_1, X_2, X_3, X_4, X_5, X_6, X_7, X_8, X_9, X_{10}, and X_{11}. These values are 1.6587, 3.84401, 1898.9064, 54.73918, 235.2941, 0.561728, 4.2871, 0.455, 80, 353501, and 85 respectively.

7.4 OPTIMIZATION OF THE MODELS FOR LINER PISTON MAINTENANCE ACTIVITY

Three models have been developed corresponding to the overhauling time of liner piston maintenance activity (z_{1P}), human energy consumed in liner piston maintenance activity (z_{2P}), and productivity of liner piston maintenance activity (z_{3P}). The ultimate objective of this work is not only developing models, but also to find out the best set of independent variables, which will result in maximization/minimization of the objective function. There will be three objective functions corresponding to the overhauling time of liner piston maintenance activity (z_{1P}), human energy consumed in liner piston maintenance activity (z_{2P}) and productivity of liner piston maintenance activity (z_{3P}). The model for the productivity of liner piston maintenance activity (z_{3P}) needs to be maximized whereas models for overhauling time of liner piston maintenance activity (z_{1P}) and human energy consumed in liner piston maintenance activity (z_{2P}) needs to be minimized. The developed models are in non-linear form. Hence, for the purpose of optimization of model, this is to be converted into linear form. This is done by taking log on both the sides of the equation. To maximize the linear function, linear programming technique is used as shown below.

For the dependent pi term (z_{1P}) we have:

$$z_{1P} = K \times (\pi_1)^a \times (\pi_2)^b \times (\pi_3)^c \times (\pi_4)^d \times (\pi_5)^e \times (\pi_6)^f \times (\pi_7)^g \times (\pi_8)^h \times (\pi_9)^i (\pi_{10})^j \times (\pi_{11})^k$$

Taking Log on both the side of this equation, we have

$$\log(z_{1P}) = \log K_1 + a\text{Log}(\pi_1) + b\text{Log}(\pi_2) + c\text{Log}(\pi_3) + d\text{Log}(\pi_4) + e\text{Log}(\pi_5) + f\text{Log}(\pi_6) + g\text{Log}(\pi_7) + h\text{Log}(\pi_8) + i\text{Log}(\pi_9) + j\text{Log}(\pi_{10}) + k\text{Log}(\pi_{11})$$

Let:

$\text{Log}(z_{1P}) = Z_1$	$\text{Log}(\pi_2) = X_2$	$\text{Log}(\pi_5) = X_5$	$\text{Log}(\pi_8) = X_8$
$\text{Log}\,K_1 = K_1$	$\text{Log}(\pi_3) = X_3$	$\text{Log}(\pi_6) = X_6$	$\text{Log}(\pi_9) = X_9$
$\text{Log}(\pi_1) = X_1$	$\text{Log}(\pi_4) = X_4$	$\text{Log}(\pi_7) = X_7$	$\text{Log}(\pi_{10}) = X_{10}$
			$\text{Log}(\pi_{11}) = X_{11}$

Then the linear model in the form of first degree of polynomial can be written as,

$$z_{1P} = K_1 + a \times X_1 + b \times X_2 + c \times X_3 + d \times X_4 + e \times X_5 + f \times X_6 + g \times X_7 + h \times X_8 + i \times X_9 + j \times X_{10} + k \times X_{11}$$

Thus the equation will be objective function for the optimization or to be very specific for minimization for the purpose of formulation of the linear programming problems. Next task is to define the constraints for the problem. The constraints can be boundaries defined for the various independent pi terms involved in the function. During collection of data, certain ranges of independent pi terms have been observed. In fact, this range has a minimum and maximum value. Therefore, there are two constraints for each independent variable.

If we denote the maximum and minimum value of dependent pi term z_{1P} by is $\pi_{1\,max}$ and π_{1min}, then the first two constraints for the problem will be obtained by taking log of these quantities and by substituting the value of multipliers of all other variables except the one under consideration equal to zero. Let the log limits be defined as C_1 (i.e. C_1= Log π_{1max}) and C_2 (i.e. C_2= Log π_{1min}).

Hence the equations of the constraints will be as follows:

$$1X_1 + 0X_2 + 0X_3 + 0X_4 + 0X_5 + 0X_6 + 0X_7 + 0X_8 + 0X_9 + 0X_{10} + 0X_{11} \leq C_1$$

$$1X_1 + 0X_2 + 0X_3 + 0X_4 + 0X_5 + 0X_6 + 0X_7 + 0X_8 + 0X_9 + 0X_{10} + 0X_{11} \geq C_2$$

The other constraints can be likewise found as follows:

$$0X_1 + 1X_2 + 0X_3 + 0X_4 + 0X_5 + 0X_6 + 0X_7 + 0X_8 + 0X_9 + 0X_{10} + 0X_{11} \leq C_3$$

$$0X_1 + 1X_2 + 0X_3 + 0X_4 + 0X_5 + 0X_6 + 0X_7 + 0X_8 + 0X_9 + 0X_{10} + 0X_{11} \geq C_4$$

$$0X_1 + 0X_2 + 1X_3 + 0X_4 + 0X_5 + 0X_6 + 0X_7 + 0X_8 + 0X_9 + 0X_{10} + 0X_{11} \leq C_5$$

$$0X_1 + 0X_2 + 1X_3 + 0X_4 + 0X_5 + 0X_6 + 0X_7 + 0X_8 + 0X_9 + 0X_{10} + 0X_{11} \geq C_6$$

$$0X_1 + 0X_2 + 0X_3 + 1X_4 + 0X_5 + 0X_6 + 0X_7 + 0X_8 + 0X_9 + 0X_{10} + 0X_{11} \leq C_7$$

$$0X_1 + 0X_2 + 0X_3 + 1X_4 + 0X_5 + 0X_6 + 0X_7 + 0X_8 + 0X_9 + 0X_{10} + 0X_{11} \geq C_8$$

$$0X_1 + 0X_2 + 0X_3 + 0X_4 + 1X_5 + 0X_6 + 0X_7 + 0X_8 + 0X_9 + 0X_{10} + 0X_{11} \leq C_9$$

$$0X_1 + 0X_2 + 0X_3 + 0X_4 + 1X_5 + 0X_6 + 0X_7 + 0X_8 + 0X_9 + 0X_{10} + 0X_{11} \geq C_{10}$$

$$0X_1 + 0X_2 + 0X_3 + 0X_4 + 0X_5 + 1X_6 + 0X_7 + 0X_8 + 0X_9 + 0X_{10} + 0X_{11} \leq C_{11}$$

$$0X_1 + 0X_2 + 0X_3 + 0X_4 + 0X_5 + 1X_6 + 0X_7 + 0X_8 + 0X_9 + 0X_{10} + 0X_{11} \geq C_{12}$$

$$0X_1 + 0X_2 + 0X_3 + 0X_4 + 0X_5 + 0X_6 + 1X_7 + 0X_8 + 0X_9 + 0X_{10} + 0X_{11} \leq C_{13}$$

$$0X_1 + 0X_2 + 0X_3 + 0X_4 + 0X_5 + 0X_6 + 1X_7 + 0X_8 + 0X_9 + 0X_{10} + 0X_{11} \geq C_{14}$$

$$0X_1 + 0X_2 + 0X_3 + 0X_4 + 0X_5 + 0X_6 + 0X_7 + 1X_8 + 0X_9 + 0X_{10} + 0X_{11} \leq C_{15}$$

$$0X_1 + 0X_2 + 0X_3 + 0X_4 + 0X_5 + 0X_6 + 0X_7 + 1X_8 + 0X_9 + 0X_{10} + 0X_{11} \geq C_{16}$$

$$0X_1 + 0X_2 + 0X_3 + 0X_4 + 0X_5 + 0X_6 + 0X_7 + 0X_8 + 1X_9 + 0X_{10} + 0X_{11} \leq C_{17}$$

$$0X_1 + 0X_2 + 0X_3 + 0X_4 + 0X_5 + 0X_6 + 0X_7 + 0X_8 + 1X_9 + 0X_{10} + 0X_{11} \geq C_{18}$$

$$0X_1 + 0X_2 + 0X_3 + 0X_4 + 0X_5 + 0X_6 + 0X_7 + 0X_8 + 0X_9 + 1X_{10} + 0X_{11} \leq C_{19}$$

$$0X_1 + 0X_2 + 0X_3 + 0X_4 + 0X_5 + 0X_6 + 0X_7 + 0X_8 + 0X_9 + 1X_{10} + 0X_{11} \geq C_{20}$$

$$0X_1 + 0X_2 + 1X_3 + 0X_4 + 0X_5 + 0X_6 + 0X_7 + 0X_8 + 0X_9 + 0X_{10} + 1X_{11} \le C_{21}$$

$$0X_1 + 0X_2 + 1X_3 + 0X_4 + 0X_5 + 0X_6 + 0X_7 + 0X_8 + 0X_9 + 0X_{10} + 1X_{11} \ge C_{22}$$

After solving this linear programming problem, we get the minimum value of the z_{1P} and the set of values of the variables to achieve this minimum value. The values of the independent pi terms can then be obtained by finding the antilog of the values of z_{1P}, X_1, X_2, X_3, X_4, X_5, X_6, X_7, X_8, X_9, X_{10}, and X_{11}. The actual vales of the multipliers and the variables are found and substituted in the above equations and the actual problem in this case can be stated below. This can now be solved as a linear programming problem using matrix analysis. Thus, the actual problem is to minimize z_{1P}, where:

$$z_{1P} = K_1 + a \times X_1 + b \times X_2 + c \times X_3 + d \times X_4 + e \times X_5 + f \times X_6 + g \times X_7 + h \times X_8 + i \times X_9 + j \times X_{10} + k \times X_{11}$$

$$z_{1P} = \log(1.0122) + (-0.0973 \times \log\pi_1) - (0.1917 \times \log\pi_2) + (0.8772 \times \log\pi_3) - (6.0098 \times \log\pi_4) + (0.0896 \times \log\pi_5) + (0.378 \times \log\pi_6) + (0.6869 \times \log\pi_7) - (0.5615 \times \log\pi_8) - (0.2282 \times \log\pi_9) + (0.0887 \times \log\pi_{10}) + (0.0838 \times \log\pi_{11})$$

$$z_{1P} = \log(1.0122) + (-0.0973 \times 1) - (0.1917 \times 2) + (0.8772 \times 3) - (6.0098 \times 4) + (0.0896 \times 5) + (0.378 \times 6) + (0.6869 \times 7) - (0.5615 \times 8) - (0.2282 \times 9) + (0.0887 \times 10) + (0.0838 \times 11)$$

Subject to following constraints:

$$1X_1 + 0X_2 + 0X_3 + 0X_4 + 0X_5 + 0X_6 + 0X_7 + 0X_8 + 0X_9 + 0X_{10} + 0X_{11} \le 1.65875$$

$$1X_1 + 0X_2 + 0X_3 + 0X_4 + 0X_5 + 0X_6 + 0X_7 + 0X_8 + 0X_9 + 0X_{10} + 0X_{11} \ge 0.531259$$

The other constraints can likewise be found as follows:

$$0X_1 + 1X_2 + 0X_3 + 0X_4 + 0X_5 + 0X_6 + 0X_7 + 0X_8 + 0X_9 + 0X_{10} + 0X_{11} \le 5.170055$$

$$0X_1 + 1X_2 + 0X_3 + 0X_4 + 0X_5 + 0X_6 + 0X_7 + 0X_8 + 0X_9 + 0X_{10} + 0X_{11} \ge 1.63055$$

$$0X_1 + 0X_2 + 1X_3 + 0X_4 + 0X_5 + 0X_6 + 0X_7 + 0X_8 + 0X_9 + 0X_{10} + 0X_{11} \le 22886690834$$

$$0X_1 + 0X_2 + 1X_3 + 0X_4 + 0X_5 + 0X_6 + 0X_7 + 0X_8 + 0X_9 + 0X_{10} + 0X_{11} \ge 22886690834$$

$$0X_1 + 0X_2 + 0X_3 + 1X_4 + 0X_5 + 0X_6 + 0X_7 + 0X_8 + 0X_9 + 0X_{10} + 0X_{11} \le 0.459878$$

$$0X_1 + 0X_2 + 0X_3 + 1X_4 + 0X_5 + 0X_6 + 0X_7 + 0X_8 + 0X_9 + 0X_{10} + 0X_{11} \ge 0.459878$$

$$0X_1 + 0X_2 + 0X_3 + 0X_4 + 1X_5 + 0X_6 + 0X_7 + 0X_8 + 0X_9 + 0X_{10} + 0X_{11} \le 116.7143$$

$$0X_1 + 0X_2 + 0X_3 + 0X_4 + 1X_5 + 0X_6 + 0X_7 + 0X_8 + 0X_9 + 0X_{10} + 0X_{11} \ge 38.90476$$

$$0X_1 + 0X_2 + 0X_3 + 0X_4 + 0X_5 + 1X_6 + 0X_7 + 0X_8 + 0X_9 + 0X_{10} + 0X_{11} \le 0.182479$$

$$0X_1 + 0X_2 + 0X_3 + 0X_4 + 0X_5 + 1X_6 + 0X_7 + 0X_8 + 0X_9 + 0X_{10} + 0X_{11} \ge 0.117284$$

$$0X_1 + 0X_2 + 0X_3 + 0X_4 + 0X_5 + 0X_6 + 1X_7 + 0X_8 + 0X_9 + 0X_{10} + 0X_{11} \le 0.213848$$

$$0X_1 + 0X_2 + 0X_3 + 0X_4 + 0X_5 + 0X_6 + 1X_7 + 0X_8 + 0X_9 + 0X_{10} + 0X_{11} \ge 0.213848$$

$$0X_1 + 0X_2 + 0X_3 + 0X_4 + 0X_5 + 0X_6 + 0X_7 + 1X_8 + 0X_9 + 0X_{10} + 0X_{11} \le 0.458$$

$$0X_1 + 0X_2 + 0X_3 + 0X_4 + 0X_5 + 0X_6 + 0X_7 + 1X_8 + 0X_9 + 0X_{10} + 0X_{11} \ge 0.26$$

$$0X_1 + 0X_2 + 0X_3 + 0X_4 + 0X_5 + 0X_6 + 0X_7 + 0X_8 + 1X_9 + 0X_{10} + 0X_{11} \le 80$$

$$0X_1 + 0X_2 + 0X_3 + 0X_4 + 0X_5 + 0X_6 + 0X_7 + 0X_8 + 1X_9 + 0X_{10} + 0X_{11} \ge 24$$

$$0X_1 + 0X_2 + 0X_3 + 0X_4 + 0X_5 + 0X_6 + 0X_7 + 0X_8 + 0X_9 + 1X_{10} + 0X_{11} \le 400695.7$$

$$0X_1 + 0X_2 + 0X_3 + 0X_4 + 0X_5 + 0X_6 + 0X_7 + 0X_8 + 0X_9 + 1X_{10} + 0X_{11} \ge 89100$$

$$0X_1 + 0X_2 + 1X_3 + 0X_4 + 0X_5 + 0X_6 + 0X_7 + 0X_8 + 0X_9 + 0X_{10} + 1X_{11} \le 89$$

$$0X_1 + 0X_2 + 1X_3 + 0X_4 + 0X_5 + 0X_6 + 0X_7 + 0X_8 + 0X_9 + 0X_{10} + 1X_{11} \ge 70$$

On solving the above problem, we get:

$$X_1 = -0.2746,$$
$$X_2 = 0.2123,$$
$$X_3 = 10.3595,$$
$$X_4 = -0.3373,$$
$$X_5 = 1.590,$$
$$X_6 = -0.9307,$$
$$X_7 = -0.6698,$$
$$X_8 = -0.5850,$$
$$X_9 = 1.3802,$$
$$X_{10} = 4.9498$$
$$X_{11} = 1.8450$$

Thus substituting these values in objective function equation we get:

$$\begin{aligned} z_{1P} = {} & \log(1.0122) - (0.0973 \times -0.2746) - (0.1917 \times 0.2123) + (0.8772 \times 10.35) \\ & -(6.0098 \times -0.3373) + (0.0896 \times 1.59) + (0.378 \times -0.9307) + (0.6869 \times -0.6698) \\ & -(0.5615 \times -0.5850) - (0.2282 \times 1.3802) + (0.0887 \times 4.9498) \times (0.0838 \times 1.8450) \end{aligned}$$

$$z_{1P} = 11.03516044$$

z_{1Pmin} = antilog(11.03516044) = 1.084327 × 10^{11} and corresponding to this, the values of z_{1Pmin} the independent pi terms are obtained by taking the antilog of X_1, X_2, X_3, X_4, X_5, X_6, X_7, X_8, X_9, X_{10}, and X_{11}. These values are 0.531259, 1.630555, 2.288669, 0.834, 0.459878, 38.90476, 0.117284, 0.213848, 0.26, 24, 89100, and 70 respectively.

Similar procedure has been adopted optimize of dependent variables z_{2P} and z_{3P} of liner piston mentioned as below:

z_{2Pmin} = antilog(3.2113) = 1626.6720 and corresponding to this, the values of z_{2Pmin} the independent pi terms are obtained by taking the antilog of X_1, X_2, X_3, X_4, X_5, X_6, X_7, X_8, X_9, X_{10}, and X_{11}. These values arc 0.531259, 1.630555, 2.288669, 0.834, 0.459878, 38.90476, 0.117284, 0.213848, 0.26, 24, 89100, and 70 respectively.

z_{3Pmax} = antilog(−1.0893106) = 0.0814121 and corresponding to this, the values of z_{3Pmax} the independent pi terms are obtained by taking the antilog of X_1, X_2, X_3, X_4, X_5, X_6, X_7, X_8, X_9, X_{10}, and X_{11}, These values are 1.6587, 5.17005, 2.288669, 0.834, 0.459878, 116.143, 0.182479, 0.213848, 0.458, 80, 400697.7, and 89 respectively.

7.5 RELIABILITY OF MODELS

Reliability is a term associated with the chances of failure. Hence reliability is used to show the performance of the model. For this, percentage error of developed model is calculated as given below:

$$\%\,\text{Error} = [(\text{Field value} - \text{Model calculated value})/\text{Field value}] \times 100$$

Once the error is calculated, then the reliability of model is calculated using following formula:

$$\text{Reliability} = 1 - \text{Mean error}$$

Where Mean error = ΣXIFI/ΣFI

Where ΣXIFI = Summation of the product for percentage of error and frequency of error occurrence ΣFI = Summation of frequency of error occurrence.

7.5.1 Reliability of Crankshaft Maintenance Activity

a. For dependent variable-overhauling time of crankshaft maintenance activity (z_{1C}):

$$\Sigma XIFI = 381.790285 \text{ and } \Sigma FI = 30$$

Mean error of overhauling time of crankshaft maintenance activity is given by:

$$\text{Mean error} = \Sigma XIFI/\Sigma FI = 381.790285/30 = 12.726342\%$$

$$\text{Mean error} = 0.12726342$$

Reliability of overhauling time crankshaft maintenance activity (z_{1C}) is given by:

$$\text{Reliability of } z_1 = R_{1C} = 1 - \text{Mean error} = 1 - 0.12726342$$

$$R_{1C} = 0.87273652$$

$$R_{1C} = 87.273\%$$

b. For dependent variable-human energy of crankshaft maintenance activity (z_{2C}):

$$\Sigma XIFI = 256.8369 \text{ and } \Sigma FI = 30$$

Mean error of overhauling human energy consumed in crankshaft maintenance activity is given by:

$$\text{Mean error} = \Sigma XIFI/\Sigma FI = 256.836998/30 = 8.561233267\%$$

$$\text{Mean error} = 0.08561233267$$

Reliability of human energy consumed in crankshaft maintenance activity (z_{2C}) is given by:

$$\text{Reliability of } z2 = R_{2C} = 1 - \text{Mean error} = 1 - 0.08561233267$$

$$R_{2C} = 0.914387667$$

$$R_{2C} = 91.4387\%$$

c. For dependent variable-productivity of crankshaft maintenance activity (z_{3C}):

$$\Sigma XIFI = 192.615254 \text{ and } \Sigma FI = 30$$

Mean error of productivity of crankshaft maintenance activity is given by,

$$\text{Mean error} = \Sigma XIFI/\Sigma FI = 192.615254/30 = 6.4205084\%$$

$$\text{Mean error} = 0.064205084$$

Reliability of productivity of crankshaft maintenance activity (z3) is given by,

$$\text{Reliability of } z3 = R_{3C} = 1 - \text{Mean error} = 1 - 0.064205084$$

$$R_{3C} = 0.935794915$$

$$R_{3C} = 93.5794915\%$$

7.5.2 Reliability of Liner Piston Maintenance Activity

a. For dependent variable-overhauling time of liner piston maintenance activity (z_{1P}):

$$\Sigma XIFI = 163.255718 \text{ and } \Sigma FI = 30$$

Mean error of overhauling time liner piston maintenance activity is given by:

$$\text{Mean error} = \Sigma XIFI/\Sigma FI = 163.255718/30 = 5.44185\%$$

$$\text{Mean error} = 0.0544185$$

Reliability of overhauling time of liner piston maintenance activity (z_{1P}) is given by:

$$\text{Reliability of } z1 = R_{1P} = 1 - \text{Mean error} = 1 - 0.0544185$$

$$R_{1P} = 0.945581427$$

$$R_{1P} = 94.5581427\%$$

b. For dependent variable-human energy consumed in liner piston maintenance activity (z_{2P}):

$$\Sigma XIFI = 209.899216 \text{ and } \Sigma FI = 30$$

Mean error of overhauling time liner piston maintenance activity is given by:

$$\text{Mean error} = \Sigma XIFI/\Sigma FI = 209.899216/30 = 6.9966405\%$$

$$\text{Mean error} = 0.069966$$

Reliability of human energy consumed in liner piston maintenance activity (z_{2P}) is given by:

$$\text{Reliability of } z2 = R_{2P} = 1 - \text{Mean error} = 1 - 0.0696640 = 0.930034$$

$$R_{2P} = 0.930034$$

$$R_{2P} = 93.0034\%$$

c. For dependent variable-productivity of liner piston maintenance activity (z_{3P})

$$\Sigma XIFI = 751.824725 \text{ and } \Sigma FI = 30$$

Mean error of overhauling time liner piston maintenance activity is given by,

$$\text{Mean error} = \Sigma XIFI/\Sigma FI = 751.824725/30 = 25.06082417\%$$

$$\text{Mean error} = 0.250608242$$

Reliability of productivity of liner piston maintenance activity (z_{3P}) is given by,

$$\text{Reliability of } z_3 = R_{3P} = 1 - \text{Mean error} = 1 - 0.250608242$$

$$R_{3P} = 0.749391758$$

$$R_{3P} = 74.9391758\%$$

7.6 SENSITIVITY ANALYSIS OF BRAKE THERMAL EFFICIENCY AND BRAKE SPECIFIC FUEL CONSUMPTION

The influences of various independent π terms have been studied by analyzing the indices of the various independent pi terms in the models. Through the technique of sensitivity analysis the change in the value of a dependent π term caused due to an introduced change in the value of individual independent π term is evaluated. In this case, the change of ±10% is introduced in the individual π terms independently (one at a time). Thus, the total range of the introduced change is 20%. The effect of this introduced change on the Percentage change value of the dependent π term is evaluated. The average value of the change in the dependent π terms due to the introduced change of 20% in each π term is shown in the Tables 7.9–7.12. These values are plotted on the on the Graph in Figures 7.7 and 7.8.

The sensitivity as evaluated is represented and discussed below.

TABLE 7.9
Sensitivity Analysis for Brake Thermal Efficiency (Z_1)

	Pi_1	Pi_2	Pi_3	Pi_4	Z_1
Mean	10.18748	0.910141	83.30429	0.01313	45.4482
10% above	11.2062	0.910141	83.30429	0.01313	46.21616146
10% below	9.16873	0.910141	83.30429	0.01313	44.61417545
% Change					0.035248613
Mean	10.18748	0.910141	83.30429	0.01313	45.44823666
10% above	10.18748	1.00116	83.30429	0.01313	42.36129577
10% below	10.18748	0.81913	83.30429	0.01313	49.12313392
% Change					−0.14878109
Mean	10.18748	0.910141	83.30429	0.01313	45.44823666
10% above	10.18748	0.910141	91.6347	0.01313	42.53526291
10% below	10.18748	0.910141	74.9739	0.01313	48.90108483
% Change					−0.14006752
Mean	10.18748	0.910141	83.30429	0.01313	45.44823666
10% above	10.18748	0.910141	83.30429	0.01444	43.54434098
10% below	10.18748	0.910141	83.30429	0.01182	47.64991728
% Change					−0.09033522

TABLE 7.10
Percentage Change in Brake Thermal Efficiency (Z1) with +/− 10% Variation of Mean Value of Each Pi Term

Sensitivity Chart

Pi Terms	Z1
Pi 1	3.52%
Pi 2	−14.87%
Pi 3	−14.00%
Pi 4	−9.03%

TABLE 7.11
Sensitivity Analysis for Brake Specific Fuel Consumption (Z_2)

	Pi_1	Pi_2	Pi_3	Pi_4	Z_1
Mean	10.18748	0.910141	83.30429	0.01313	8.180616082
10% above	11.2062	0.910141	83.30429	0.01313	8.101781284
10% below	9.16873	0.910141	83.30429	0.01313	8.268656919
% Change					–0.02039891
Mean	10.18748	0.910141	83.30429	0.01313	8.180616082
10% above	10.18748	1.00116	83.30429	0.01313	8.750858069
10% below	10.18748	0.81913	83.30429	0.01313	7.5933858
% Change					0.141489621
Mean	10.18748	0.910141	83.30429	0.01313	8.180616082
10% above	10.18748	0.910141	91.6347	0.01313	7.701221353
10% below	10.18748	0.910141	74.9739	0.01313	8.745365353
% Change					–0.12763635
Mean	10.18748	0.910141	83.30429	0.01313	8.180616082
10% above	10.18748	0.910141	83.30429	0.01444	7.246120322
10% below	10.18748	0.910141	83.30429	0.01182	9.354521335
% Change					–0.25773132

TABLE 7.12
Percentage Change in Brake Specific Fuel Consumption (Z_2) with +/– 10% Variation of Mean Value of Each Pi Term

Sensitivity Chart

Pi Terms	Z_2
Pi_1	–2.03%
Pi_2	14.14%
Pi_3	–12.76%
Pi_4	–25.77%

7.6.1 Effect of Introduced Change on the Dependent π Term: Brake Thermal Efficiency

In this model, when a total change of 20% is introduced in the value of independent π term π_1, a change of 3.52% occurs in the value of Z_1, (i.e. brake thermal efficiency; computed from the model). The change brought in the value of Z_1, because of the change in the value of the other independent π terms namely π_2 is – 14.57%. Similarly, the change of about –14.00%, and –9.03% takes place in the value of Z_1, because of the change in the values of π_3 and π_4 respectively.

It can be seen that the highest change takes place in Z_1, because of the π term π_2. Thus, π_2 related to fuel consumption and engine load is most sensitive πterm. The sequence of the various π terms in the descending order of sensitivity is π_3, π_4, and π_1.

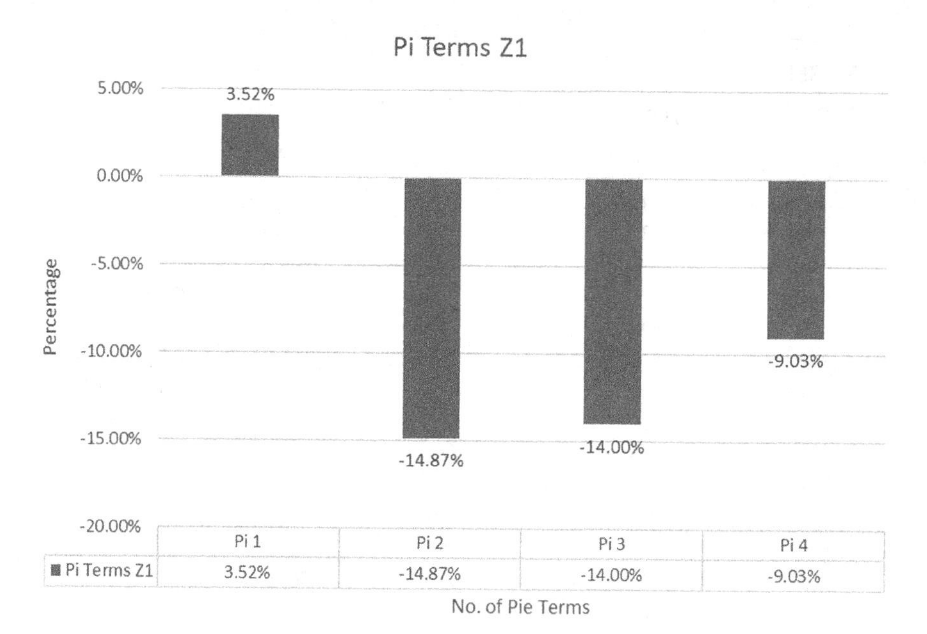

FIGURE 7.7 Percentage change in Pi terms variation for dependent term Z_1 (brake thermal efficiency).

Percentage change in all the four independent Pi terms to evaluate the effect of this changes on the dependent variable brake thermal efficiency.

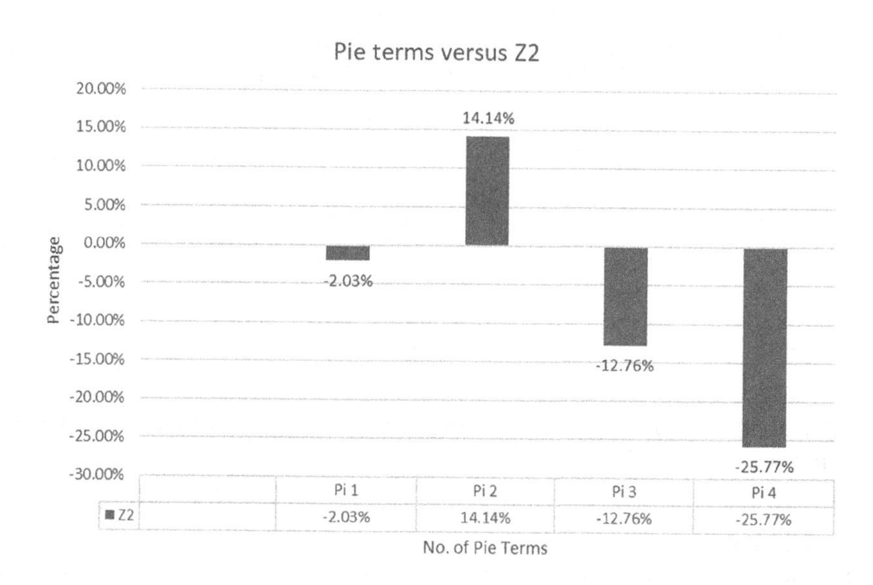

FIGURE 7.8 Percentage change in pi terms variation for dependent term Z_2 (brake specific fuel consumption).

Percentage change in all the six independent Pi terms to evaluate the effect of this changes on the dependent variable brake specific fuel consumption.

The sensitivity as evaluated is represented and discussed below.

7.6.2 Effect of Introduced Change on the Dependent π Term: Brake Thermal Efficiency

In this model, when a total change of 20% is introduced in the value of independent π term π_1, a change of −2.03% occurs in the value of Z_1 (i.e. brake thermal efficiency; computed from the model). The change brought in the value of Z_1, because of the change in the value of the other independent π terms, namely π_2, is 14.14%. Similarly, the change of about −12.76%, and −25.77% takes place in the value of Z_1, because of the change in the values of π_3 and π_4 respectively.

It can be seen that the highest change takes place in Z_1, because of the π term π_2. Thus, π_4 related to engine specification is most sensitive πterm. The sequence of the various π terms in the descending order of sensitivity is π_3, π_4, and π_1.

7.7 SENSITIVITY ANALYSIS OF TURBINE SPEED AND TURBINE POWER

The effect of this introduced change on the percentage change value of the dependent π term is evaluated. The average value of the change in the dependent π terms due to the introduced change of 20% in each π term is shown in the below Table 7.13. These values are plotted on the on the Graph in Figures 7.9 and 7.10 as shown below.

7.7.1 Effect of Introduced Change on the Dependent π Term: Turbine Speed

In this model, when a total range of change of 20% is introduced in the value of independent π term π_1, a change of + 18.3722% occurs in the value of Z_1 (i.e. Speed of Turbine; computed from the model). The change brought in the value of Z_1, because of the change in the value of the other independent π term π2 is +0.8506%. Similarly, the change of about +0.9504%, +6.0840%, +12.3994%, and +10.0165% takes place in the value of Z_1, because of the change in the values of π_3, π_4, π_5, and π_6 respectively.

It can be seen that the highest change takes place in Z_1, because of the π term π_1, whereas the least change takes place due to π_2. Thus, π_1 is most sensitive πterm and π_2 term is the least sensitive π term. The sequence of the various π terms in the descending order of sensitivity is π_5, π_6, π_5, π_4, and π_3.

7.7.2 Effect of Introduced Change on the Dependent π Term: Power Developed

In this model, when a total range of change of 20% is introduced in the value of independent π term π_1, a change of + 81.9168% occurs in the value of Z_1 (i.e. power developed; computed from the model). The change brought in the value of Z_1, because of the change in the value of the other independent π term π2 is +3.5188%. Similarly, the changes of about +5.7337%, +17.5160%, +138.4902%, and +23.1725% take place in the value of Z_1, because of the changes in the values of π_3, π_4, π_5, and π_6 respectively. It can be seen that the highest change takes place in Z_1, because of the π term π_1, whereas the least change takes place due to π_2. Thus, π_1 is most sensitive π term, and π_2 is the least sensitive π term. The sequence of the various π terms in the descending order of sensitivity is π_5, π_6, π_5, π_4, and π_3.

TABLE 7.13
Sensitivity Analysis for Turbine Speed and Turbine Power

	Independent Pi Terms (Varied by ±10%)						% of Change Effect on Dependent Pi Terms	
	Pi_1	Pi_2	Pi_3	Pi_4	Pi_5	Pi_6	PD_1	PD_2
Average	1.548E–05	0.0622222	0.18	21.3787	3.543396	9990.096	195.801844	4.090981819
10%	1.703E–05	0.0622222	0.18	21.3787	3.543396	9990.096	179.533416	2.82030927
–10%	1.393E–05	0.0622222	0.18	21.3787	3.543396	9990.096	215.506647	6.17151204
		% Change					18.3722635	81.91683362
Average	1.548E–05	0.0622222	0.18	21.3787	3.543396	9990.096	195.801844	4.090981819
10%	1.548E–05	0.0684444	0.18	21.3787	3.543396	9990.096	196.59471	4.159986926
–10%	1.548E–05	0.056	0.18	21.3787	3.543396	9990.096	194.929092	4.016031494
		% Change					0.85066505	3.518848008
Average	1.548E–05	0.0622222	0.18	21.3787	3.543396	9990.096	195.801844	4.090981819
10%	1.548E–05	0.0622222	0.198	21.3787	3.543396	9990.096	196.68842	4.204070562
–10%	1.548E–05	0.0622222	0.162	21.3787	3.543396	9990.096	194.82643	3.9695048
		% Change					0.95095614	5.733727804
Average	1.548E–05	0.0622222	0.18	21.3787	3.543396	9990.096	195.801844	4.090981819
10%	1.548E–05	0.0622222	0.18	23.51657	3.543396	9990.096	201.550369	4.447039475
–10%	1.548E–05	0.0622222	0.18	19.24083	3.543396	9990.096	189.637754	3.730459168
		% Change					6.08401546	17.51609611
Average	1.548E–05	0.0622222	0.18	21.3787	3.543396	9990.096	195.801844	4.090981819
10%	1.548E–05	0.0622222	0.18	21.3787	3.897736	9990.096	184.644102	7.699157073
–10%	1.548E–05	0.0622222	0.18	21.3787	3.189057	9990.096	208.92244	2.033546249
		% Change					12.3994427	138.4902469
Average	1.548E–05	0.0622222	0.18	21.3787	3.543396	9990.096	195.801844	4.090981819
10%	1.548E–05	0.0622222	0.18	21.3787	3.543396	10989.11	186.730662	3.667820634
–10%	1.548E–05	0.0622222	0.18	21.3787	3.543396	8991.087	206.343257	4.615803697
		% Change					10.0165523	23.17250736

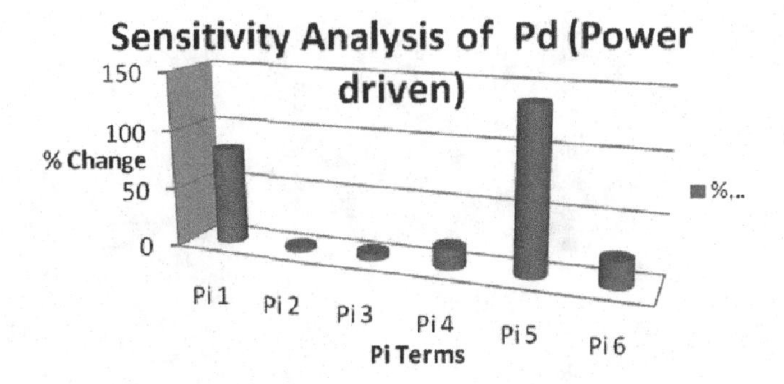

FIGURE 7.9 Percentage change in turbine speed with variation of Pi term.

Percentage change in all the six independent pi terms to evaluate the effect of this changes on the dependent variable Turbine Speed.

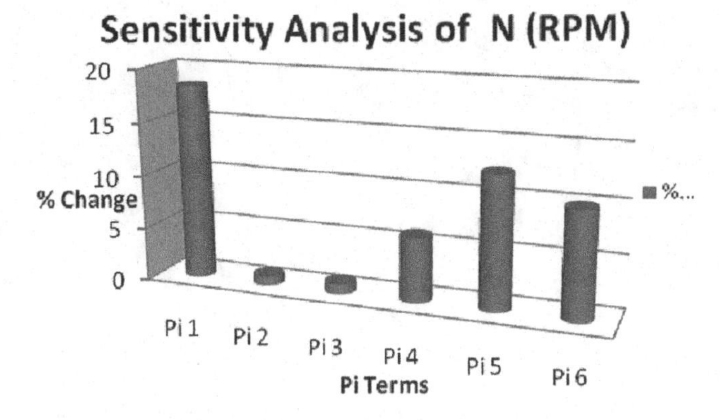

FIGURE 7.10 Percentage change in power developed with variation of pi.

Percentage change in all the six independent pi terms to evaluate the effect of this changes on the dependent variable Power Developed.

8 Interpretation of Mathematical Models

8.1 MODELS DEVELOPED FOR DEPENDENT VARIABLES OF CRANKSHAFT MAINTENANCE ACTIVITY

The exact forms of models for dependent variables, overhauling time of crankshaft maintenance activity, human energy consumed in crankshaft maintenance activity, and productivity of crankshaft maintenance activity are as follows:

$$\begin{aligned} z_{1C} &= \pi_{D1} \\ &= 1.7198\,(\pi_1)^{0.0873}(\pi_2)^{0.1026}(\pi_3)^{2.8137}(\pi_4)^{0.2115}(\pi_5)^{-0.2101}(\pi_6)^{-0.0025}(\pi_7)^{3.5409}(\pi_8)^{-0.4451} \\ &\quad (\pi_9)^{-0.125}(\pi_{10})^{-0.0609}(\pi_{11})^{0.1131} \end{aligned} \tag{8.1}$$

$$\begin{aligned} z_{2C} &= \pi_{D2} \\ &= 1.2203(\pi_1)^{0.2769}(\pi_2)^{0.018}(\pi_3)^{-0.2583}(\pi_4)^{3.3045}(\pi_5)^{0.0597}(\pi_6)^{-0.0314}(\pi_7)^{-1.2341}(\pi_8)^{0.0415} \\ &\quad (\pi_9)^{0.0103}(\pi_{10})^{0.0002}(\pi_{11})^{-0.3935} \end{aligned} \tag{8.2}$$

$$\begin{aligned} z_{3C} &= \pi_{D3} \\ &= 1.6837(\pi_1)^{-0.0873}(\pi_2)^{-0.1026}(\pi_3)^{-0.1837}(\pi_4)^{-1.2275}(\pi_5)^{0.2101}(\pi_6)^{0.0025}(\pi_7)^{0.9916}(\pi_8)^{0.4451} \\ &\quad (\pi_9)^{0.125}(\pi_{10})^{0.0609}(\pi_{11})^{-0.1131} \end{aligned} \tag{8.3}$$

In the above equations, (z_{1C}) relates to response variable for overhauling time of crankshaft maintenance activity, (z_{2C}) relates to response variable for human energy consumed in crankshaft maintenance activity, and (z_{3C}) relates to response variable for productivity of crankshaft maintenance activity.

8.1.1 Interpretation of Model of Crankshaft Maintenance Activity

The indices of the models of crankshaft maintenance activity are the indicators of how the phenomenon is affected by the interaction of various independent pi terms in the model. Interpretation of the model is reported in terms of several aspects: (a) interpretation of curve fitting constant K; (b) order of influence of various inputs (causes) on outputs (effects); and (c) Relative influence of causes on effect.

a. Interpretation of curve fitting constant K_1 for z_{1C}
   The value of curve fitting constant in z_{1C} is 1.7198. This collectively represents the combined effect of all extraneous variables. Further, as it is positive, this indicates that there are good numbers of causes, which have increasing influence on overhauling time of crankshaft maintenance activity (z_{1C}).

8.1.1.1 Analysis of the Model for Dependent Pi Term – Overhauling Time of Crankshaft Maintenance Activity (z_{1C})

1. The absolute index of π_7 is the highest at 3.5409. Thus, the term related to workstation of crankshaft maintenance activity is the most influencing π term in this model. The value of

DOI: 10.1201/9781003318699-8

this index is positive indicating that overhauling time of crankshaft maintenance activity is directly proportional to workstation of crankshaft maintenance activity. It indicates that increase in length and width, and decrease in height of frame, can increase the overhauling time of crankshaft maintenance activity (z_{1C}).

2. The absolute index of π_8 is the lowest at −0.4451. Thus, the term related to temperature is the least influencing π term in this model. The value of this index is negative indicating that overhauling time of crankshaft maintenance activity is inversely proportional to term related to temperature (π_8) at crankshaft maintenance workstation. It indicates that increase in temperature can reduce the overhauling time of crankshaft maintenance activity (z_{1C}).
3. The sequence of influence of other independent π terms present in this model is π_3, π_4, π_{11}, π_2, π_1, π_6, π_{10}, π_9, and π_5, having absolute indices as 2.8137, 0.2115, 0.1131, 0.1026, 0.0873, −0.0025, −0.0609, −0.125 and −0.2101, respectively.

The overhauling time of crankshaft maintenance activity (z_{1C}) is directly proportional to the terms related to specification of crankshaft data (π_3) with parameters including length of Allen bolt, length of saddle bolt, and diameter of bearing of crankpin. As the length of saddle bolt, Allen bolts and diameter of crankpin bearing increases the overhauling time of crankshaft maintenance increases with the index as 2.8137.

The overhauling time of crankshaft maintenance activity (z_{1C}) is directly proportional to the term related to specification of tools used for maintenance activity (π_4) with parameters including length of spanner and length of barring rod. As the length of spanner and barring rod increases the overhauling time of crankshaft maintenance increases with the index as 0.2115.

The overhauling time of crankshaft maintenance activity (z_{1C}) is directly proportional to the term related to noise at workstation (π_{11}). As the noise at workstation increases the overhauling time of crankshaft maintenance activity increases with the index as 0.1131.

The overhauling time of crankshaft maintenance activity (z_{1C}) is directly proportional to the term related to workers data (π_2) with parameters including the age, enthusiasm, and general health status of workers with the index as 0.1026. Increase in age, increase in health problems, and decrease in enthusiasm and work experience can increase the overhauling time of crankshaft maintenance activity (z_{1C}).

The overhauling time of crankshaft maintenance activity (z_{1C}) is directly proportional to the term related to anthropometric data (π_1) with parameters including stature, elbow height with the index as 0.0873. It indicates that increase in stature of worker, Elbow height of worker, fingertip height, and decrease in shoulder height of worker, eye level height of workers can increase the overhauling time of crankshaft maintenance activity (z_{1C}).

Similarly, the overhauling time of crankshaft maintenance activity (z_{1C}) is inversely proportional to the term related to axial clearance (π_6), with the index as −0.0025. It indicates that increase in axial clearance and bolt elongation can reduce the overhauling time of crankshaft maintenance activity (z_{1C}).

The overhauling time of crankshaft maintenance activity (z_{1C}) is inversely proportional to the term related to illumination (π_{10}) at workstation, with the index as −0.0609. It indicates that increase in illumination can reduce the overhauling time of crankshaft maintenance activity (z_{1C}).

The overhauling time of crankshaft maintenance activity (z_{1C}) is inversely proportional to the term related to humidity (π_9) at workstation, with the index as −0.125. It indicates that increase in humidity can reduce the overhauling time of crankshaft maintenance activity (z_{1C}).

The overhauling time of crankshaft maintenance activity (z_{1C}) is inversely proportional to the term related to specification of solvent (π_5) at workstation with the index as −0.2101. It indicates that increase in solvents can reduce the overhauling time of crankshaft maintenance activity (z_{1C}).

b. Interpretation of curve fitting constant K_2 for z_{2C}

The value of curve fitting constant in (z_{2C}) is 1.2203. This collectively represents the combined effect of all extraneous variables. Further, as it is positive, this indicates that there is a good number of causes which have increasing influence on human energy consumed in crankshaft maintenance activity (z_{2C}).

8.1.1.2 Analysis of the Model for Dependent Pi Term – Human Energy Consumed in Crankshaft Maintenance Activity (z_{2C})

1. The absolute index of π_4 is the highest at 3.3045. Thus the term related to tools used for crankshaft maintenance activity is the most influencing π term in this model. The value of this index is positive indicating that human energy consumed in crankshaft maintenance activity is directly proportional to tools of crankshaft maintenance activity with parameters including length of spanner, and length of barring rod. As the length of spanner and barring rod increases, the so does the human energy consumed in the crankshaft maintenance activity (z_{2C}).
2. The absolute index of π_7 is the lowest at −1.2341. Thus the term related to workstation of crankshaft maintenance activity is the least influencing π term in this model. The value of this index is negative, indicating that human energy consumed in crankshaft maintenance activity is inversely proportional to the term related to workstation of crankshaft maintenance. It indicates that increase in height of frame can reduce the human energy consumed in the crankshaft maintenance activity (z_{2C}).
3. The sequence of influence of other independent π terms are π_1, π_5, π_8, π_2, π_9, π_{10}, π_6, π_3, and π_{11}, having absolute indices as 0.2709, 0.0597, 0.0415, 0.018, 0.0103, 0.0002, −0.0314, −0.2583, and −0.3935, respectively.

The human energy consumed in crankshaft maintenance activity (z_{2C}) is directly proportional to the term related to anthropometric data (π_1) with parameters including stature and elbow height with the index as 0.2769. It indicates that increase in stature, elbow height, fingertip height, and decrease in shoulder height, and eye level height of worker can increase the human energy consumed in crankshaft maintenance activity (z_{2C}).

The human energy consumed in crankshaft maintenance activity (z_{2C}) is directly proportional to the term related to solvent (π_5) at workstation with the index as 0.0597. It indicates that increase in solvents can raise the human energy consumed in crankshaft maintenance activity (z_{2C}).

The human energy consumed in crankshaft maintenance activity (z_{2C}) is directly proportional to the term related to temperature (π_8) at workstation with the index as 0.0415. It indicates that increase in temperature can increase the human energy consumed in crankshaft maintenance activity (z_{2C}).

The human energy consumed in crankshaft maintenance activity (z_{2C}) is directly proportional to the term related to workers data (π_2) with parameters including age, enthusiasm, and general health status of workers with the index as 0.018. It indicates that increase in age, increase in health problems, and decrease in enthusiasm and work experience can increase the human energy consumed in crankshaft maintenance activity (z_{2C}).

The human energy consumed in crankshaft maintenance activity (z_{2C}) is directly proportional to the term related to humidity (π_9) at workstation with the index as 0.0103. It indicates that increase in humidity can increase the human energy consumed in crankshaft maintenance activity (z_{2C}).

The human energy consumed in crankshaft maintenance activity (z_{2C}) is directly proportional to the terms related to illumination (π_{10}) at workstation with the index as 0.0002. It indicates that increase in illumination can increase the human energy consumed in crankshaft maintenance activity (z_{2C}).

Similarly, human energy consumed in crankshaft maintenance activity (z_{2C}) is inversely proportional to the term related to axial clearance (π_6), with the index as −0.0314. It indicates that increase in axial clearance and bolt elongation can reduce the human energy consumed in crankshaft maintenance activity (z_{2C}).

The human energy consumed in crankshaft maintenance activity (z_{2C}) is indirectly proportional to the term related to specification of crankshaft data (π_3) with parameters including length of saddle bolt, Allen bolt, and diameter of bearing of crankpin. As the length of saddle bolt, Allen bolts and diameter of crankpin bearing increases, the human energy consumed in crankshaft maintenance activity (z_{2C}) reduces with the index as −0.2583.

The human energy consumed in crankshaft maintenance activity (z_{2C}) is indirectly proportional to the term related to noise at workstation (π_{11}). As the noise at workstation increases, the human energy consumed in crankshaft maintenance activity (z_{2C}) reduces with the index as −0.3935.

c. Interpretation of curve fitting constant K_3 for z_{3C}
The value of curve fitting constant in z_3 is 1.6837. This collectively represents the combined effect of all extraneous variables. Further, as it is positive, this indicates that there are good number of causes, which have increasing influence on productivity of crankshaft maintenance activity (z_{3C}).

8.1.1.3 Analysis of the Model for Dependent Pi Term – Productivity in Crankshaft Maintenance Activity (z_{3C})

1. The absolute index of π_7 is the highest at 0.9916. Thus the term related to workstation of crankshaft maintenance activity is the most influencing π term in this model. The value of this index is positive indicating that productivity of crankshaft maintenance activity (z_{3C}) is directly proportional to workstation of crankshaft maintenance activity with parameters such as length of frame. It indicates that increase in height of frame can increase the productivity of the crankshaft maintenance activity (z_{3C}).
2. The absolute index of π_4 is the lowest at −1.2275. Thus the term related to tools used for crankshaft maintenance activity is the least influencing π term in this model. The value of this index is negative indicating that productivity of crankshaft maintenance activity is directly proportional to tools of crankshaft maintenance activity with parameters including length of spanner, length of barring rod. As the length of spanner and barring rod increases, the productivity of the crankshaft maintenance activity (z_{3C}) decreases.
3. The sequence of influence of other independent π terms are π_8, π_5, π_9, π_{10}, π_6, π_1, π_2, π_{11}, and π_3, having absolute indices as 0.4451, 0.2101, 0.125, 0.0609, 0.0025, −0.0873, −0.1026, −0.1131, and −0.1837, respectively.

The productivity of the crankshaft maintenance activity (z_{3C}) is directly proportional to the term related to temperature (π_8) at workstation with the index as 0.4451. It indicates that increase in temperature can increase productivity of the crankshaft maintenance activity (z_{3C}).

The productivity of the crankshaft maintenance activity (z_{3C}) is directly proportional to the term related to solvent (π_5) at workstation with the index as 0.2101. It indicates that increase in solvents can increase the productivity of the crankshaft maintenance activity (z_{3C}).

The productivity of the crankshaft maintenance activity (z_{3C}) is directly proportional to the term related to humidity (π_9) at workstation with the index as 0.125. It indicates that increase in humidity can increase the productivity of the crankshaft maintenance activity (z_{3C}).

The productivity (z_{3C}) of crankshaft maintenance activity is directly proportional to the term related to illumination (π_{10}) at workstation with the index as 0.0609. It indicates that increase in illumination can increase the productivity of the crankshaft maintenance activity (z_{3C}).

The productivity of the crankshaft maintenance activity (z_{3C}) is directly proportional to the term related to axial clearance (π_6) with the index as 0.0025. It indicates that increase in axial clearance and bolt elongation can increase productivity of the crankshaft maintenance activity (z_{3C}).

Similarly, the productivity of the crankshaft maintenance activity (z_{3C}) is indirectly proportional to the term related to anthropometric data (π_1) with parameters including stature, elbow height with the index as −0.0873. It indicates that increase in stature, elbow height, fingertip height, and decrease in shoulder height, eye level height of workers can reduce the productivity of the crankshaft maintenance activity (z_{3C}).

The productivity of the crankshaft maintenance activity (z_{3C}) is indirectly proportional to the term related to workers data (π_2) with parameters including the age, enthusiasm, general health status of workers with the index as −0.1026. It indicates that increase in age, increase in health problems, decrease in enthusiasm and work experience of workers can decrease the productivity of the crankshaft maintenance activity (z_{3C}).

The productivity of the crankshaft maintenance activity (z_{3C}) is indirectly proportional to the term related to noise at workstation (π_{11}). As the noise at workstation increases, the productivity of the crankshaft maintenance activity (z_{3C}) reduces with the index as −0.1131.

The productivity of the crankshaft maintenance activity (z_{3C}) is indirectly proportional to the term related to specification of crankshaft data (π_3) with parameters including length of saddle bolt, length of Allen bolt, diameter of bearing of crankpin. As the length of saddle bolt and length of Allen bolt and diameter of crankpin bearing increases, the productivity of the crankshaft maintenance activity (z_{3C}) reduces with the index as −0.1837.

8.1.2 Analysis of Performance of Models by ANN Simulation of Crankshaft Maintenance Activity

The models have been formulated mathematically as well as by using ANN. We have three sets of values: computed by field values, computed by mathematical models, and those of ANN. All these values match well. This is justified by calculating their respective mean values and standard errors of performance. The comparison of values of dependent pi term obtained by field calculated values, mathematical models values, and ANN-based values are indicated in the Table 8.1.

TABLE 8.1
Comparison of Field Calculated Vales, Model Based Values, and ANN Based Values of Crankshaft Maintenance Activity

	Dependent Pi Term of Overhauling Time of Crankshaft Maintenance Activity (z_{1C})			Dependent Pi Term Human Energy Consumed in Crankshaft Maintenance Activity (z_{2C})			Dependent Pi Term Productivity of Crankshaft Maintenance Activity (z_{3C})		
Ser	Field	Model	ANN	Field	Model	ANN	Field	Model	ANN
1	1.8185×10^{11}	1.5592×10^{11}	1.9439×10^{11}	55.074	46.884	51.962	0.0697	0.0547	0.06518
2	1.8766×10^{11}	1.5212×10^{11}	1.6414×10^{11}	45.420	45.264	48.340	0.0588	0.0561	0.06726
3	2.0639×10^{11}	1.6386×10^{11}	2.1767×10^{11}	42.954	47.014	47.507	0.078	0.0521	0.07398
4	1.4866×10^{11}	1.3997×10^{11}	1.4569×10^{11}	43.496	34.471	49.930	0.0522	0.0610	0.05328
5	2.4824×10^{11}	2.139×10^{11}	2.3375×10^{11}	39.122	42.088	47.741	0.0838	0.0399	0.08898
6	2.4475×10^{11}	1.7524×10^{11}	2.1843×10^{11}	39.335	41.198	43.700	0.0783	0.0487	0.08773
7	1.6054×10^{11}	1.3777×10^{11}	1.8066×10^{11}	43.106	45.487	52.634	0.0648	0.0620	0.05754
8	1.6739×10^{11}	1.435×10^{11}	1.9018×10^{11}	35.325	43.887	39.882	0.0682	0.0595	0.06
9	1.6532×10^{11}	1.5186×10^{11}	2.0776×10^{11}	38.322	42.750	41.457	0.0745	0.0562	0.05925
10	1.5240×10^{11}	1.5253×10^{11}	1.9824×10^{11}	33.506	43.212	40.499	0.0711	0.0560	0.05463
11	1.6002×10^{11}	1.3818×10^{11}	1.9436×10^{11}	28.707	35.778	39.275	0.0697	0.0618	0.05736
12	1.1559×10^{11}	1.1162×10^{11}	1.6113×10^{11}	50.656	45.116	52.684	0.0578	0.0765	0.04143
13	1.5499×10^{11}	1.5576×10^{11}	1.7723×10^{11}	45.732	44.291	50.913	0.0635	0.0548	0.05555
14	1.6945×10^{11}	1.4397×10^{11}	1.752×10^{11}	31.043	41.678	33.627	0.0628	0.0593	0.06074
15	1.6106×10^{11}	1.3708×10^{11}	1.5686×10^{11}	26.025	40.549	36.539	0.0562	0.0623	0.05773
16	1.5021×10^{11}	1.4764×10^{11}	1.5782×10^{11}	46.720	41.731	45.474	0.0566	0.0578	0.05384
17	1.8172×10^{11}	1.4255×10^{11}	1.7455×10^{11}	40.369	43.083	46.949	0.0626	0.0599	0.06513
18	1.7785×10^{11}	1.4291×10^{11}	1.7983×10^{11}	45.418	39.745	46.859	0.0645	0.0597	0.06375
19	1.7113×10^{11}	1.4301×10^{11}	1.7967×10^{11}	53.215	39.611	52.369	0.0644	0.0597	0.06134

(Continued)

TABLE 8.1 (Continued)

	Dependent Pi Term of Overhauling Time of Crankshaft Maintenance Activity (z_{1C})			Dependent Pi Term Human Energy Consumed in Crankshaft Maintenance Activity (z_{2C})			Dependent Pi Term Productivity of Crankshaft Maintenance Activity (z_{3C})		
Ser	Field	Model	ANN	Field	Model	ANN	Field	Model	ANN
20	1.8598×10^{11}	1.538×10^{11}	1.7329×10^{11}	34.803	39.085	43.951	0.0621	0.0555	0.06666
21	1.8624×10^{11}	1.5668×10^{11}	1.9152×10^{11}	53.050	43.482	53.789	0.0687	0.0545	0.06675
22	1.5822×10^{11}	1.5478×10^{11}	1.8859×10^{11}	58.354	50.147	48.680	0.0676	0.0551	0.05671
23	1.5563×10^{11}	1.4008×10^{11}	1.6258×10^{11}	38.132	44.692	39.170	0.0583	0.0609	0.05578
24	1.8728×10^{11}	1.6527×10^{11}	1.9363×10^{11}	34.244	38.450	37.177	0.0694	0.0516	0.06713
25	1.8986×10^{11}	1.4452×10^{11}	2.0706×10^{11}	29.606	41.265	29.862	0.0742	0.0591	0.06805
26	1.7591×10^{11}	1.5688×10^{11}	1.9784×10^{11}	33.509	48.319	40.089	0.0709	0.0544	0.06305
27	1.7927×10^{11}	1.4413×10^{11}	2.0036×10^{11}	39.159	45.030	40.754	0.0718	0.0592	0.06425
28	1.8366×10^{11}	1.4897×10^{11}	1.7611×10^{11}	34.500	39.619	35.636	0.0631	0.0573	0.06583
29	1.6855×10^{11}	1.8455×10^{11}	1.8781×10^{11}	36.352	37.415	39.847	0.0673	0.0462	0.06041
30	1.8392×10^{11}	1.5311×10^{11}	1.5188×10^{11}	31.264	39.062	44.080	0.0544	0.0557	0.06592

TABLE 8.2
Comparison between Observed and Computed Values of Dependent Pi Terms of Crankshaft Piston Maintenance Operation

Dependent Pi Term	Overhauling Time (z_{1C})	Human Energy Consumed (z_{2C})	Productivity(z_{3C})
Mean field	1.7532e+011	40.2178	0.0628
Mean ANN	1.8461e+011	44.0464	0.0662
Mean model	1.5174e+011	42.3473	0.0570
Mean absolute error performance function	1.8013e+010	4.8203	0.0065
Mean squared error performance function	4.8121e+020	35.4194	6.1831e–005

From these comparisons, phenomena response by a conventional approach and ANN simulation, it seems that the curve obtained by dependent pi terms for crankshaft maintenance activity – overhauling time of crankshaft maintenance activity (z_{1C}), human energy consumed in crankshaft maintenance activity (z_{2C}) and productivity of crankshaft maintenance activity (z_{3C}) –overlap due to fewer errors, which is on the positive side, and gives an accurate relationship between ANN simulation and field data (Table 8.2).

From the above tables, it seems that mathematical models and ANN developed using MATLAB can be successfully used for computation of dependent pi terms for a given set of independent pi terms.

8.2 MODELS DEVELOPED FOR DEPENDENT VARIABLES OF LINER PISTON MAINTENANCE ACTIVITY

The exact forms of models for dependent variables, overhauling time of liner piston maintenance activity, human energy consumed in liner piston maintenance activity, and productivity of liner piston maintenance activity are as follows:

$$\begin{aligned} z_{1P} &= \pi_{D1} \\ &= 1.0122(\pi_1)^{-0.0973}(\pi_2)^{-0.1917}(\pi_3)^{0.8772}(\pi_4)^{-6.0098}(\pi_5)^{0.0896}(\pi_6)^{0.378}(\pi_7)^{0.6869}(\pi_8)^{-0.5615} \\ &\quad (\pi_9)^{-0.2282}(\pi_{10})^{0.0887}(\pi_{11})^{0.0838} \end{aligned} \tag{8.4}$$

$$z_{2P} = \pi_{D2}$$
$$= 1.7874(\pi_1)^{0.355}(\pi_2)^{0.3448}(\pi_3)^{0.1681}(\pi_4)^{4.4021}(\pi_5)^{-0.0249}(\pi_6)^{-0.131}(\pi_7)^{-0.3313}(\pi_8)^{-0.1018}$$
$$(\pi_9)^{-0.1479}(\pi_{10})^{0.0643}(\pi_{11})^{0.2221} \tag{8.5}$$

$$z_{3P} = \pi_{D3}$$
$$= 1.1148(\pi_1)^{0.0973}(\pi_2)^{0.1917}(\pi_3)^{0.0188}(\pi_4)^{-2.5291}(\pi_5)^{-0.0896}(\pi_6)^{-0.378}(\pi_7)^{2.971}(\pi_8)^{0.5615}$$
$$(\pi_9)^{0.2282}(\pi_{10})^{-0.0887}(\pi_{11})^{-0.0838} \tag{8.6}$$

In the above equations, (z_{1P}) relates to response variable of overhauling time of liner piston maintenance activity, (z_{2P}) relates to response variable for human energy consumed in liner piston maintenance activity, and (z_{3P}) relates to response variable for productivity of liner piston maintenance activity.

8.2.1 Interpretation of Models of Liner Piston Maintenance Activity

The indices of the models of liner piston maintenance activity are the indicators of how the phenomenon is getting affected because of the interaction of various independent pi terms in model. Interpretation of model is being reported in terms of several aspects: (a) interpretation of curve fitting constant K; (b) order of influence of various inputs (causes) on outputs (effects); and (c) Relative influence of causes on effect.

a. Interpretation of curve fitting constant K_1 for z_{1P}

The value of curve fitting constant in z_{1P} is 1.0122. This collectively represents the combined effect of all extraneous variables. Further, as it is positive, this indicates that there are good number of causes, which have increasing influence on time of liner piston maintenance activity (z_{1P}).

1. The absolute index of π_3 is the highest at 0.8772. Thus the term related specification of liner piston is the most influencing π term in this model. The value of this index is positive indicating that overhauling time of liner piston maintenance activity (z_{1P}) is directly proportional to liner piston specifications with parameters including length and diameter of liner, length of piston pin, and piston stroke length. It indicates that increase in length of piston pin, piston stroke length, and increase in length and diameter of liner can increase the overhauling time liner piston maintenance activity (z_{1P}).
2. The absolute index of π_4 is the lowest at −6.0098. Thus the term related to specification tools used for liner piston maintenance activity is the least influencing π term in this model. The value of this index is negative indicating that overhauling time of liner piston maintenance activity (z_{1P}) is inversely proportional to tools used for liner piston overhauling with parameters including length and diameter of bracket expander, length and diameter of socket tool, and length and diameter of liner puller. It indicates that increase in length and diameter of bracket expander, length and diameter of socket tool, and length and diameter of liner puller can reduce the overhauling time liner piston maintenance activity (z_{1P}).
3. The sequence of influence of other independent pi terms are π_7, π_6, π_5, π_{10}, π_{11}, π_1, π_2, π_9, and π_8, having absolute indices as 0.6869, 0.378, 0.0896, 0.0887, 0.0838, −0.0973, −0.1917, −0.2282 and −0.5615, respectively.

The overhauling time of liner piston maintenance activity (z_{1P}) is directly proportional to the term related to workstation data (π_7), with parameters including height and breath of frame with the index as 0.6869. It indicates that increase in height and breath of frame can increase overhauling time of liner piston maintenance activity (z_{1P}).

The overhauling time of liner piston maintenance activity (z_{1P}) is directly proportional to the term related to axial clearance (π_6) with the index as 0.378. It indicates that increase in axial clearance of small end bearing and bolt elongation can increase overhauling time of liner piston maintenance activity (z_{1P}).

The overhauling time of liner piston maintenance activity (z_{1P}) is directly proportional to the term related to solvent (π_5) at workstation with the index as 0.0896. It indicates that increase in solvents can increase overhauling time of liner piston maintenance activity (z_{1P}).

The overhauling time of liner piston maintenance activity (z_{1P}) is directly proportional to the term related to illumination (π_{10}) at workstation with the index as 0.0887. It indicates that increase in illumination can increase the overhauling time of liner piston maintenance activity (z_{1P}).

The overhauling time of liner piston maintenance activity (z_{1P}) is directly proportional to the term related to noise at workstation (π_{11}). As the noise at workstation increases, the overhauling time of liner piston maintenance activity (z_{1P}) increases with the index as 0.0838.

Similarly, the overhauling time of liner piston maintenance activity (z_{1P}) is indirectly proportional to the term related to anthropometric data (π_1) with parameters including stature, elbow height, fingertip height, shoulder height and eye level height, with the index as –0.0973. It indicates that increase in stature, elbow height, fingertip height, and decrease in shoulder height and eye level height of workers can reduce the overhauling time of liner piston maintenance activity (z_{1P}).

The overhauling time of liner piston maintenance activity (z_{1P}) is indirectly proportional to the term related to workers data (π_2) with parameters including the age, enthusiasm, general health status of workers with the index as –0.1917. It indicates that decrease in age, decrease in health problems, increase in enthusiasm and work experience can decrease the overhauling time of liner piston maintenance activity (z_{1P}).

The overhauling time of liner piston maintenance activity (z_{1P}) is indirectly proportional to the term related to humidity (π_9) at workstation, with the index as –0.2282. It indicates that increase in humidity can reduce the overhauling time of liner piston maintenance activity (z_{1P})

The overhauling time of liner piston maintenance activity (z_{1P}) is indirectly proportional to the term related to temperature (π_8) at workstation, with the index as –0.5615. It indicates that increase in temperature can reduce overhauling time of liner piston maintenance activity (z_{1P}).

b. Interpretation of curve fitting constant K_2 for z_{2P}

The value of curve fitting constant in z_{2P} is 1.7874. This collectively represents the combined effect of all extraneous variables. Further, as it is positive, this indicates that there are good numbers of causes, which have increasing influence on human energy consumed in liner piston maintenance activity (z_{2P}).

1. The absolute index of π_4 is the highest at 4.4021. Thus the term related to tools used for liner piston maintenance activity is the most influencing π term in this model. The value of this index is positive, indicating that human energy consumed in liner piston maintenance activity (z_{2P}) is directly proportional to tools used for liner piston overhauling with parameters including length and diameter of bracket expander, length, and diameter of socket tool, and length and diameter of liner puller. It indicates that increase in length and diameter of bracket expander, length, and diameter of socket tool, length, and diameter of liner puller can increase the human energy consumed in liner piston maintenance activity (z_{2P}).
2. The absolute index of π_7 is the lowest at –0.3313. Thus the term related to workstation data is the least influencing π term in this model. The value of this index is negative indicating that human energy consumed in liner piston maintenance activity (z_{2P}) is inversely proportional to liner piston specifications with parameters including height and breath of frame. It indicates that increase in height and breath of frame can reduce the human energy consumed in liner piston maintenance activity (z_{2P}).
3. The sequence of influence of other independent pi terms are π_1, π_2, π_{11}, π_3, π_{10}, π_5, π_8, π_6, and π_9, having absolute indices as 0.355, 0.3448, 0.2221, 0.1681, 0.0643, –0.0249, –0.1018, –0.131, and –0.1479, respectively.

The human energy consumed in liner piston maintenance activity (z_{2P}) is directly proportional to the term related to anthropometric data (π_1), with parameters including stature, elbow height with the index as 0.355. It indicates that increase in stature, elbow height, fingertip height, and decrease in shoulder, eye level height of worker can increase the human energy consumed in liner piston maintenance activity (z_{2P}).

The human energy consumed in liner piston maintenance activity (z_{2P}) is directly proportional to the term related to workers data (π_2) with parameters including the age, enthusiasm, and general health status of workers with the index as 0.3448. It indicates that increase in age, increase in health problems, decrease in enthusiasm and work experience can increase the human energy consumed in liner piston maintenance activity (z_{2P}).

The human energy consumed in liner piston maintenance activity (z_{2P}) is directly proportional to the term related to noise at workstation (π_{11}). As the noise at workstation increases, the human energy consumed in liner piston maintenance activity (z_{2P}) also increases with the index as 0.2221.

The human energy consumed in liner piston maintenance activity (z_{2P}) is directly proportional to the term related to specification of liner piston (π_3) at workstation, with the index as 0.1681, and parameters including length and diameter of liner, length of piston pin, and piston stroke length. It indicates that increase in length of piston pin, piston stroke length and increase in length and diameter of liner can increase the human energy consumed in liner piston maintenance activity (z_{2P}).

The human energy consumed in liner piston maintenance activity (z_{2P}) is directly proportional to the term related to illumination (π_{10}) at workstation with the index as 0.0643. It indicates that increase in illumination can increase the human energy consumed in liner piston maintenance activity (z_{2P}).

Similarly, the human energy consumed in liner piston maintenance activity (z_{2P}) is indirectly proportional to the term related to solvent (π_5) at workstation with the index as −0.0249. It indicates that increase in solvents can reduce the human energy consumed in liner piston maintenance activity (z_{2P}).

The human energy consumed in liner piston maintenance activity (z_{2P}) is indirectly proportional to the term related to temperature (π_8) at workstation with the index as −0.1018. It indicates that increase in temperature can reduce the human energy consumed in liner piston maintenance activity (z_{2P})

The human energy consumed in liner piston maintenance activity (z_{2P}) is indirectly proportional to the term related to axial clearance (π_6) with the index as −0.131. It indicates that increase in axial clearance of small end bearing and bolt elongation can reduce the human energy consumed in liner piston maintenance activity (z_{2P}).

The human energy consumed in liner piston maintenance activity (z_{2P}) is indirectly proportional to the term related to humidity (π_9) at workstation with the index as −0.1479. It indicates that increase in humidity can reduce the human energy consumed in liner piston maintenance activity (z_{2P}).

c. Interpretation of curve fitting constant K_3 for z_{3P}

The value of curve fitting constant in (z_{3P}) is 1.1148. This collectively represents the combined effect of all extraneous variables. Further, as it is positive, this indicates that there are good number of causes, which have increasing influence on productivity of liner piston maintenance activity (z_{3P}).

1. The absolute index of π_7 is the highest at 2.7919. Thus the term related to workstation data is the most influencing π term in this model. The value of this index is positive indicating that productivity of liner piston maintenance activity (z_{3P}) is directly proportional to liner piston specifications, with parameters including height and breath of frame. It indicates that increase in height and breath of frame can increase the productivity of liner piston maintenance activity (z_{3P}).

2. The absolute index of $\pi4$ is the lowest at −2.5291. Thus the term related to tools used for liner piston maintenance activity is the least influencing π term in this model. The value of this index is negative indicating that productivity of liner piston maintenance activity (z_{3P}) is inversely proportional to tools used for liner piston overhauling with parameters including length and diameter of bracket expander, length and diameter of socket tool, and length and diameter of liner puller. It indicates that increase in length and diameter of bracket expander, length and diameter of socket tool, and length and diameter of liner puller can reduce the productivity of liner piston maintenance activity (z_{3P}).
3. The sequence of influence of other independent pi terms are π_8, π_9, π_2, π_1, π_3, π_{11}, π_{10}, π_5, and π_6 having absolute indices as 0.5615, 0.2282, 0.1917, 0.0973, 0.0188, −0.0838, −0.0887, −0.0896, and −0.378, respectively.

The productivity of liner piston maintenance activity (z_{3P}) is directly proportional to the term related to temperature (π_8) at workstation with the index as 0.5615. It indicates that increase in temperature can increase the productivity of liner piston maintenance activity (z_{3P}).

The productivity of liner piston maintenance activity (z_{3P}) is directly proportional to the term related to humidity (π_9) at workstation with the index as 0.2282. It indicates that increase in humidity can increase productivity of liner piston maintenance activity (z_{3P}).

The productivity of liner piston maintenance activity (z_{3P}) is directly proportional to the term related to workers data (π_2) with parameters including the age, enthusiasm, and general health status of workers with the index as 0.1917. It indicates that increase in age, increase in health problems, decrease in enthusiasm and work experience of workers can increase the productivity of liner piston maintenance activity (z_{3P}).

The productivity of liner piston maintenance activity (z_{3P}) is directly proportional to the term related to anthropometric data (π_1), with parameters including stature, elbow height, fingertip height, shoulder height, and eye level height, with the index as 0.0973. It indicates that increase in stature, elbow height fingertip height and decrease in shoulder height, and eye level height of workers can increase the productivity of liner piston maintenance activity (z_{3P}).

The productivity of liner piston maintenance activity (z_{3P}) is directly proportional to the term related to specification of liner piston (π_3) at workstation, with the index as 0.0188 with parameters including length and diameter of liner, length of piston pin, piston stroke length. It indicates that increase in length of piston pin, piston stroke length, and increase in length and diameter of liner can increase the productivity of liner piston maintenance activity (z_{3P}).

Similarly, the productivity of liner piston maintenance activity (z_{3P}) is inversely proportional to the term related to noise at workstation (π_{11}). As the noise at workstation increases, the productivity of liner piston maintenance activity (z_{3P}) reduces with the index as −0.0838.

The productivity of liner piston maintenance activity (z_{3P}) is indirectly proportional to the term related to illumination (π_{10}) at workstation, with the index as −0.0887. It indicates that increase in illumination can reduce the productivity of liner piston maintenance activity (z_{3P}).

The productivity of liner piston maintenance activity (z_{3P}) is indirectly proportional to the term related to solvent (π_5) at workstation, with the index as −0.0896. It indicates that increase solvents can reduce the productivity of liner piston maintenance activity (z_{3P}).

The productivity of liner piston maintenance activity (z_{3P}) is indirectly proportional to the term related to axial clearance (π_6), with the index as −0.378. It indicates that increase in axial clearance of small end bearing and bolt elongation can reduce the productivity of liner piston maintenance activity (z_{3P}).

8.2.2 Analysis of Performance of Models by ANN Simulation of Liner Piston Maintenance Activity

The models have been formulated mathematically as well as by using ANN. We have three sets of values: computed by field observation, computed by mathematical models, and those of ANN. All these values match well. This is justified by calculating their respective mean values and standard errors of performance. The comparison of values of dependent pi term obtained by field calculated values, mathematical models values, and ANN-based values are indicated in Table 8.3.

From these comparisons, phenomena response by a conventional approach and ANN simulation, it seems that the curve obtained by dependent pi terms for liner piston maintenance activity – overhauling time of liner piston maintenance activity (z_{1P}), human energy consumed in liner maintenance activity (z_{2P}) and productivity of liner maintenance activity (z_{3P}) –overlap, due to fewer errors, which is on the positive side, and gives an accurate relationship between ANN simulation and field data (Table 8.4).

From the above tables, it seems that mathematical models and ANN developed using MATLAB can be successfully used for computation of dependent pi terms for a given set of independent pi terms.

TABLE 8.3
Comparison of Field Calculated Vales, Model Based Values, and ANN-based Values of Liner Piston Maintenance Activity

	Dependent Pi Term Overhauling Time of Liner Piston Maintenance Activity (z_{1P})			Dependent Pi Term Human Energy Consumed in Liner Piston Maintenance Activity (z_{2P})			Dependent Pi Term Productivity of Liner Piston Maintenance Activity (z_{3P})		
	Field	Model	ANN	Field	Model	ANN	Field	Model	ANN
1	8.956×10^{10}	8.292×10^{10}	10.039×10^{10}	36.948	33.5859	34.281	0.0875	0.0701	0.0948
2	9.431×10^{10}	8.799×10^{10}	14.156×10^{10}	46.703	42.3948	44.246	0.0898	0.066	0.1083
3	8.731×10^{10}	9.335×10^{10}	12.768×10^{10}	40.509	36.3640	43.933	0.0864	0.0623	0.1022
4	9.156×10^{10}	8.765×10^{10}	11.036×10^{10}	35.852	33.1269	38.486	0.0885	0.0663	0.0963
5	8.703×10^{10}	8.315×10^{10}	10.365×10^{10}	48.076	44.0842	43.396	0.0863	0.0699	0.0958
6	8.342×10^{10}	8.920×10^{10}	10.059×10^{10}	45.233	36.2659	40.374	0.0845	0.0652	0.0919
7	9.056×10^{10}	9.034×10^{10}	9.769×10^{10}	31.887	31.6475	35.745	0.0880	0.0643	0.0908
8	8.999×10^{10}	8.775×10^{10}	10.757×10^{10}	47.736	39.16944	44.458	0.0877	0.0662	0.095
9	9.755×10^{10}	8.089×10^{10}	10.676×10^{10}	35.440	36.38274	43.857	0.0913	0.0718	0.0929
10	9.563×10^{10}	7.692×10^{10}	13.042×10^{10}	35.798	42.58796	45.548	0.0904	0.0756	0.1057
11	9.214×10^{10}	7.670×10^{10}	9.751×10^{10}	22.407	24.43367	32.454	0.0888	0.0758	0.0929
12	9.257×10^{10}	9.724×10^{10}	11.642×10^{10}	29.116	28.70447	37.005	0.0890	0.0598	0.0982
13	7.671×10^{10}	7.940×10^{10}	13.474×10^{10}	42.507	35.85690	41.270	0.0810	0.0732	0.1051
14	8.288×10^{10}	8.800×10^{10}	13.734×10^{10}	29.704	29.00556	40.839	0.0842	0.066	0.1074
15	8.452×10^{10}	8.359×10^{10}	13.375×10^{10}	38.854	31.47235	40.164	0.0850	0.0695	0.1077
16	8.057×10^{10}	8.867×10^{10}	12.642×10^{10}	35.962	29.68863	38.457	0.0830	0.0655	0.104
17	8.084×10^{10}	7.293×10^{10}	12.019×10^{10}	40.679	34.94850	39.590	0.0831	0.0797	0.102
18	8.219×10^{10}	8.707×10^{10}	14.885×10^{10}	26.040	32.79451	43.704	0.0838	0.0667	0.1109
19	8.480×10^{10}	8.609×10^{10}	10.318×10^{10}	36.554	29.37651	39.343	0.0852	0.0675	0.0933
20	7.671×10^{10}	8.887×10^{10}	12.434×10^{10}	44.581	32.9072	42.927	0.0810	0.0654	0.1023
21	7.790×10^{10}	8.651×10^{10}	10.02×10^{10}	36.487	37.9480	36.382	0.0816	0.0672	0.0944
22	8.370×10^{10}	7.694×10^{10}	9.682×10^{10}	36.486	36.7986	34.519	0.0846	0.0755	0.0908
23	7.592×10^{10}	8.354×10^{10}	10.239×10^{10}	37.525	37.6237	40.734	0.0806	0.0696	0.0927
24	9.666×10^{10}	8.718×10^{10}	8.450×10^{10}	32.845	27.3869	27.346	0.0909	0.0667	0.0863
25	2.208×10^{10}	10.625×10^{10}	12.782×10^{10}	32.927	28.9748	42.257	0.1375	0.0547	0.1039
26	2.353×10^{10}	10.482×10^{10}	11.263×10^{10}	36.704	33.6636	39.331	0.1419	0.0554	0.0961
27	9.028×10^{10}	9.733×10^{10}	9.019×10^{10}	34.207	32.6940	33.676	0.0879	0.0597	0.0896
28	2.181×10^{10}	9.986×10^{10}	7.673×10^{10}	33.057	29.6553	29.524	0.1366	0.0582	0.0821
29	9.301×10^{10}	10.314×10^{10}	8.028×10^{10}	29.044	27.8825	28.265	0.0892	0.0563	0.0819
30	8.383×10^{10}	8.741×10^{10}	7.135×10^{10}	36.381	30.1070	32.078	0.0847	0.0665	0.0779

TABLE 8.4
Comparison between Observed and Computed Values of Dependent Pi Terms of Liner Piston Maintenance Operation

Dependent pi Term	Overhauling Time (z_{1P})	Human Energy Consumed (z_{2P})	Productivity (z_{3P})
Mean field	1.0055e+011	36.5421	0.0914
Mean ANN	1.1041e+011	38.4734	0.0964
Mean model	0.8805e+011	33.5846	0.0666
Mean absolute error performance function	3.6168e+010	4.5073	0.0153
Mean squared error performance function	2.4051e+021	35.3167	3.8629e–004

8.3 ANALYSIS OF THE MATHEMATICAL MODEL FOR THE DEPENDENT PI TERM BRAKE THERMAL EFFICIENCY

Z1 is the brake thermal efficiency. The formulated model based on the experimental data is as follows:

$$Z_1 = 85.1138 \times (\Pi_1)^{0.1758} \times (\Pi_2)^{-0.7380} \times (\Pi_3)^{-0.6950} \times (\Pi_4)^{-0.4490}$$

Where:

Z1 = Brake thermal efficiency
π_1 = Pi terms related to the blend formation and time involvement
π_2 = Pi terms related to the fuel consumption and engine load
π_3 = Pi terms related to the fuel characteristic
π_4 = Pi terms related to the engine specifications

The absolute index of π_1 is highest at 0.1758. Thus, the term related to the blend formation and time involved in the experimental. The value of the index is positive, indicating that the brake thermal efficiency is directly proportional to terms related to the blend formation and time involved.

The termπ_2 has a negative index –0.7380, which is inversely proportional to brake thermal efficiency. The π_2 term is brake thermal efficiency increase with the increase in mass of fuel and blend formation with decrease in stroke length and aniline point.

The term π_3 has a negative index –0.6950, which is inversely proportional to brake thermal efficiency. The π_3 term is brake thermal efficiency increase with the decrease in aniline point and diesel index with increase in cetane number.

The term π_4 has a negative index –0.4490, which is inversely proportional to brake thermal efficiency. The π_4 term is brake thermal efficiency increase with the decrease instroke length, cubic capacity, flue tank capacity and engine speed (Table 8.5).

8.4 ANALYSIS OF THE MATHEMATICAL MODEL FOR THE DEPENDENT PI TERM BRAKE SPECIFIC FUEL CONSUMPTION

Z2 is the brake specific fuel consumption. The formulated model based on the experimental data is as follows:

$$Z_2 = 0.7348 \times (\Pi_1)^{-0.1016} \times (\Pi_2)^{0.7070} \times (\Pi_3)^{-0.6336} \times (\Pi_4)^{-1.2727}$$

TABLE 8.5
Comparison of Experimentally Calculated Values and Mathematical Model Equation Base Values

Ser	Calculated from Mathematical Model	Experimental Observation	Error = Experimental Observation – Calculated Value
1	25.5516	25.8273	0.2757
2	29.9799	29.2810	–0.6989
3	31.8660	32.9520	1.0860
4	33.0555	34.7332	1.6777
5	35.3126	35.4921	0.1795
6	40.9147	40.8440	–0.0707
7	47.6204	44.2315	–3.3889
8	50.6068	49.7231	–0.8837
9	52.6469	53.2748	0.6278
10	56.9381	58.3856	1.4475
11	54.6252	57.8613	3.2361
12	63.2968	61.1010	–2.1959
13	66.4347	63.9941	–2.4406
14	69.2473	69.4127	0.1654
15	74.6129	74.2919	–0.3210

Where:

Z_2 = Brake specific fuel consumption
π_1 = Pi terms related to the blend formation and time involvement
π_2 = Pi terms related to the fuel consumption and engine load
π_3 = Pi terms related to the fuel characteristic
π_4 = Pi terms related to the engine specifications

The term π_2 has a positive index 0.7070 which is directly proportional to brake thermal efficiency. The π_2 term is brake specific fuel consumption decrease with the increase in mass of fuel and blend formation with decrease in stroke length and aniline point.

The absolute index of π_1 is –0.1016. Thus the term related to the blend formation and time involved in the experimental. The value of the index is negative, indicating that the brake specific fuel consumption is inversely proportional to terms related to the blend formation and time involved.

The term π_3 has a negative index –0.6336, which is inversely proportional to brake specific fuel consumption. The π_3 term is brake specific fuel consumption increase with the decrease in aniline point and diesel index with increase in cetane number.

The term π_4 has a negative index –1.2727, which is inversely proportional to brake specific fuel consumption. The $\pi_4$terms is brake specific fuel consumption increase with the decrease in stroke length, cubic capacity, flue tank capacity and engine speed (Table 8.6).

TABLE 8.6
Comparison of Experimentally Calculated Values and Mathematical Model Equation Base Values

Ser	Calculated from Mathematical Model	Experimental Observation	Error = Experimental Observation – Calculated Value
1	13.7323	13.9388	0.2064
2	12.2896	12.2947	0.0051
3	11.5461	10.9250	−0.6211
4	10.6962	10.3647	−0.3315
5	9.9682	10.1431	0.1749
6	8.6128	8.8140	0.2012
7	7.7439	8.1390	0.3951
8	7.2762	7.2401	−0.0361
9	6.7294	6.7574	0.0280
10	6.2270	6.1659	−0.0611
11	6.5045	6.2218	−0.2827
12	5.8633	5.8919	0.0286
13	5.5489	5.6255	0.0766
14	5.1262	5.1864	0.0602
15	4.7537	4.8458	0.0921

8.5 MODELS DEVELOPED FOR DEPENDENT VARIABLES TURBINE SPEED

The exact forms of models obtained for the dependent variables turbine speed in solar updraft tower is as under:

$$(Z_1) = 0.5705 \times (\pi_1)^{-0.9101} \times (\pi_2)^{0.0424} \times (\pi_3)^{0.0474} \times (\pi_4)^{0.3036*} (\pi_5)^{-0.6156} \times (\pi_6)^{-0.4977}$$

In the above equations (Z_1) relates to response variable for turbine speed of solar updraft tower. Interpretation of the model:

Several aspects are reported by interpretation of models: (1) outputs (effects) influenced by various inputs (causes); (2) effect influenced by causes; and (3) curve fitting constant K is interpreted.

a. Interpretation of curve fitting constant (K_1) for Z_1

   The value of curve fitting constant in this model for (Z_1) is 0.5705. This collectively represents the combined effect of all extraneous variables such as leakages, heat transfer, the divergent top chimney, thermal mass associated with the ground area etc. Further, as it is positive, this indicates that these causes have increasing influence on the turbine speed (Z_1).

   The constant in this model is 0.5707. This value is less than one, hence it has no magnification effect in the value computed from the product of the various terms of the model.

8.5.1 Analysis of the Model for Dependent π Term Z_1

i. The absolute index of π_4 is highest at 0.3036. The factor π_4 is related to ambient condition and diameter of collector which is the most influencing π term in this model. The value of this index is positive indicating ambient condition has strong impact on turbine speed. π_{D1} is directly varying with respect to π_4. As the index is high with increasing the temperature and air velocity, increases turbine speed during the experimentation. Further turbine speed increases with the increase in the diameter of solar collector.

The absolute index of π_2 is lowest at 0.0424. Thus π, the term related to parameters such as solar chimney. Chimney height is directly proportional to the turbine speed π_{D1}, with the increase height of chimney increases the turbine speed.

ii. The influence of the other independent pi terms present in this model is π_1, having absolute index of –0.9101. The indices of π_3, π_5, and π_6 are 0.0424, –0.6156, and –0.4977, respectively. The negative indices indicate need for improvement. The negative indices indicate that π_{D1} varies inversely with respect to π_1, π_5, and π_6.

The index of π_1 is negative, indicating that the turbine speed (Z_1) varies, inversely proportional to the term related to solar collector [π_1] with the index as –0.9101, indicating that with the raise in collector roof height and collector thickness, the turbine speed of solar updraft tower reduces.

The index of π_5 is negative, indicating that the turbine speed (Z_1) varies inversely proportional to the term related to solar heating [π_5], with the index as –0.6156, indicating that with the increase in air inlet area, the turbine speed of solar updraft tower reduces.

The index of π_6 is negative, indicating that the turbine speed (Z_1) varies inversely proportional to the term related to thermal conductivity [π_6], with the index as –0.4977. The turbine speed of solar updraft tower increases with the rise in thermal conductivity.

8.6 ANALYSIS OF PERFORMANCE OF THE MODELS OF POWER DEVELOPED

The exact forms of models obtained for the dependent variables turbine speed in solar updraft tower is as follows:

$$(Z) = 7.1796\times10^{-19}\times(\pi_1)^{-3.9024}\times(\pi_2)^{0.1755}\times(\pi_3)^{0.2861}\times(\pi_4)^{0.8756}\times(\pi_5)^{6.6344}\times(\pi_6)^{-1.1456} \quad (8.7)$$

In the above equations (Z_2) relates to response variable for power developed of solar updraft tower.

8.6.1 Interpretation of the Model

Interpretation of the model is reported in terms of several aspects: (1) order of influence of various inputs (causes) on outputs (effects); (2) relative influence of causes on effect; and (3) interpretation of the curve fitting constant K.

a. Interpretation of curve fitting constant (K_1) for Z_2

The value of curve fitting constant in this model for (Z_2) is 7.1796×10^{-19}. This collectively represents the combined effect of all extraneous variables such as leakages, heat transfer, the divergent top chimney, thermal mass associated with the ground area etc. Further, as it is positive, this indicates that these causes have increasing influence on the turbine speed (Z_2).

The constant in this model is 7.1796×10^{-19}. This value is less than one, hence it has no magnification effect in the value computed from the product of the various terms of the model.

8.6.2 Analysis of the Model for Dependent π Term Z_2

i. The absolute index of π_5 is highest at 6.6344. The factor π_5 is related to solar heating and diameter of collector which is the most influencing π term in this model. The value of this index is positive indicating solar heating has strong impact on power developed. π_{D2} is directly varying with respect to π_4. As the index is high with increasing the temperature and heating duration, increases power developed during the experimentation.

ii. The absolute index of π_2 is lowest at 0.1755. Thus π, the term related to parameters such as solar chimney. As chimney height is directly proportional to the power developed π_{D2}, with the increase height of chimney, the power developed also increases.

iii. The influence of the other independent pi terms present in this model is π_1, having absolute index of −3.9024. The indices of π_3, π_4, and π_6 are 0.2861, 0.8756, and −1.1456, respectively. The negative indices indicate need for improvement. The negative indices indicate that π_{D2} varies inversely with respect to π_6.

The index of π_1 is negative indicating that the power developed (Z_2) varies inversely proportional to the term related to solar collector [π_1] with the index as −3.9024, indicating that with the raise in collector roof height and collector thickness, the power developed of solar updraft tower reduces.

The power developed (Z_2) is directly proportional to the term related to the ambient condition [π_4] with the index as 0.8756, The value of this index is positive indicating ambient condition has strong impact on turbine speed. π_{D2} is directly varying with respect to π_4. As the index is high with increasing the temperature and air velocity, so does power developed increase. Further power developed increases with the increase in the diameter of solar collector.

The index of π_6 is negative indicating that the power developed (Z_2) varies inversely proportional to the term related to thermal conductivity [π_6], with the index as −1.1456. The power developed of solar updraft tower increases with the rise in thermal conductivity.

8.6.2.1 Analysis of Performance of the Models

The models have been formulated mathematically as well as using the ANN. The two sets of values of two dependent π terms values are computed by experimental observations, values computed by mathematical models, and the values obtained by ANN. All these values match well. This is justified by calculating their respective mean values and standard errors of estimation. The comparison of dependent π terms obtained by experimental calculated values; equation-based values and ANN-based values are indicated in the Table 8.7.

TABLE 8.7
Comparison of Experimental Calculated Values, Equation Base Values, and ANN-based Values

	Dependent Variable Turbine Speed			Dependent Variable Power Developed		
Ser	Experimental	Mathematical Model	ANN	Experimental	Mathematical Model	ANN
01	152.3962	155.3322	150.5109	1.5000	1.6003	1.7205
02	157.1374	155.2094	153.2339	1.6667	1.6182	1.5659
03	161.2013	161.9593	156.5123	1.8333	1.8571	1.9017
04	159.1693	159.5207	155.3095	1.7500	1.7420	1.7284
05	162.5559	166.8146	159.6308	1.9167	2.0440	2.0660
06	161.2013	163.8501	157.5171	1.8333	1.9060	1.9964
07	159.1693	160.15	156.0304	1.7500	1.7884	1.9290
08	159.1693	160.8255	155.8496	1.7500	1.8298	1.7457
09	169.3291	169.1511	166.9503	2.0833	2.1497	1.9494
10	162.5559	167.2207	161.8905	1.9167	2.0743	1.7901
11	142.2364	143.5051	142.9686	4.1667	4.5656	4.1066
12	138.8498	141.8347	142.8476	3.8889	4.4390	4.0030
13	149.0096	148.0432	145.6899	5.6667	5.1090	5.1385
14	142.2364	143.5051	142.9686	4.1667	4.5656	4.1066
15	149.0096	148.4231	148.2144	5.6667	5.1886	5.6198
16	142.2364	143.9181	143.656	4.1667	4.6455	4.2282
17	140.8818	151.0812	147.6495	4.1667	5.3715	5.1765
18	138.8498	141.887	143.4833	3.8889	4.5311	4.3345
19	159.1693	156.832	156.1698	6.3333	6.1632	6.0372
20	155.7827	154.9865	156.3251	6.0000	5.9400	6.2812

Bibliography

Babaleye O. Ahmed, *Design and Thermodynamic Analysis of Solar Updraft Tower*, Project Report, June 2011.

S.S. Al-Azawie, S. Hassan and M.F. Zammeri, "Experimental and Numerical Study on Ground Material Absorptivity for Solar Chimney Power Applications", *WIT Transactions on Ecology and the Environment*, 186, 219–230, 2014.

Mohammed Awwad Al-Dabbas, "A Performance Analysis of Solar Chimney Thermal Power Systems", *Thermal Science*, 15 (3), 619–642, 2011.

Mohammed Awwad Al-Dabbas, "The First Pilot Demonstration: Solar Updraft Tower Power Plant in Jordan", *International Journal of Sustainable Energy*, 31 (6), 399–410, 2012.

A. Angstrom, "Solar and Terrestrial Radiation", *Quarterly Journal of the Royal Meteorological Society*, 50, 121–126, 1924.

I. Asiltürk and M. Çunkaş, "Modeling and Prediction of Surface Roughness in Turning Operations Using Artificial Neural Network and Multiple Regression Method", *Expert Systems with Applications*, 38, 5826–5832, 2011. doi: 10.1016/j.eswa.2010.11.041

A. Atmanli, "Comparative Analyses of Diesel-waste Oil Biodiesel and Propanol, n-Butanol or 1-Pentanol Blends in a Diesel Engine", *Fuel*, 176, 209–215, 2016. doi: 10.1016/j.fuel.2016.02.076

A. Atmanli, E. Ileri, B. Yuksel and N. Yilmaz, "Extensive Analyses of Diesel-Vegetable Oil-n-Butanol Ternary Blends in a Diesel Engine", *Applied Energy*, 145, 155–162, 2015. doi: 10.1016/j.apenergy.2015.01.071

Vineet Banodha and Praveen Patel, "Risk Assessment in Maintenance Work at Diesel Locomotive Workshop", *International Journal on Emerging Technologies*, 5 (1), 59–63, 2014.

R.K. Bansal, *Fluid Mechanics and Hydraulic Machines*, Laxmi Publications (P) Ltd., 2005.

Ralph M. Barnes, *Motion and Time Study Design and Measurement of Work*, John Wiley & Sons, 7th Edition, 202–205.

Basic Research Needs for Solar Energy Utilization, *Report of the Basic Energy Sciences Workshop on Solar Energy Utilization*, April 18–21. Office of Science, US Department of Energy, 2005.

A. Bejalwar and P. Belkhode, "Analysis of Experimental Setup of a Small Solar Chimney Power Plant", *Procedia Manufacturing*, 20, 481–486, 2018. doi: 10.1016/j.promfg.2018.02.071

P. Belkhode, C. Sakhale and A. Bejalwar, "Evaluation of the Experimental Data to Determine the Performance of a Solar Chimney Power Plant", *Materials Today: Proceedings*, 27, 102–106, 2020. doi: 10.1016/J.MATPR.2019.09.006

P.N. Belkhode, "Mathematical Modelling of Liner Piston Maintenance Activity Using Field Data to Minimize Overhauling Time and Human Energy Consumption", *Journal of the Institution of Engineers (India): Series C*, 99, 701–709, 2018.

P.N. Belkhode, "Analysis and Interpretation of Steering Geometry of Automobile Using Artificial Neural Network Simulation", *Engineering*, 11, 231–239, 2019.

P.N. Belkhode, "Development of Mathematical Model and Artificial Neural Network Simulation to Predict the Performance of Manual Loading Operation of Underground Mines", *Journal of Materials Research and Technology*, 8, 2309–2315, 2019.

P.N. Belkhode, V.N. Ganvir, A.C. Shende and S.D. Shelare, "Utilization of Waste Transformer Oil as a Fuel in Diesel Engine", *Materials Today: Proceedings*, 2021. doi: 10.1016/j.matpr.2021.02.008

P.N. Belkhode, S.D. Shelare, C.N. Sakhale, R. Kumar, S. Shanmugan, M.E.M. Soudagar and M.A. Mujtaba, "Performance Analysis of Roof Collector Used in the Solar Updraft Tower", *Sustainable Energy Technologies and Assessments*, 48, 101619, 2021. doi: 10.1016/J.SETA.2021.101619

M.D.S. Bernardes, A. Voß and G. Weinrebe, "Thermal and Technical Analyses of Solar Chimneys", *Solar Energy*, 75 (6), 511–524, 2003.

E. Bilgen and J. Rheault, "Solar Chimney Power Plants for High Latitudes", *Solar Energy*, 79 (5), 449–458, 2005.

Tushigjargal Bold, "Solar Updraft Tower: A Chimney that Generates Electricity", *Tunza Eco-Generation*, 37, 27–38, 2017.

A. James Bowery, "Brief Proforma for a Solar Updraft Tower Algae Biosphere", July 8, 2006.

Saurabh Singh Chandrawat and Doulat T. Gianchandani, "Reduction of the Time Required for POH of Diesel Locomotives Using Unit Replacement", *IOSR Journal of Mechanical and Civil Engineering (IOSR-JMCE)*, 3 (2), 13–19, September–October 2012.

P.J. Cottam, P. Duffour, P. Lindstrand and P. Fromme, "Effect of Canopy Profile on Solar Thermal Chimney Performance", *Solar Energy*, 129, 286–296, 2016.

Y.J. Dai, H.B. Huang and R.Z. Wang, "Case Study of Solar Chimney Power Plants in Northwestern Regions of China", *Renewable Energy*, 28, 1295–1304, 2003.

Biman Das and Robert M. Grady, "Industrial Workplace Layout Design-An Application of Engineering Anthropometry", *Ergonomics*, 26 (5), 433–447, 1983.

Tapas K. Das and Sudeep Sarkar, "Optimal Preventive Maintenance in a Production: Inventory System", *IIE Transactions*, 31, 537–551, 1999.

Fabio De Felice and Antonella Petrillo, "Methodological Approach for Performing Human Reliability and Error Analysis in Railway Transportation System", *International Journal of Engineering and Technology*, 3 (5), 341–353, 2011.

Rommert Dekker, "Integrating Optimization, Priority Setting, Planning and Combing of Maintenance Activities", *European Journal of Operational Research*, 82, 225–240, 1995.

B.S. Dhillon and Y. Liu, "Human Error in Maintenance: A Review", *Journal of Quality in Maintenance Engineering*, 12 (1), 21–36, 2006.

I. Dincer and M.A. Rosen, "A Worldwide Perspective on Energy, Environment and Sustainable Development", *International Journal of Energy Research*, 22 (15), 1305–1321, 1998.

Jillian Dorrian and Gregory D. Roach, "Simulated Train Driving: Fatigue, Self-Awareness and Cognitive Disengagement", *Elsevier Applied Ergonomics*, 38, 155–166, 2007.

Marco Aurelio dos Santos Bernardes, Theodor W. Von Backstrom and Detlev G. Kroger, "Analysis of Some Available Heat Transfer Coefficients Applicable to Solar Chimney Power Plant Collectors", *Solar Energy*, 83, 264–275, 2009.

Eastman Kodak Co. Ltd., "Chapter V Environment", *Ergonomic Design for People at Work*, Van Nustrand Reinhold, New York, 1983a.

Eastman Kodak Co. Ltd., "Section VIA Appendix – A Anthropometric Data", *Ergonomic Design for People at Work*, Van Nustrand Reinhold, New York, 1983b.

Eastman Kodak Co. Ltd., "Sitting Standing Seat Stand Work Place", *Ergonomic Design for People at Work*, Van Nostrand Reinhold, New York, 1983c.

Eastman Kodak Co. Ltd., *"WORK PLACE" Ergonomic Design for People at Work*, Van Nostrans Reinhold, New York, 1983d.

Amin Mohamed El-Ghonemy, "Solar Chimney Power Plant with Collector", *IOSR Journal of Electronics and Communication Engineering (IOSR-JECE)*, 11 (2), 28–35, March–April 2016.

Andre G. Ferreira, Cristiana B. Maia, Marcio F.B. Cortez and Ramon M. Valle, "Technical Feasibility Assessment of a Solar Chimney for Food Drying", *Solar Energy*, 82, 198–205, 2008.

T.P. Fluri, J.P. Pretorius, C. Van Dyk, T.W. Von Backstrom, D.G. Kroger and G.P.A.G. Van Zijl, "Cost Analysis of Solar Chimney Power Plants", *Solar Energy*, 83, 246–256, 2009.

T. Fulder, P. Pizmoht, A. Polanjar and M. Leber, "Ergonomically Designed Workstation Based on Simulation of Worker's Movements", *International Journal of Simulation, Model*, 4 (1), 27–34, 2005.

A.J. Gannon and T.W. Von Backström, "Solar Chimney Cycle Analysis with System Loss and Solar Collector Performance", *Journal of Solar Energy Engineering*, 122 (3), 133–137, 2000.

N. Geuder, F. Trieb, C. Schillings, R. Meyer, and V. Quaschning, "Comparison of Different Methods for Measuring Solar Irradiation Data", *3rd International Conference on Experiences with Automatic Weather Stations*, Torremolinos, Spain, February 2003.

P. Gopalkrishnan and A.K. Banerji, *Maintenance & Spare Parts Management*, Prentice Hall India Pvt. Ltd., New Delhi, 2015.

Iwona Grabarek, "Ergonomic Diagnosis of the Driver's Workplace in an Electric Locomotive", *International Journal of Occupational Safety and Ergonomics*, 8 (2), 225–242, 2002.

Wiesław Grzybowski, "A Method of Ergonomic Workplace Evaluation for Assessing Occupational Risks at Workplaces", *International Journal of Occupational Safety and Ergonomics*, 7 (2), 223–237, 2001.

W. Haaf, "Solar Chimneys, Part II: Preliminary Test Results from the Manzanares Pilot Plant", *International Journal of Solar Energy*, 2, 141–161, 1984.

W. Haaf, K. Friedrich, G. Mayr and J. Schlaich, "Solar Chimneys, Part I: Principle and Construction of the Pilot Plant in Manzanares", *International Journal of Solar Energy*, 2, 3–20, 1983.

Mohammad O. Hamdan, "Analysis of a Solar Chimney Power Plant in the Arabian Gulf Region", *Renewable Energy*, 36, 1–6, 2010.

Salman H. Hammadi, *Solar Updraft Tower Power Plant with Thermal Storage*, Research Gate, January 2008.

Magdy Bassily Hanna, Tarek Abdel-Malak Mekhai, Omar Mohamed Dahab, Mohamed Fathy Cidek Esmail and Ahmed Rekaby Abdel-Rahman, "Experimental and Numerical Investigation of the Solar Chimney Power Plant's Turbine", *Open Journal of Fluid Dynamics*, 6, 332–342, 2016.

Reinhard Harte, Markus Tschersich, Rüdiger Höffer and Tarek Mekhail, "Design and Construction of a Prototype Solar Updraft Chimney in Aswan/Egypt", *Acta Polytechnica*, 57 (3), 167–181, 2017.

Reinhard Harte and Gideon P.A.G. Van Zijl, "Structural Stability of Concrete Wind Turbines and Solar Chimney Towers Exposed to Dynamic Wind Action", *Journal of Wind Engineering and Industrial Aerodynamics*, 95, 1079–1096, 2007.

R.A. Hedderwick, "Performance Evaluation of a Solar Chimney Power Plant", Master's thesis, University of Stellenbosch, Stellenbosch, South Africa, 2001.

M. Jamil Ahmad and G.N. Tiwari, "Solar Radiation Models – Review", *International Journal of Energy and Environment*, 1 (3), 513–532, 2010.

Harold E. Johnson and Piero P. Bonissone, "Expert System for Diesel Electric Locomotive Repair", *The Journal of Fourth Application and Research*, 1 (1), 1983.

Shadi Kalasha, Wajih Naimeh and Salman Aji, "Experimental Investigation of a Pilot Sloped Solar Updraft Power Plant Prototype Performance Throughout a Year", *Energy Procedia*, 50, 627–633, 2014.

Stamatics V. Kartalopous, *Understanding Neural Networks and Fuzzy Logic*, Prentice-Hall of India Pvt. Ltd., New Delhi, 2004.

A.B. Kasaeian, Sh. Molana, K. Rahmani and D. Wen, *A Review on Solar Chimney Systems*, Elsevier, September, 2016.

O.P. Khanna, *Work Study Motion & Time Study*, Dhanpat Rai Publications, 1992.

H.H. Kimbal, "Variations in Total and Luminous Solar Radiation with Geographical Position in the United States", *Monthly Weather Review*, 47, 769, 1919.

A. Koonsrisuk, S. Lorente and A. Bejan, "Constructal Solar Chimney Configuration", *International Journal of Heat and Mass Transfer*, 53, 327–333, 2010.

Jozef Kotus and Malgorzata Szwarc, "Problems of Railway Noise – A Case Study", *International Journal of Occupational Safety and Ergonomics (JOSE)*, 17 (3), 309–325, 2011.

S.C.W. Krauter, "Solar Electric Power Generation Photovoltaic Energy Systems", Springer.com, 2009.

R.J.K. Krisst, "Energy Transfer System", *Alternative Sources of Energy*, 63, 8–11, 1983.

D.G. Kröger and D. Blaine, "Analysis of the Driving Potential of a Solar Chimney Power Plant", *South African Institution of Mechanical Engineering, R & D Journal*, 15, 85–94, 1999.

H. Kulunk, "A Prototype Solar Convection Chimney Operated Under IZMIT Conditions", in T.N. Veziroglu, editor, *Proceedings of Seventh MICAES*, p. 162, 1985.

S. Kunar and G. Ghosh, "Measurement and Evaluation of Reliability, Availability and Maintainability of a Diesel Locomotive Engine", *IOSR Journal of Mechanical and Civil Engineering (IOSR-JMCE)*, 8 (1), 31–46, July–August 2013.

Asaf Lerin, Gur Mosheior and Assaf Sarigz, "Scheduling a Maintenance Activity on Parallel Identical Machines", Department of Statistics, The Hebrew University, July 2008.

Haotian Liu, Justin Weibel and Eckhard Groll, "Performance Analysis of an Updraft Tower System for Dry Cooling in Large-Scale Power Plants", November 9, 2017.

F. Lupi, C. Borri, R. Harte, W.B. Krätzig and H. Niemann, "Facing Technological Challenges of Solar Updraft Power Plants", *Journal of Sound and Vibration*, 334, 57–84, 2015.

Cristiana B. Maia, André G. Ferreira, Ramón M. Valle, Márcio F.B. Cortez, "Theoretical Evaluation of the Influence of Geometric Parameters and Materials on the Behavior of the Airflow in a Solar Chimney", *Computers & Fluids*, 38, 625–636, 2009.

Neeraj Mehla, Rahul Makade and N.S. Thakur, "Experimental Analysis of A Velocity Field Using Variable Chimney Diameter For Solar Updraft Tower", *IJEST*, April, 2011.

T. Mekhail, A. Rekaby, M. Fathy, M. Bassily and R. Harte, "Experimental and Theoretical Performance of Mini Solar Chimney Power Plant", *Journal of Clean Energy Technologies*, 5 (4), 294–298, July 2017.

R. Mofidian, A. Barati, M. Jahanshahi and M. Hassan, "Optimization on Thermal Treatment Synthesis of Lactoferrin Nanoparticles via Taguchi Design Method", *SN Applied Sciences*, 2019. doi: 10.1007/s42452-019-1353-z

Arash Motaghedi-Larijani and Hamid Reza Haddad, "A New Single Machine Scheduling Problem with Setup Time, Job Deterioration and Maintenance Costs", *International Journal of Management Science and Engineering Management*, 6 (4), 284–291, 2011.

Gerald Muller, "Low Pressure Solar Thermal Converter", *Renewable Energy*, 35, 318–321, 2010.

L.B. Mullett, "The Solar Chimney – Overall Eficiency, Design and Performance", *International Journal of Ambient Energy*, 8, 35–40, 1987.

K.F.H. Murrell, "Design of Seating", *Ergonomics (Man in His Working Environment)*, Chapman and Hall, London, New York, 1986.

K.F.H. Muwell, "Nature of Ergonomics", *Ergonomics*, Chapman and Hall, London, New York, 25–28, 1956.

Daryl R. Myers, "Solar Radiation Modeling and Measurements for Renewable Energy Applications: Data and Model Quality", *Energy*, 30, 1517–1531, 2005.

M.M. Padki and S.A. Sherif, "Fluid Dynamics of Solar Chimneys", *Proceedings of the ASME Winter Annual Meeting*, 43–46, 1988.

H. Pastohr, O. Kornadt and K. Gürlebeck, "Numerical and Analytical Calculations of the Temperature and Flow Field in the Upwind Power Plant", *International Journal of Energy Research*, 28, 495–510, 2004.

N. Pasumarthi, and S.A. Sherif, 'Experimental and Theoretical Performance of a Demonstration Solar Chimney Model – Part I: Mathematical Model Development", *International Journal of Energy Research*, 22, 277–288, 1998.

N. Pasurmarthi and S.A. Sherif, "Performance of a Demonstration Solar Chimney Model for Power Generation", *Proceeding of the 1997 35th Heat Transfer and Fluid*, Sacrmento, CA, 203–240, 1997.

Yuriy Posudin, "Measurement of Solar Radiation", Book Chapter 12, 1–13, August 11, 2014.

J.P. Pretorius, "Solar Tower Power Plant Performance Characteristics", Master's thesis, University of Stellenbosch, Stellenbosch, South Africa, 2004.

Johannes Petrus Pretorius, "Optimization and Control of a Large-scale Solar Chimney Power Plant", Department of Mechanical Engineering University of Stellenbosch, Matieland, South Africa, 2007.

Shiv Pratap Raghuvanshi et al., "Carbon Dioxide Emissions from Coal Based Power Generation in India", *Journal Energy Conversion and Management*, 47, 421–441, 2006.

S.S. Rao, *Optimization Theory and Application*, Wiley Eastern Ltd., New Delhi, 1994.

M.A. Rosen, "The Role of Energy Efficiency in Sustainable Development", *Technical Society*, 15 (4), 21–26, 1996.

E.P. Sakonidou, T.D. Karapantsios, A.I. Balouktsis, and D. Chassapis, "Modeling of the Optimum Tilt of a Solar Chimney for Maximum Air Flow", *Solar Energy*, 82, 80–94, 2008.

M. Sawka, Solar Chimney—Untersuchungen zur Strukturintegritat des Stahlbetonturms, Bergische Universitat Wuppertal, 2004.

H. Schenck Jr., *Theories of Engineering Experimentation*, Mc-Graw Hill, 1961.

H. Schenck Jr., *Theories of Engineering Experimentation*, 1st Edition, McGraw Hill Inc., 1967. Report Load Generation Balance Report 2016–2017, Brought Out Annually by Central Electricity Authority, New Delhi, May 2016.

Schlaich Bergermann Solar GmbH, Stuttgart, Solar Updraft Tower, Project Report, October 2011.

J. Schlaich, "World Energy Demand, Population Explosion and Pollution: Could Solar Energy Utilisation Become a Solution?" *The Structural Engineer*, 69 (10), 189–192, 1991.

J. Schlaich, *The Solar Chimney: Electricity from the Sun*, Deutsche Verlags-Anstalt, Stuttgart, 1994.

J. Schlaich, "Solar Updraft Towers", *Slovak Journal of Civil Engineering*, 03, 39–42, 2009.

Manajit Sengupta, "Measurement and Modeling of Solar Radiation", Workshop, April, 01, 13–19, 2016.

S.D. Shelare, K.R. Aglawe and P.N. Belkhode "A Review on Twisted Tape Inserts for Enhancing the Heat Tansfer", *Materials Today: Proceedings*, 2021. doi: 10.1016/j.matpr.2021.09.012

Ashaf A. Shikdar and Mohamed A. AI-Hqadhrami, "Smart Workstation Design: Ergonomics and Methods Engineering Approach", *International Journal of Industrial and System Enginerring*, 2 (4), 67–69, 2007.

Sushil Kumar Shriwastawa, *Industrial Maintenance Management*, S. Chand & Company Pvt. Ltd., New Delhi, 2013.

S. Simsek, "Effects of Biodiesel Obtained from Canola, Sefflower Oils and Waste Oils on the Engine Performance and Exhaust Emissions", *Fuel*, 265, 117026, 2020. doi: https://doi.org/10.1016/j.fuel.2020.117026

Ashutosh Vikram Singh and Preetam Singh Suryavanshi, "A Generalized Overview of Solar Updraft Towers", *Imperial International Journal of Eco-friendly Technologies*, 1 (1), 31–37, 2016.

Divya Singh and Seema Kwatra, "Psychological Effects of Shift Work on the Lives of Railway Employees: An Ergonomics Intervention", *International Journal of Advanced Engineering Research and Studies*, I (II), 99–105, January–March, 2012.

S. Singh, R. Kumar, A. Barabadi and S. Kumar, "Human Error Quantification of Railway Maintenance Tasks of Disc Brake Unit", *Proceedings of the World Congress on Engineering*, Vol II, WCE 2014, July 2–4, 2014, London, 2014.

S.N. Sivanandam, S. Sumathi and S.N. Deepa, *Introduction to Neutral Networks using Matlab 6.0*, Tata Mcgraw-Hill Publishing Company Limited, New Delhi, 2005.

Sushil Kumar Srivastava, *Maintenance Engineering (Principles, Practices and Manageent)*, S. Chand & Company Ltd., New Delhi, 2007.

M. Tingzhen, L. Wei, X. Guoling, X. Yanbin, G. Xuhu and P. Yuan, "Numerical Simulation of the Solar Chimney Power Plant Systems Coupled with Turbine", *Renewable Energy*, 33 (5), 897–905, 2008.

Jeffrey H.Y. Too and C.S. Nor Azwadi, "A Brief Review on Solar Updraft Power Plant", *Journal of Advanced Review on Scientific Research*, 18 (1), 1–25, 2016.

F. Trieb, O. Langniÿ and H. Klaiÿ, "Solar Electricity Generation – A Comparative View of Technologies, Costs and Environmental Impact", *Solar Energy*, 59 (1–3), 88–99, 1997.

T.W. Von Backström and A.J. Gannon, "The Solar Chimney Air Standard Thermodynamic Cycle", *South African Institution of Mechanical Engineering, R & D Journal*, 16 (1), 16–24, 2000.

Theodor W. von Backstrom and Thomas P. Fluri, "Maximum Fluid Power Condition in Solar Chimney Power Plants – An Analytical Approach", *Solar Energy*, 80, 1417–1423, 2006.

John Wilson and Lucy Mitchell, "Understanding of Mental Workload in the Railways", *Rail Human Factors Supporting the Integrated Railways*, Ashgat, Aldershot, 309–318, 2005.

Malima Isabelle Wolf, "Solar Updraft Towers: Their Role in Remote On-Site Generation", Project Report, April 29, 2008.

J.K. Yang, J. Li, B. Xiao, J.J. Li, J.F. Zhang and Z.Y. Ma, "A Novel Technology of Solar Chimney for Power Generation", *Acta Energinae Solaris Sinica*, 24 (4), 565–570, 2003.

N. Yilmaz, A. Atmanli and F.M. Vigil, "Quaternary Blends of Diesel, Biodiesel, Higher Alcohols and Vegetable Oil in a Compression Ignition Engine", *Fuel*, 212, 462–469, 2018. doi: 10.1016/j.fuel.2017.10.050

A. Zandian and M. Ashjaee, "The Thermal Efficiency Improvement of a Steam Rankine Cycle by Innovative Design of a Hybrid Cooling Tower and a Solar Chimney Concept", *Renewable Energy*, 51, 465–473, 2013.

Chuan-li Zhao and Heng-yong Tang, "Single Machine Scheduling with General Job-Dependent Aging Effect and Maintenance Activities to Minimize Makespan", *Applied Mathematical Modelling*, 34, 837–841, 2010.

Y. Zheng, T.Z. Ming, Z. Zhou, X.F. Yu, H.Y. Wang, Y. Pan and W. Liu, "Unsteady Numerical Simulation of Solar Chimney Power Plant System with Energy Storage Layer", *Journal of the Energy Institute*, 83 (2), 86–92, 2010.

Joshua Mah Jing Zhi, Melissa Law Jia Li and Ho Kwan Yu, "Experiments with Solar Updraft Tower Models", January 2016.

Xinping Zhou and Yangyang Xu, "Solar Updraft Tower Power Generation", June 22, 2014.

Xinping Zhou, Jiakuan Yang, Bo Xiao and Xiaoyan Shi, "Special Climate around a Commercial Solar Chimney Power Plant", *Journal of Energy Engineering*, 134, 25–37, March 2008.

Y. Zhou, X.H. Liu and Q. Li, "Unsteady Conjugate Numerical Simulation of the Solar Chimney Power Plant System with Vertical Heat Collector", *Physical and Numerical Simulation of Material Processing Vi, Pts 1 and 2*, J. Niu and G.T. Zhou, editors, Trans Tech Publications Ltd., Stafa-Zurich, pp. 535–540, 2012.

Index

A

Alternative, 1, 14, 16, 32, 38, 74
Analysis, 78, 85–87, 89, 90, 100, 102
Analytical, 2, 13
Anthropometry, 2
Artificial, 3, 113, 115, 117, 119, 122, 157
Assembly, 8, 15

B

Blending, 32, 106
Brake, 1, 11, 32, 33, 37, 74, 75, 77, 108, 188, 190, 206, 212
Buckingham, 9, 11, 41

C

Chimney, 1, 38–40, 41, 71, 73
Civil, 8
Clearance, 17, 25, 28, 48, 55, 59, 66, 83, 169, 196, 201
Collector, 1, 6, 38, 39, 40
Complexity, 9, 73
Conductivity, 210
Consumption, 8, 11, 13, 32, 33, 37, 38, 74
Costs, 3, 38
Crankshaft, 13–18

D

Deficiency, 2
Diesel Blends, 3
Dimensionless, 3, 7, 11, 13, 14, 18

E

Efficiency, 1, 11, 15, 32
Elbow, 36, 38, 48, 184–186, 190, 197, 198
Engine, 1, 7, 15
Enhancing, 1, 2
Enthusiasm, 6, 9, 16, 18, 25
Environment, 1–3, 5, 15, 32, 38
Ergonomic, 1, 2, 3
Expenditure, 3
Experimentation, 2, 3, 7, 8, 13, 15
Extraneous, 6–8, 14, 15, 22, 39, 49, 195, 197
Eye Level, 197, 198, 202

F

Field, 7, 13
Fingertip, 197, 198, 202
Fossil, 1, 32

G

Generators, 38, 71

H

Heat Transfer, 208, 209
Humidity, 3, 6, 9, 15, 16, 18, 20, 21, 25, 38, 40, 41, 47

I

Illumination, 202, 205
Index, 33, 195, 196
Industrial, 1–3, 13, 15
Instrumentation, 13, 47, 49, 58, 73, 74
Intensity, 3
Interpretation, 109
Investigation, 7, 8, 14, 47

L

Leakages, 109
Liner Piston, 3, 11, 15, 16
Loading, 1
Locomotive, 11, 15
Locoshed, 3, 13, 15, 20, 49, 54, 57, 60

M

Magnification, 108, 109
Management, 1
Manually, 1
Manufacturing, 8
Materials, 1, 39, 71
Mathematical, 3, 6
Matlab, 24, 32, 46, 87, 90, 93, 97
Maximum, 178
Mean Error, 185, 186, 187
Mechanics, 7, 9
Mechanization, 1
Minimum, 178
Modelling, 1, 5, 15, 21, 32

O

Objective, 178, 180
Operations, 1
Operators, 1, 2, 17, 24, 49
Optimization, 3, 15, 16, 20, 177
Optimum, 3, 16
Overhauling, 16, 19, 25

P

Performance, 1–3, 8, 13, 18
Physical, 2, 5, 7–9, 11, 13
Planning, 7, 8, 13
Poor Health, 1
Postural, 2
Power, 164, 192, 193, 209

Prediction, 119, 123
Production, 11, 32
Productivity, 1–5, 8
Pulse Rate, 6, 47, 49, 58, 70

Q

Qualitatively, 8, 16

R

Radiation, 1, 38, 71, 73
Reliability, 38, 185, 187

S

Scientific, 7
Sensitivity Analysis, 165
Shoulder, 197, 198, 202
Simulation, 113, 131
Solar, 1, 3, 11, 38–41
Stroke Length, 207

T

Technological, 1
Test Envelope, 13
Test Points, 13
Test Sequence, 13
Theoretical, 7, 14, 16
Thermal, 1, 11, 32, 33, 37, 39, 77, 107, 149, 187, 190
Transformer, 32, 74, 96
Turbine, 1, 38–41, 71, 73, 109, 111

V

Variables, 1, 3, 6

W

Work Measurement, 1
Workstation, 1, 3, 15, 25, 47

Printed in the United States
by Baker & Taylor Publisher Services